AF581724

Hyperspectral Imaging for Fine to Medium Scale Applications in Environmental Sciences

Hyperspectral Imaging for Fine to Medium Scale Applications in Environmental Sciences

Editors

Michael Vohland
András Jung

MDPI • Basel • Beijing • Wuhan • Barcelona • Belgrade • Manchester • Tokyo • Cluj • Tianjin

Editors
Michael Vohland
Leipzig University
Germany

András Jung
Eötvös Loránd University
Hungary

Editorial Office
MDPI
St. Alban-Anlage 66
4052 Basel, Switzerland

This is a reprint of articles from the Special Issue published online in the open access journal *Remote Sensing* (ISSN 2072-4292) (available at: https://www.mdpi.com/journal/remotesensing/special_issues/hyperspectral_imaging).

For citation purposes, cite each article independently as indicated on the article page online and as indicated below:

LastName, A.A.; LastName, B.B.; LastName, C.C. Article Title. *Journal Name* **Year**, *Volume Number*, Page Range.

ISBN 978-3-0365-0878-8 (Hbk)
ISBN 978-3-0365-0879-5 (PDF)

Contents

About the Editors

Michael Vohland (Dr., Prof.): Full Professor for Geography with a main focus on Geoinformatics and Remote Sensing at Leipzig University (Germany). He studied Physical Geography with Geobotany and Remote Sensing as minor subjects and received his PhD (Dr. rer. nat.) at the University of Trier (Germany) in 2003. His PhD thesis focused on the use of optical remote sensing data to optimize mesoscale precipitation–runoff modelling. His current research activities include the analysis of multi- and hyperspectral remote sensing data for geoecosystem research, the application of imaging and non-imaging spectroscopy for soil and vegetation parameters, and the coupling of remote sensing data (reflective, thermal) with modelling approaches (e.g., hydrological process modelling) at different spatial scales. He is one of the scientific founding members of the Remote Sensing Centre for Earth System Research (Leipzig University, UFZ—Helmholtz-Centre for Environmental Research) established in 2020, and he currently acts as one of the co-chairs heading the "Multi- and Hyperspectral: Methodology" working group of the German Society for Photogrammetry, Remote Sensing and Geoinformation (DGPF).

András Jung is associate professor at the Institute of Cartography and Geoinformatics, Faculty of Informatics, Eötvös Loránd University (Hungary). He obtained his engineering degree from the Szent István University (SZIE) in Budapest in 2001. In 2004, he completed his second degree at the Budapest University of Technology and Economics (BME) in Geodesy and Geoinformatics. In 2006 he earned his PhD from Corvinus University of Budapest in the field of hyperspectral remote sensing and plant examination methods. He had longer stays at the Humboldt University of Berlin, at the University of Agricultural Sciences (Vienna) and at the University of Halle. Since 2006 he has been researching and teaching at universities in Germany (2006-2010, 2019-2020: Martin Luther University Halle-Wittenberg, University of Ulm, and University of Leipzig). He also acts as one of the co-chairs heading the "Multi- and Hyperspectral: Methodology" working group of the German Society for Photogrammetry, Remote Sensing and Geoinformation (DGPF). His research focuses on field spectroscopy and hyperspectral remote sensing, with special regard to scientific and industrial development of spectral imaging sensors.

Editorial

Hyperspectral Imaging for Fine to Medium Scale Applications in Environmental Sciences

Michael Vohland [1,2,*] and András Jung [3]

1 Geoinformatics and Remote Sensing, Institute for Geography, Leipzig University, Johannisallee 19a, 04103 Leipzig, Germany
2 Remote Sensing Centre for Earth System Research, Leipzig University, Talstr. 35, 04103 Leipzig, Germany
3 Institute for Cartography and Geoinformatics, Eötvös Lorand University, Pázmány P. Sétány 1/C., 1117 Budapest, Hungary; jung@inf.elte.hu
* Correspondence: michael.vohland@uni-leipzig.de; Tel.: +49-341-97-32798

Received: 2 September 2020; Accepted: 6 September 2020; Published: 11 September 2020

Hyperspectral imaging (HSI) combines conventional imaging and spectroscopic techniques in a way of spatially organized spectroscopy. Technical developments in the last three decades have brought the capacity of HSI to provide spectrally, spatially and temporally detailed data. The latter crucially relates to rapid data acquisition, favoured by hyperspectral snapshot technologies, i.e., no scanning as, e.g., push broom scanning as one conventionally remote sensing technique is needed for obtaining 3D image cubes. Furthermore, the development of miniaturized hyperspectral sensors has fostered their application with lightweight unmanned aerial vehicle (UAV) platforms [1–3]. HSI sensor technology with 3D reconstruction capacities is currently available [4]. Among HSI, hyperspectral microscopy imaging is another emerging field facilitating new applications [5–7].

Beyond this background, the aim of this Special Issue (SI) is to present a selection of innovative applications of HSI in the environmental and earth sciences, with a focus on the fine- to the medium-scale ranging from the microscale to field- and airborne data acquisition and analysis. The SI comprises a total of nine papers in various thematic fields, which can be organized into the following categories: geology/mineral exploration (one published paper), digital soil mapping (one), the mapping and characterization of vegetation (two) and the sensing of water bodies (including under-ice and underwater applications) (three); two rather methodically/technically oriented contributions focus on the optimized processing of UAV data and on the design and test of a receiver for simultaneous hyperspectral and differential laser absorption spectrometry (LAS) measurements.

In geological field studies, almost vertical-oriented outcrops may be mapped and characterised most properly by tripod-mounted close-range imaging instruments [8]. In this context, the study of Lorenz et al. [9] presents an adapted workflow for outcrop sensing by including atmospheric and topographic corrections, which are markedly beneficial for close- to long-range observations covering different sensing distances and viewing perspectives. For two different datasets, both acquired with an AisaFENIX push broom scanner (SPECIM, Spectral Imagig Ltd., Oulu, Finland), HSI mapping products were integrated with 3D photogrammetric data to create "hyperclouds", i.e., geometrically correct representations of the hyperspectral data cube.

Airborne hyperspectral imaging has been used in many studies to quantify soil variables, but soil studies with UAV data are still rare (see, for example, the recent review in [10]). The SI contribution of Hu et al. [11] aims at filling one gap in the UAV-based mapping of soil salinity. For this purpose, data were acquired from a UAV platform with a hyperspectral camera (Rikola Ltd., Oulu, Finland), providing data at a spatial resolution of 0.1 m and covering the 0.50–0.89 μm wavelength region with 62 spectral bands. With these data, random forest regression was used to estimate the electrical conductivity (EC) values and to generate EC maps for fields with different vegetation cover conditions, located in the region of Aksu, Western Xinjiang, China.

Different soil types were selected by Salazar et al. [12] for hyperspectral measurements from different distances to test a newly developed multichannel receiver. The configuration of this receiver allows the range-resolved collection of hyperspectral data in the 350–2500 nm range, combined with LAS measurements in the 820–850 nm wavelength region. Acquired test data indicated consistent hyperspectral measurements, independent of the range to the target. Envisioned applications include the rapid classification of soils, rocks, minerals and vegetation for ecological or agronomic research or the monitoring of earth construction sites as, for example, mine tailings.

Two SI papers focus on the forest ecosystems of different ecofloristic zones. Issues such as forest health, productivity and ecosystem services are often discussed in the context of forest diversity [13,14] and motivate researchers to seek out new inventory methods with the required spatial details. Recent developments in remote sensing technologies and image processing techniques thus extend the toolbox of forest researchers and managers [15].

Based on airborne HSI data, acquired with NEO Hyspex VNIR 1600 and NEO Hyspex SWIR 320m-e (Norsk Elektro Optikk AS, Skedsmokorset, Norway), Knauer et al. [16] evaluated the benefits of combining state-of-the-art classification techniques by turning them into an ensemble classifier, implemented for the discrimination of, in total, 15 forest tree species. The study was performed for forests of the temperate zone of the Northern hemisphere, located in Saxony Anhalt and Thuringia (Germany). The obtained results indicated that even the best available classifiers could be further improved by incorporating them into a multiple classifier system and using a specific (precision-weighted) voting strategy. Furthermore, MCLDA (multiclass linear discriminant analysis) was proposed for the image data analysis, as it performed best among different spectral dimensionality reduction methods.

The second forest-related contribution of Cao et al. [17] dealt with salt-tolerant mangroves, distributed to intertidal regions along tropical and subtropical coastlines. Over the past 50 years, global mangrove resources have rapidly decreased due to human interference and natural causes; for their monitoring and management, remote sensing techniques have been widely used [17,18].

Cao et al. [17] used a snapshot hyperspectral imager (UHD 185, Cubert GmbH, Ulm, Germany) to capture field reflectance data covering the spectral range of 450–998 nm with 138 spectral bands. They tested different hyperspectral information extraction methods to investigate the applicability of field snapshot HSI for the identification of mangrove species and to determine the spectral wavebands relevant for an effective classification. As an outcome, the authors underlined the potential of close-range HSI as a tool in monitoring mangrove forests at the species level.

Three SI contributions dealt with applications in water bodies, each with a different focus. In polar marine ecosystems, sea ice-associated algae are an essential feature characterised by a high spatiotemporal variability. The algal biomass is typically concentrated in the bottom ice layers and at the ice-water interfaces, thus not detectable with classical airborne and/or satellite remote sensing techniques [19]. Cimoli et al. have given an extensive overview about adapted capturing techniques, including spectral under-ice measurements and the use of unmanned underwater vehicles as sensing platforms [20]. In the current SI contribution [19], they coupled an AISA Kestrel 10 push broom sensor (SPECIM, Specim Spectral Imaging Ltd., Oulu, Finland) with a standard digital RGB camera and trialled this system at Cape Evans, Antarctica. For a ~20-m-long transect, ultra-high-resolution HSI data were used to quantify per-pixel algal biomass and pigments at the ice-water interface; RGB imagery was processed with digital photogrammetry to capture the under-ice structure and topography.

The use of aboveground remote sensing data of inland waters suffers from some marked limitations. The water-leaving radiation is largely affected by refraction at the water surface and atmospheric absorption and scattering. Therefore, an accurate atmospheric correction is a critical issue for the precise quantification of optically active substances (OAS) in the water column, especially from space [21,22]. For airborne hyperspectral image data with a pixel size of 2 m (AISA DUAL imaging system; SPECIM, Spectral Imaging Ltd., Oulu, Finland), Pyo et al. [23] tested different atmospheric correction approaches for their influence on the retrieval of phycocyanin (PC) and chlorophyll-a (Chl-a) for the water body of the Baekje Reservoir (Geum River, South Korea). Based on different bio-optical retrieval algorithms, the distribution maps of PC and Chl-a were generated to indicate risk regions for cyanobacterial blooms.

A different approach was followed by Seidel et al. [24] to quantify OAS (Chl-a and coloured dissolved organic matter) for a suite of freshwater lakes with different trophic levels, all located in Central Germany. Hyperspectral data for the OAS retrieval were acquired at various depths of each water column by means of a submersible hyperspectral camera (UHD 285, Cubert GmbH, Ulm, Germany), incorporated in a waterproof casing and equipped with a portable halogen lamp. Different from aboveground remote sensing methods, these measurements allowed for the monitoring of the vertical distribution of OAS in the water column; hence, they potentially bridge the gap between point sensors that provide continuous measurements at and below the water surface and spatially continuous remote sensing observations, e.g., from satellites or UAV platforms.

For the latter, the fast retrieval of high-quality and geometrically accurate mosaics of image data is still a challenge. Angel et al. [25] reviewed that existing techniques of mosaicking UAV images are often time-consuming and complex, so that there is a general need to accelerate and automate this procedure. Following this paradigm, they implemented a fully automated workflow to produce geo-rectified and mosaicked hyperspectral UAV images with an optimized co-registration strategy based on a small number of ground control points. The performance of the automated approach was evaluated by comparing its computational effort with that of other available approaches and by determining the standard metrics of spatial accuracy.

Author Contributions: The two authors contributed equally to all aspects of this editorial. All authors have read and agreed to the published version of the manuscript.

Funding: This research received no external funding.

Acknowledgments: We would like to thank all authors who have contributed to this volume. Our special thanks also go to the anonymous reviewers for providing their valuable comments to help the authors improve their manuscripts. Last, but not least, we also would like to thank the Remote Sensing editorial team for its substantial support during the complete process of compiling and editing this volume. We especially thank Nelson Peng, without whom this SI could not have been realized.

Conflicts of Interest: The authors declare no conflict of interest.

References

1. Stuart, M.B.; McGonigle, A.J.S.; Willmott, J.R. Hyperspectral Imaging in Environmental Monitoring: A Review of Recent Developments and Technological Advances in Compact Field Deployable Systems. *Sensors* **2019**, *19*, 3071. [CrossRef] [PubMed]
2. Qin, J. Hyperspectral imaging instruments. In *Hyperspectral Imaging for Food Quality Analysis and Control*; Sun, D.-W., Ed.; Academic Press: San Diego, CA, USA, 2010; pp. 129–172.
3. Saari, H.; Aallos, V.-V.; Akujärvi, A.; Antila, T.; Holmlund, C.; Kantojärvi, U.; Mäkynen, J.; Ollila, J. Novel Miniaturized Hyperspectral Sensor for UAV and Space Applications. In *Proc. SPIE 7474, Sensors, Systems, and Next-Generation Satellites XIII, 74741M*; SPIE Remote Sensing: Berlin, Germany, 2009.
4. Renhorn, I.G.E.; Axelsson, L. High spatial resolution hyperspectral camera based on exponentially variable filter. *Opt. Eng.* **2019**, *58*, 103106. [CrossRef]
5. Pu, H.; Lin, L.; Sun, D.-W. Principles of Hyperspectral Microscope Imaging Techniques and Their Applications in Food Quality and Safety Detection: A Review. *Compr. Rev. Food Sci. Food Saf.* **2019**, *18*, 853–866. [CrossRef]
6. Bi, X.; Zhu, S.; Xian, J.; Wei, L.; Yin, H.; Li, M.; Chen, Z. Multihyperspectral Microscopic Imaging for the Precise Identification of Pollen. *Anal. Lett.* **2018**, *51*, 2295–2303. [CrossRef]
7. Wei, L.; Su, K.; Zhu, S.; Yin, H.; Li, Z.; Chen, Z.; Li, M. Identification of microalgae by hyperspectral microscopic imaging system. *Spectrosc. Lett.* **2017**, *50*, 59–63. [CrossRef]
8. Kurz, T.H.; Buckley, S.J.; Howell, J.A. Close-range hyperspectral imaging for geological field studies: Workflow and methods. *Int. J. Remote Sens.* **2013**, *34*, 1798–1822. [CrossRef]

9. Lorenz, S.; Salehi, S.; Kirsch, M.; Zimmermann, R.; Unger, G.; Vest Sørensen, E.; Gloaguen, R. Radiometric Correction and 3D Integration of Long-Range Ground-Based Hyperspectral Imagery for Mineral Exploration of Vertical Outcrops. *Remote Sens.* **2018**, *10*, 176. [CrossRef]
10. Lausch, A.; Baade, J.; Bannehr, L.; Borg, E.; Bumberger, J.; Chabrilliat, S.; Dietrich, P.; Gerighausen, H.; Glässer, C.; Hacker, J.M.; et al. Linking Remote Sensing and Geodiversity and Their Traits Relevant to Biodiversity—Part I: Soil Characteristics. *Remote Sens.* **2019**, *11*, 2356. [CrossRef]
11. Hu, J.; Peng, J.; Zhou, Y.; Xu, D.; Zhao, R.; Jiang, Q.; Fu, T.; Wang, F.; Shi, Z. Quantitative Estimation of Soil Salinity Using UAV-Borne Hyperspectral and Satellite Multispectral Images. *Remote Sens.* **2019**, *11*, 736. [CrossRef]
12. Salazar, S.E.; Coffman, R.A. Multi-Channel Optical Receiver for Ground-Based Topographic Hyperspectral Remote Sensing. *Remote Sens.* **2019**, *11*, 578. [CrossRef]
13. Brockerhoff, E.G.; Barbaro, L.; Castagneyrol, B.; Forrester, D.I.; Gardiner, B.; González-Olabarria, J.R.; Lyver, P.O.; Meurisse, N.; Oxbrough, A.; Taki, H.; et al. Forest biodiversity, ecosystem functioning and the provision of ecosystem services. *Biodivers. Conserv.* **2017**, *26*, 3005–3035. [CrossRef]
14. Ammer, C. Diversity and forest productivity in a changing climate. *New Phytol.* **2019**, *221*, 50–66. [CrossRef] [PubMed]
15. White, J.C.; Coops, N.C.; Wulder, M.A.; Vastaranta, M.; Hilker, T.; Tompalski, P. Remote Sensing Technologies for Enhancing Forest Inventories: A Review. *Can. J. Remote Sens.* **2016**, *42*, 619–641. [CrossRef]
16. Knauer, U.; Von Rekowski, C.S.; Stecklina, M.; Krokotsch, T.; Pham Minh, T.; Hauffe, V.; Kilias, D.; Ehrhardt, I.; Sagischewski, H.; Chmara, S.; et al. Tree species classification based on hybrid ensembles of a convolutional neural network (CNN) and random forest classifiers. *Remote Sens.* **2019**, *11*, 2788. [CrossRef]
17. Cao, J.; Liu, K.; Liu, L.; Zhu, Y.; Li, J.; He, Z. Identifying mangrove species using field close-range snapshot hyperspectral imaging and machine-learning techniques. *Remote Sens.* **2018**, *10*, 2047. [CrossRef]
18. Duke, N.C.; Meynecke, J.-O.; Dittmann, S.; Ellison, A.M.; Anger, K.; Berger, U.; Cannicci, S.; Diele, K.; Ewel, K.C.; Field, C.D. A world without mangroves? *Science* **2007**, *317*, 41–42. [CrossRef]
19. Cimoli, E.; Meiners, K.M.; Lucieer, A.; Lucieer, V. An Under-Ice Hyperspectral and RGB Imaging System to Capture Fine-Scale Biophysical Properties of Sea Ice. *Remote Sens.* **2019**, *11*, 2860. [CrossRef]
20. Cimoli, E.; Meiners, K.M.; Lund-Hansen, L.C.; Lucieer, V. Spatial variability in sea-ice algal biomass: An under-ice remote sensing perspective. *Adv. Polar Sci.* **2017**, *28*, 268–296.
21. Dörnhöfer, K.; Oppelt, N. Remote sensing for lake research and monitoring–Recent advances. *Ecol. Indic.* **2016**, *64*, 105–122. [CrossRef]
22. Mouw, C.B.; Greb, S.; Aurin, D.; DiGiacomo, P.M.; Lee, Z.; Twardowski, M.; Binding, C.; Hu, C.; Ma, R.; Moore, T.; et al. Aquatic color radiometry remote sensing of coastal and inland waters: Challenges and recommendations for future satellite missions. *Remote Sens. Environ.* **2015**, *160*, 15–30. [CrossRef]
23. Pyo, J.C.; Ligaray, M.; Kwon, Y.S.; Ahn, M.-H.; Kim, K.; Lee, H.; Kang, T.; Cho, S.B.; Park, Y.; Cho, K.H. High-Spatial Resolution Monitoring of Phycocyanin and Chlorophyll-a Using Airborne Hyperspectral Imagery. *Remote Sens.* **2018**, *10*, 1180. [CrossRef]
24. Seidel, M.; Hutengs, C.; Oertel, F.; Schwefel, D.; Jung, A.; Vohland, M. Underwater Use of a Hyperspectral Camera to Estimate Optically Active Substances in the Water Column of Freshwater Lakes. *Remote Sens.* **2020**, *12*, 1745. [CrossRef]
25. Angel, Y.; Turner, D.; Parkes, S.; Malbeteau, Y.; Lucieer, A.; McCabe, M.F. Automated Georectification and Mosaicking of UAV-Based Hyperspectral Imagery from Push-Broom Sensors. *Remote Sens.* **2020**, *12*, 34. [CrossRef]

Article

Radiometric Correction and 3D Integration of Long-Range Ground-Based Hyperspectral Imagery for Mineral Exploration of Vertical Outcrops

Sandra Lorenz [1,*], Sara Salehi [2,3], Moritz Kirsch [1], Robert Zimmermann [1], Gabriel Unger [1], Erik Vest Sørensen [2] and Richard Gloaguen [1]

1 Helmholtz-Zentrum Dresden-Rossendorf, Helmholtz Institute Freiberg for Resource Technology, Division "Exploration Technology", Chemnitzer Straße 40, 09599 Freiberg, Germany; m.kirsch@hzdr.de (M.K.); r.zimmermann@hzdr.de (R.Z.); g.unger@hzdr.de (G.U.); r.gloaguen@hzdr.de (R.G.)
2 Department of Petrology and Economic Geology, Geological Survey of Denmark and Greenland, 1350 Copenhagen K, Denmark; ssal@geus.dk (S.S.); evs@geus.dk (E.V.S.)
3 Department of Geosciences and Natural Resource Management, University of Copenhagen, 1165 Copenhagen K, Denmark
* Correspondence: s.lorenz@hzdr.de; Tel.: +49-351-260-4487

Received: 22 December 2017; Accepted: 24 January 2018; Published: 26 January 2018

Abstract: Recently, ground-based hyperspectral imaging has come to the fore, supporting the arduous task of mapping near-vertical, difficult-to-access geological outcrops. The application of outcrop sensing within a range of one to several hundred metres, including geometric corrections and integration with accurate terrestrial laser scanning models, is already developing rapidly. However, there are few studies dealing with ground-based imaging of distant targets (i.e., in the range of several kilometres) such as mountain ridges, cliffs, and pit walls. In particular, the extreme influence of atmospheric effects and topography-induced illumination differences have remained an unmet challenge on the spectral data. These effects cannot be corrected by means of common correction tools for nadir satellite or airborne data. Thus, this article presents an adapted workflow to overcome the challenges of long-range outcrop sensing, including straightforward atmospheric and topographic corrections. Using two datasets with different characteristics, we demonstrate the application of the workflow and highlight the importance of the presented corrections for a reliable geological interpretation. The achieved spectral mapping products are integrated with 3D photogrammetric data to create large-scale now-called "hyperclouds", i.e., geometrically correct representations of the hyperspectral datacube. The presented workflow opens up a new range of application possibilities of hyperspectral imagery by significantly enlarging the scale of ground-based measurements.

Keywords: hyperspectral; topographic correction; atmospheric correction; radiometric correction; long-range; long-distance; Structure from Motion (SfM); photogrammetry; mineral mapping; minimum wavelength mapping; Maarmorilik; Riotinto

---

1. Introduction

Hyperspectral imaging has been increasingly used to support mineral exploration and geological mapping campaigns. The obtained spectral signatures provide detailed information about the composition of rocks and the occurrence of economic minerals. The hyperspectral instruments are conventionally operated with a nadir viewing angle, comprising different scales of area coverage and spatial resolution by operation on satellite [1,2], airplane [3–6] or drone [7]. Depending on the acquisition altitude, a varying influence of the atmosphere between sensor and target, as well as illumination differences due to topography, can be observed in the acquired spectral imagery. Numerous approaches have been introduced in an attempt to overcome these effects: Atmospheric influences are either corrected by atmospheric modelling using radiative transfer models (e.g., [8–10]), the use of ground targets with known or assumed spectra (empirical line calibration [11], flat field correction [12], dark object subtraction [13]), or a combination of both [14]. Whereas radiative transfer models rely on the correct input of a set of external parameters and are mainly used for satellite and airborne data, the use of ground targets, dark objects, or flat fields provides a much more straightforward approach. However, these methods require a spatial resolution high enough to resolve spectrally uniform reference target(s) and/or a reasonable knowledge on the spectra of those materials present, and are therefore mainly used for drone- or airborne data with low acquisition altitudes (e.g., [7,15]).

In the last few years, a ground-based approach of using hyperspectral sensors for geological applications has emerged. A tripod-mounted device can be used to rapidly acquire spectrally and spatially highly resolved data of near-vertical geological outcrops, i.e., spatial orientations that are not (or hardly) observable by nadir-faced instruments. Near-vertical outcrops may comprise steep mountain slopes, water-faced cliffs, open pit mine walls, and road cuts. Particularly in arctic or humid regions, where snow and ice, lichens, or dense vegetation cover the Earth's surface, the investigation of such natural or artificial cuts through the strata might be the only possibility to obtain spectral information of the local geology. Currently, ground-based hyperspectral sensors for geological applications are nearly exclusively used for targets at distances between one to several hundred metres (e.g., [16–18]). Within this range, the spatial resolution varies between centimetre and decimetre scale, enough to resolve even small-scale mineral compounds and fault systems. Another significant benefit of close-distance measurements is the negligible influence of the atmosphere, which potentially voids the need for an elaborate radiometric correction. Instead, an empirical line approach using reference targets with the same orientation, distance, and illumination conditions as the geological target is sufficient for the conversion to reflectance. However, observing a geological target at close range is not always feasible or reasonable. In particular, larger and vertically oriented targets such as steep mountain slopes, sea- or lake-faced cliffs, and walls of large open pit mines are often only fully visible from an opposing location such as a neighbouring mountain [19], pit level, shore, or even a boat [20]. The distance between the sensor and the target of interest can then easily exceed the close-range and extend to several kilometres. These distances not only lead to major atmospheric distortions, but also prevent the logistical setup of visible reference targets for radiometric correction as well as ground control points for image georeferencing. Additionally, owing to the much larger scale of the observed surface and the ground-based viewing perspective, pixels within one scene can represent a range of different distances and orientations, leading to highly variable radiometric distortions. For those reasons, correction methods established for nadir acquisitions are not applicable or need to be intensely modified to account for the special conditions of long-range ground-based sensing.

In this paper, we meet these additional challenges and present a novel workflow that allows the creation of fully corrected long-range ground-based hyperspectral image data for geological applications. In addition to sensor-induced geometric distortion corrections, the workflow now includes a new approach for the radiometric correction of long-range ground-based data as well as a topographic correction algorithm based on integration with 3D surface data using automatic matching algorithms. We also describe a detailed methodology for producing 3D hyperclouds, i.e., geometrically correct representations of the hyperspectral datacube, for the display of generated spectral mapping products. The methods presented will be included in the open source Mineral Exploration Python Hyperspectral Toolbox MEPHySTo [7]. We demonstrate the methodology in two areas that differ in geology, climate, and scientific objectives. The first area is located in an arctic environment, where two hyperspectral scans acquired from different points of view are used to detect and map mineralogical variations in the composition of the Mârmorilik Formation marbles in West Greenland. The single result map is integrated with photogrammetry data to provide spatial context and a 3D view that can be integrated into 3D modelling. The second dataset was acquired at the now-abandoned open pit mine Corta Atalaya near Minas de Riotinto, Spain. The Spanish dataset demonstrates the applicability of the corrected dataset for alteration zone mapping of a massive sulphide deposit under hot and dusty conditions as well as the integratability of datasets acquired at different times.

2. Areas of Investigation

2.1. Nunngarut Peninsula, Maarmorilik, Greenland

The first study area is located in central West Greenland, within the regions of Uummannaq Fjord and Karrat Isfjord (Figure 1). The investigated area covers large parts of the Nunngarut Peninsula at the Qaamarujuk fjord, where the former mining town of Maarmorilik is located. The nearby Black Angle Pb–Zn deposit is separated from the Nunngarut Peninsula by the smaller Affarlikassaa fjord. The study area belongs to the Mârmorilik Formation, a 1600 m thick carbonate-dominated rock sequence representing the southernmost stratigraphy of the Paleoproterozoic Karrat Group [21]. It was deposited between 2.1 and 1.9 Ga in an epicontinental marginal basin as platform carbonates [21], nonconformably overlies a suite of strong deformed Archean orthogneisses, and is overlain by flysch-type metasedimentary rocks of the Nûkavsak Formation [22].

The Mârmorilik Formation is dominated by dolomite-rich marbles in the lower part and calcite-rich marbles in the upper part. Locally, interbedded horizons of quartzites, tremolite-rich marbles and possible metamorphosed evaporites in the form of anhydrite occur [21,23]. The Black Angel Mississippi-Valley-Type (MVT) Pb–Zn deposit is emplaced within the Mârmorilik Formation [22,24], causing an overprint of the marbles by basal brines. The whole succession of Archean basement and the Karrat Group was strongly folded and thrusted by the Nagssugtoqidian–Rinkian orogenesis. During this orogenesis, the Mârmorilik Formation underwent at least three phases of deformation [19], leading to recrystallisation and metamorphism under high greenschist to amphibolite facies conditions [25]. The Mârmorilik Formation is interpreted to be the lateral equivalent to the Qaarsukassak Formation [26], and together they form a several hundred square kilometre large prospective region for zinc mineralisation [19,27].

2.2. Corta Atalaya, Riotinto, Spain

Corta Atalaya, near Minas de Riotinto in the province of Huelva (southern Spain), is, with a size of 1200 × 900 m and a maximal depth of 365 m, one of the most famous open pits of the Riotinto mining district (Figure 1). The Volcanogenic Massive Sulphide (VMS) mineralisation of Riotinto is associated with the Iberian Pyrite Belt (IPB), which is considered to host the largest concentration of massive sulphides in the Earth's crust [28]. The IPB is located in a north-vergent fold and thrust belt of late Variscan age [29] extending from east of Setubal, Portugal, to north of Seville, Spain, and has been extensively mined for copper, manganese, iron, and gold since the Bronze Age. At Riotinto, the lithostratigraphic succession can be divided into three units (from bottom to top): (i) phyllites and quartzites; (ii) slates, basalt sills, felsic volcanics (rhyolites and dacites); and (iii) the so-called Culm series (greywackes and slates). The stratabound, VMS lenses are located within felsic volcanics of Upper Devonian to Lower Carboniferous ages [28]. Zones of chloritic and argillitic alteration are associated with the massive sulphide mineralisation. Stockwork zones occur underneath the lenses in the vicinity of faults [28]. A gossan usually forms in the cap-rock above. The deposit of Riotinto itself is situated in the hinge of an E–W-trending anticline with an east-plunging fold axis. Corta Atalaya is located on the southern flank of this so-called Riotinto anticline. Stockwork and massive ore bodies are associated with E–W-striking thrusts. A set of later NW–SE-oriented transverse faults offsets the Riotinto anticline. The most prominent of these faults, the Falla Eduardo, displaces the massive sulphide body San Dionisio about 150 m to the south and finds its continuation in the Filón Sur ore body east of Corta Atalaya [28]. The massive sulphide body San Dionisio, which was exploited in Corta Atalaya, originally had reserves of 100 million tonnes. Originally, the mine was dedicated to the extraction of iron and copper sulphides (mainly pyrite with smaller amounts of chalcopyrite).

The initial objective was to extract copper from copper sulphides, but, subsequently, the sulphur contained in pyrite was used for the manufacturing of sulphuric acid until final closure of the open pit in 1991 [28].

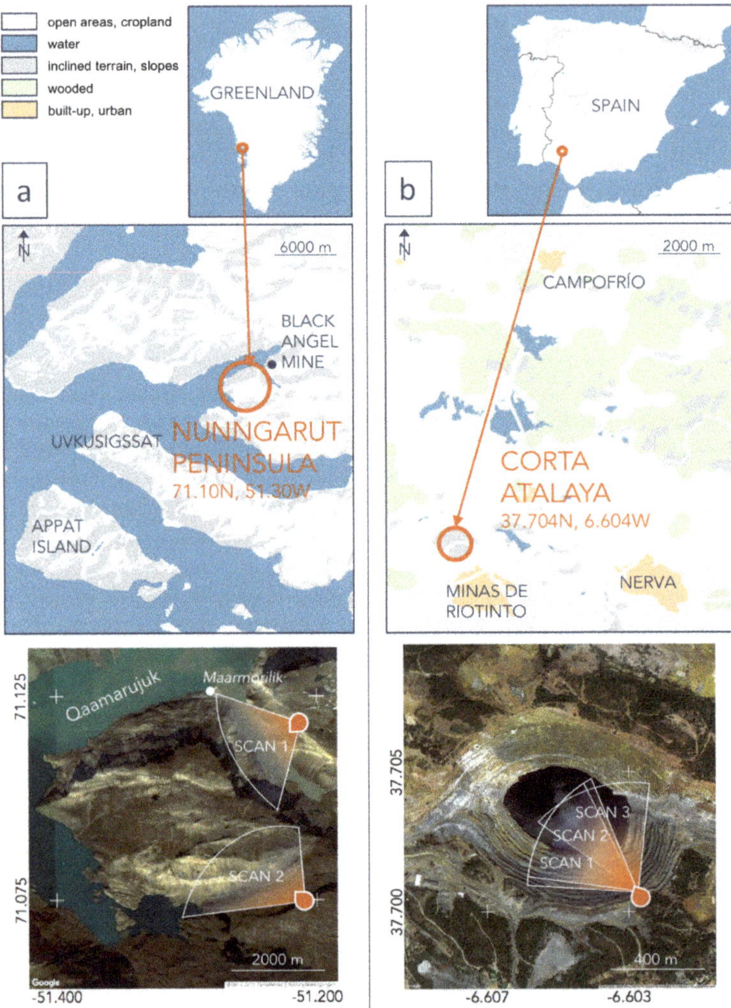

Figure 1. Location of the two investigated sites and schematic coverage of the acquired AisaFENIX hyperspectral imagery at: (**a**) Nunngarut Peninsula, Maarmorilik, Greenland; and (**b**) Corta Atalaya open pit, Minas de Rio Tinto, Spain.

3. Data Acquisition

3.1. Hyperspectral Imagery

The hyperspectral image (HSI) data was acquired using a SPECIM AisaFENIX push-broom scanner. The scanner has 384 swath pixels with 624 spectral bands each, covering the visible and near-infrared (VNIR) to short-wave infrared (SWIR) range between 380 and 2500 nm. The spectral resolution (Full Width at Half Maximum—FWHM) varies between 3.5 nm for the VNIR and 12 nm in the SWIR at a spectral sampling distance of about 1.5 nm (VNIR) and 5 nm (SWIR), respectively. By mounting the instrument on a rotary stage, a continuous hyperspectral image with a vertical field of view (FOV) of 32.3° and a maximum scanning angle of 130° could be acquired in one measurement. During the measurements, the GPS position of the camera, acquisition time, and general viewing direction (from here on referred to as 'camera angle') of the scan were recorded. A Spectralon SRS-99 white panel was set up near the camera within the FOV and with a similar general orientation as the imaged outcrop.

3.2. Photogrammetry Data/3D Data

Images for reconstruction of surface geometry were recorded using precalibrated RGB and hyperspectral cameras. In the case of Maarmorilik, a Nikon D800E with a 35 mm 1.4 Zeiss lens was used from a helicopter. The 3D pointcloud of Corta Atalaya was based on fusion of drone-borne images from a Rikola Hyperspectral Imager (red band) and a Canon EOS M with EF-M 22 mm f/2 STM lens (as grey-scale image). Camera positions were obtained from an attached GPS device, whereas the imaging geometry was reconstructed using a Structure from Motion (SfM) and MultiView Stereo (MVS) workflow. Prior to the photogrammetry workflow, image distortions were removed.

3.3. Validation Sampling

Samples of the main lithologies were taken for a validation of the correction workflow and of the mineral mapping results. Sample locations were recorded using a handheld GPS device. Spectra of representative fresh and altered rock surfaces were acquired in situ using a portable Spectral Evolution PSR-3500 spectro-radiometer using a contact probe (8 mm spot size) with an internal, artificial light source. Its spectral resolution is 3.5 nm (1.5 nm sampling interval) in VNIR and 7 nm (2.5 nm sampling interval) in the SWIR, resulting in 1024 channels in the spectral range from 350 to 2500 nm. Radiance values were converted to reflectance using a calibrated PTFE panel with >99% reflectance in VNIR and >95% in SWIR (either Spectralon SRS-99 or Zenith Polymer). Each spectral record consisted of 10 individual measurements, which were taken consecutively and then averaged.

4. Processing Workflow

4.1. Preprocessing of Hyperspectral Raw Data

The acquired raw hyperspectral datasets are first converted to At-Sensor-Radiance using dark-current subtraction followed by image normalisation and multiplication of sensor- and band-specific radiometric calibration data (Figure 2). In a second step, two geometric corrections of sensor-specific optical distortions need to be applied. The first effect is a distortion along the FOV comparable to the distortion of fish-eye lenses. This leads to an increasing shortening of the image from the centre to the upper and lower image boundaries. The second effect can be described as slit bending and refers to a curved recording of the currently scanned (straight) line. Both effects can be removed by applying correction values for each pixel in the FOV. The required parameters are included in a lookup table provided by the manufacturer of the sensor. In the case that several scans of the same scene have been acquired with the same settings, a stacking and averaging of those scenes can be performed at this point. By image stacking, the signal-to-noise ratio can be increased, reducing possible temporal illumination variations due to changing cloud cover.

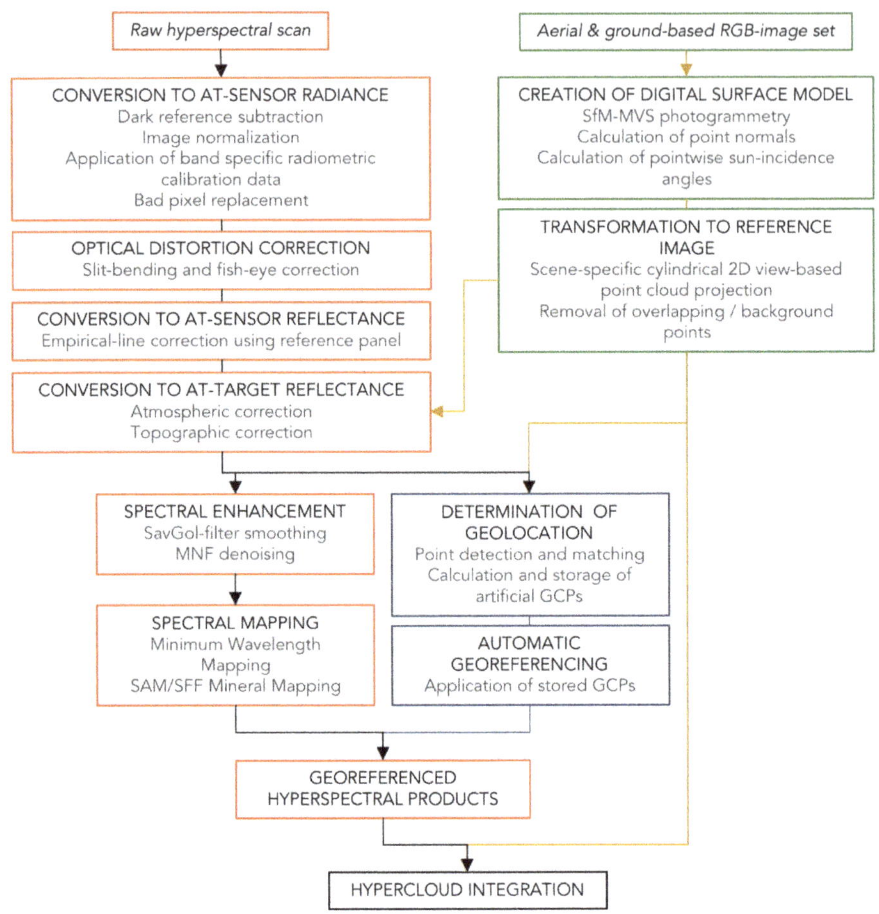

Figure 2. Schematic workflow for the correction, processing, and 3D integration of long-range ground-based hyperspectral imagery.

4.2. Radiometric Correction of Hyperspectral Radiance Data

Subsequent to the transformation of the raw hyperspectral data into radiance, a conversion to at-sensor reflectance needs to be applied, which can be achieved using a white reference panel placed near the sensor. This Spectralon (SRS-99) reference target is close to an ideal Lambertian reflector with >99% reflectance in the VNIR and >95% in the SWIR. Its exact reflectance spectrum is known and can be used for an empirical line correction of the radiance data. Hereby, a linear regression between the image radiance values and the reference reflectance values is calculated and applied for each band.

Depending on the imaging distance and the climatic conditions, the resulting at-sensor reflectance image may still feature atmospheric distortions (see Figure 3). In contrast to air- or spaceborne data, the scene-specific intermediate atmospheric layer can be assumed to have a uniform composition with only negligible variations. Nevertheless, the amount of atmospheric influence varies for each pixel and depends mainly on the distance between sensor and target, but can be also influenced by local variations, e.g., differing intensities of upwelling water vapour.

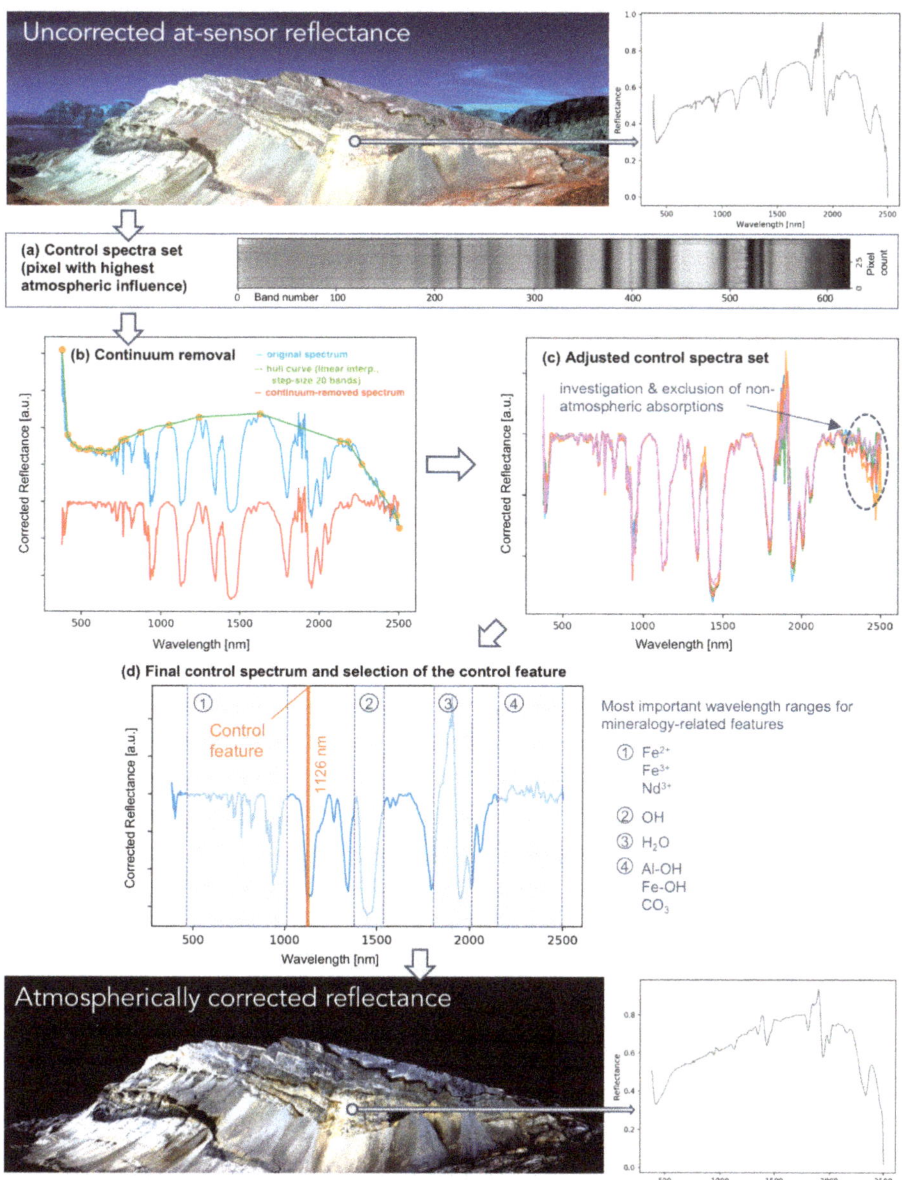

Figure 3. Atmospheric correction workflow on the example of the Maarmorilik marble cliffs (Nunngarut, Scan 2). Hyperspectral images are displayed using spectral true colour representative bands (R: 640 nm G: 550 nm B: 470 nm). See text for a detailed description. (**a**) Control spectra set; (**b**) continuum removal; (**c**) adjusted control spectra set; (**d**) final control spectrum and selection of the control feature.

Given these circumstances, we attempt to perform a radiometric correction to remove atmospheric distortions using a single atmospheric correction spectrum for each scene. The intensity of correction needs to be varied according to the amount of atmospheric distortion. For the correction approach to be robust and independent from additional parameters or knowledge about the composition of the influencing atmospheric layer, the atmospheric correction spectrum is derived directly and automatically from the hyperspectral image itself. Hereby, the correction spectrum is a comprehensive representation of all scene-abundant spectrally influencing atmospheric components, which may encompass atmospheric dust, water vapour, and other atmospheric gases. The correction spectrum is neither selective nor restricted to defined components and is thus applicable for any atmospheric setting.

Owing to the assumed constant composition of the atmosphere over the scene, the depths of all atmosphere-related features should change equally if the atmospheric influence is altered. This approach allows us to evaluate the amount of atmospheric influence for each pixel by the depth of only one atmospheric absorption feature and eliminates the need for atmospheric models, additional calibration targets, and distance measurements. The now-called control feature must necessarily be both common in all possibly occurring atmospheric compositions and strong enough to be detectable even for low atmospheric influence. Additionally, it should not overlap with any characteristic mineralogy-related features to avoid interference and miscorrections. The absorption band we found to fulfill these conditions best is situated at 1126 nm (Figure 3d) and is related to atmospheric water vapour [14].

The atmospheric correction workflow consists of several steps, which can also be retraced in Figure 3:

1. Masking of sky-related pixels: All image pixels representing sky and sky reflected by mirroring surfaces such as water are masked out automatically from the reflectance image using a ratio between the image bands located at 410 and 890 nm. These wavelength positions are set to encompass two ends of the extreme decline in VNIR reflectance that is specific for sky-related spectra. This characteristic shape leads to a usually very distinct ratio difference between sky and non-sky pixels. In our examples, the masking threshold was most successful in a ratio range between 1.0 and 2.0.
2. Determination and processing of possible correction spectra: The depth of the control feature at 1126 nm is calculated for all remaining pixels. All pixel spectra with a control feature depth within 80–100% of the maximum are extracted as a control spectrum set (Figure 3a), which will be used to determine the final atmospheric correction spectrum. A continuum removal and an equalisation of the control feature depth are applied on each spectrum of the control set separately. The respective continuum hull is calculated using a linear interpolation of stepwise acquired maxima all over the respective spectrum (Figure 3b). The moving window for the continuum hull calculation can either be set to a fixed step size or restricted to specific stored wavelength ranges that are located outside or at the edge of known atmospheric absorption windows.
3. Exclusion of nonatmospheric features: Some spectra of the resulting equalised control spectra set may still contain additional nonatmospheric absorptions. These features should be excluded from the correction spectrum to avoid a weakening or deletion of important mineralogical features during the atmospheric correction process. In contrast to atmospheric features, nonatmospheric absorptions occur with differing intensities and only in a spectral subset of the control spectra (Figure 3c,d). They can be excluded from the control spectrum set by maintaining only the highest of all spectral values for each wavelength. The used threshold can be varied manually if needed.
4. Calculation and application of the final control spectrum: The remaining spectral information is averaged for each wavelength to reduce possible noise. The outcome of the whole procedure provides a single continuum-removed correction spectrum containing solely the characteristic atmospheric contribution of the analysed hyperspectral image (Figure 3d). The atmospheric correction itself is performed pixelwise. For each pixel, the intensity of the correction spectrum needs to be adjusted to both depth and reflectance value of the control feature in the pixel spectrum. The correction itself is then achieved by a simple division of the pixel spectrum by the

adjusted correction spectrum. The original reflectance intensities are maintained in the corrected image spectra during that process.

The processing time for the automatic correction of a hyperspectral scan with the spatial and spectral dimensions as in our examples is less than one minute. Thus, the method is extremely time- and effort-saving and can be easily integrated into a batch-processing workflow.

Depending on the Signal-to-Noise ratio (SNR) of the processed dataset, a subsequent Minimum Noise Fraction (MNF) smoothing can be advantageous. MNF smoothing entails a transformation of the image into MNF space, a rejection of bands with low SNR, and a subsequent back-transformation into the original image space [30]. The number of MNF bands to be rejected can be determined by looking at the eigenvalue function of the calculated MNF bands, which reaches a plateau after a sharp increase and suggests a rejection if the asymptotic eigenvalue function approaches a linear function [31].

4.3. SfM-MVS Photogrammetry

The Digital Surface Model is derived from aerial and ground-based images using the Structure-from-Motion MultiView Stereo (SfM-MVS) algorithms in Agisoft Photoscan Professional 1.2.5. SfM-MVS is a low-cost, user-friendly workflow combining photogrammetric techniques, 3D computer vision, and conventional surveying techniques. It solves the equations for camera pose and scene geometry automatically using a highly redundant bundle adjustment [32,33]. A typical SfM-MVS workflow towards a final surface model consists of the following eight steps [33,34]:

1. Detection of characteristic image points;
2. Automatic point matching using a homologous transformation;
3. Keypoint filtering—this step is crucial for model accuracy and validation of later results [35];
4. Iterative bundle adjustment to reconstruct the image acquisition geometry and internal camera parameters;
5. Scaling and georeferencing of the intrinsic coordinate system to available reference points (GCPs) or camera coordinates and optimisation of the resulting sparse cloud;
6. Applying MultiView Stereo algorithms (dense matching) to compute the dense cloud—the resulting dense cloud is the basis for the geometric correction of the hyperspectral data;
7. Interpolation of the dense cloud by, e.g., Meshing or Inverse Distance Weighting (IDW), to retrieve a Digital Surface Model (DSM);
8. Texturising of the 3D model.

4.4. Calculation of Sun Incidence Angles for Topographic Correction

Knowledge of the sun incidence angle for each pixel of the hyperspectral image is crucial for its topographic correction. In contrast to nadir data, vertical outcrop scans can have multiple pixels located at any given latitude/longitude coordinate position, which can be only spatially differentiated by their elevation values. Therefore, common tools for the calculation of slope, aspect, and sun incidence angle of Digital Elevation Models (DEM) cannot be applied here. Instead, we calculate the sun incidence angle for each individual point of the point cloud generated in Section 4.3 as the angle between the point normal and the sun vector (Figure 4a). The point normals were either calculated during the point cloud construction or can be computed retroactively using a triangulation of neighboring points. The sun vector is characterised by

$$sunvec = \begin{pmatrix} \cos(SE) * \sin(AZ) \\ \cos(SE) * \cos(AZ) \\ \sin(AZ) \end{pmatrix} \tag{1}$$

with *SE* being the sun elevation angle and *AZ* the sun azimuth at the given date, time, and position of the acquisition. The calculated sun incidence angles are stored as additional point properties in the point cloud file and retained in all following processing steps.

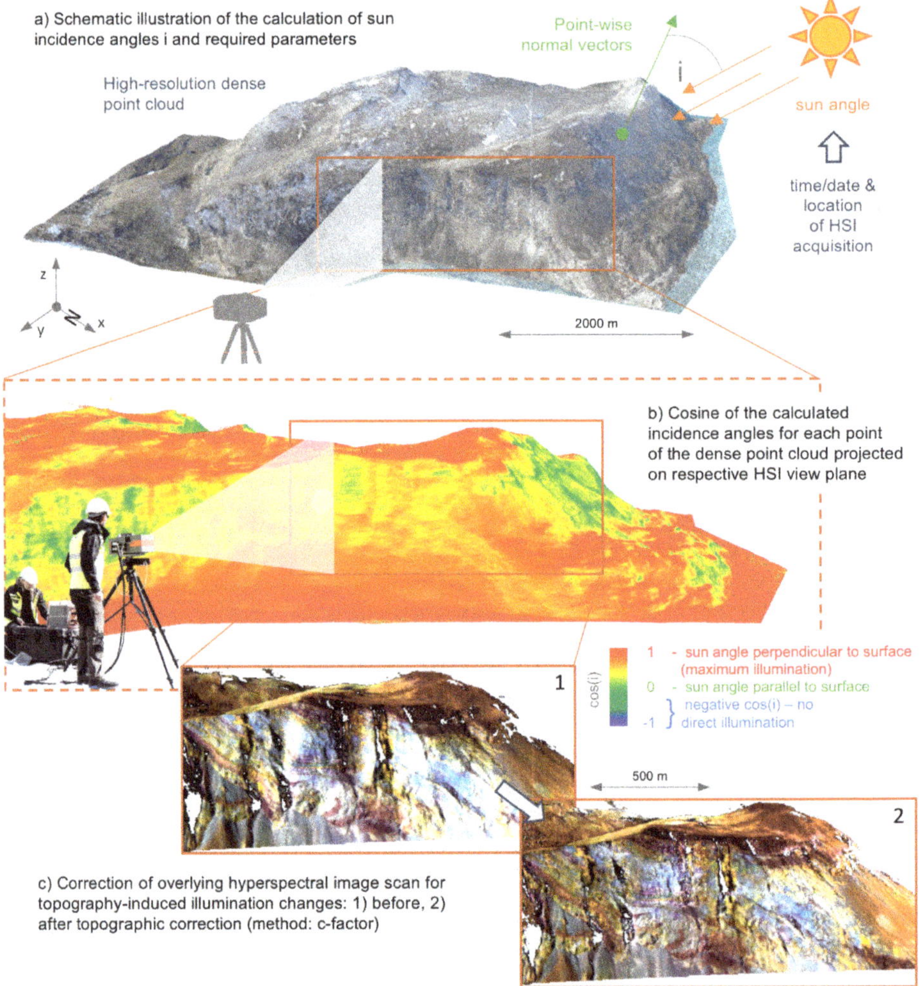

Figure 4. Topographic correction of vertical hyperspectral image (HSI) (Nunngarut, Scan 1). (**a**) Schematic illustration of the calculation of sun incidence angles i and required parameters; (**b**) cosine of the calculated incidence angles for each point of the dense point cloud projected on respective HSI view plane; (**c**) correction of overlying hyperspectral image scan for topography-induced illumination changes: (1) before, (2) after topographic correction (method: c-factor).

4.5. Projection of Pointcloud and HSI Matching

An integration of 2D hyperspectral data and 3D point cloud data is needed for topographic correction and final creation of the 3D hypercloud. In order to facilitate automatic matching and reduce distortion in the subsequent wrapping process, the point cloud is projected onto a 2D surface in a way that resembles the view of the hyperspectral camera during image acquisition. It is crucial here that through the entire process of ensuing transformations the original coordinates of each point of the cloud are stored as additional parameters. Due to the push-broom character of the sensor, a simple orthographic projection of the point cloud onto a plane is not suitable. Instead, the point cloud is first transformed so that the camera position is set as the new origin and the camera viewing angle is set along the y-axis of the coordinate system by

$$Transformed\ points = Original\ points - Camera\ Position * (-Camera\ Angle). \tag{2}$$

The spatial relation between point cloud, camera angle, and camera position in the transformed coordinate system is displayed in Figure 5.

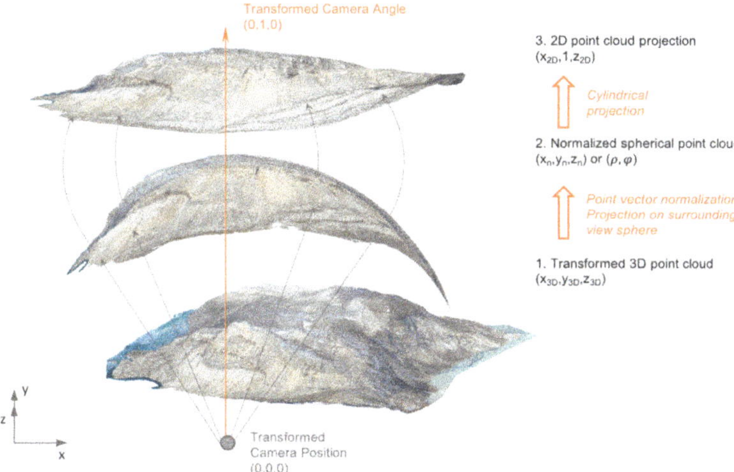

Figure 5. Schematic workflow of the point cloud transformation and projection to create a 2D image resembling the panoramic view of a push-broom hyperspectral imager (Nunngarut, Scan 2).

Each point coordinate of the transformed point cloud now corresponds to the vector $\vec{v}$ between the transformed camera position at (0,0,0) and the point at (x_{3D}, y_{3D}, z_{3D}). If we assume that the camera FOV is a subset of a virtual surrounding view sphere with the center at the camera position, the point cloud can be projected onto that sphere by normalizing each point vector by

$$\begin{gathered} (x_n, y_n, z_n) = \frac{\vec{v}}{|\vec{v}|} \\ \text{with } \vec{v} = \begin{pmatrix} x_{3D} \\ y_{3D} \\ z_{3D} \end{pmatrix}; \end{gathered} \tag{3}$$

see also Figure 5b.

The projected point cloud is now unfolded onto a 2D plane using a cylindrical projection with

$$\begin{gathered} x_{2D} = \rho \text{ with } \rho = \tan^{-1}(y_n/x_n), \\ y_{2D} = 1, \\ z_{2D} = \tan\varphi \text{ with } \varphi = \pi/2 - \tan^{-1}\left(\sqrt{x_n{}^2 + y_n{}^2}/z_n\right), \end{gathered} \tag{4}$$

with x_{2D} and y_{2D} being the Cartesian coordinates of the created 2D image, and with x_n, y_n, and z_n or ρ and φ being the Cartesian or spherical coordinates of the normalised 3D point cloud, respectively (Figure 5c). The angle at which the cylinder is cut for the projection can be set by an additional parameter.

The projection into 2D space considers all of the points in the true line of sight of the hyperspectral camera, which includes points hidden behind points in the foreground (front points), such as the backside of a mountain (back points). This leads to artefacts within the created 2D image (see Figure 6a) and would adversely affect subsequent processing steps. Using a maximum threshold for the original spatial distance between neighbouring points, the adverse back points can be removed. To ensure a fast processing even for huge point clouds, a moving window is used to process several points at once. For each applied window, the contained point with the closest distance to the camera position is found. This distance can be calculated from the original coordination of the point cloud, which is still saved as additional point parameters. Hereby, it is advantageous to use only the original coordination axis that was closest to the original camera angle. While neighbouring front points show a similar location with generally from decimetres to a few metres difference (depending on the spatial accuracy of the data), back points mostly feature locations far off, with distances of several tens to hundreds of metres from the camera-closest front point. According to this, the threshold is set and all resulting back points are deleted (Figure 6b). Due to the nature of this workflow, a smaller window size guarantees a higher accuracy, but also a higher computation time.

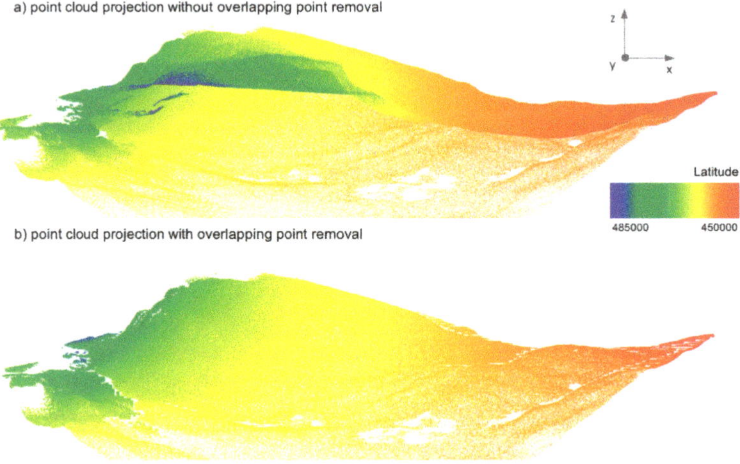

Figure 6. Effect of the overlapping point removal on the quality of the 2D point cloud projection image on the example of Nunngarut, Scan 2. The original x-coordination of the points is illustrated by a colour gradient. (**a**) Point cloud projection without overlapping point removal; (**b**) point cloud projection with overlapping point removal.

After the deletion of the interfering back points, the remaining front points are interpolated into a raster with a spatial resolution similar to or slightly higher than the spatial resolution of the hyperspectral data. Apart from RGB colour information, this ortho-image has four additional bands containing the original point cloud coordinates and the calculated sun incidence angles. The created RGB raster can now be used for an automatic co-registration of the hyperspectral image. The matching workflow used for the co-registration will be part of the MEPHySTo toolbox presented in Jakob et al. [7] and is also successfully adapted and used for the integration of vessel-based hyperspectral data and 3D point clouds in an accompanying paper [20]. The workflow is based on the SIFT (Scale-invariant feature transform) algorithm [36], which, from both images, extracts local features or keypoints that are invariant to translation, rotation, and scale and partly invariant to affine or 3D projection and illumination changes. Using the FLANN (Fast Library for Approximate Nearest Neighbors) matching algorithm library [37], correlating point pairs between both keypoint sets are found. The best-matching point pairs are used as control points for a polynomial warping of the hyperspectral image to fit on the RGB raster. After the co-registration, each overlapping point of both datasets features high-resolution spectral data, geographic position, and elevation, as well as the sun incidence angle at the time of the acquisition.

4.6. Topographic Correction of Referenced HSI

The topographic correction is similar to the approach described in Jakob et al. [7]. The main difference is the calculation of pixel-specific sun incidence angles, which is described above in Section 4.4. The calculated angles can now be used to apply a topographic correction algorithm. The c-factor method returned the best correction results of all the methods implemented in the toolbox and achieved a very smooth and accurate correction even for high illumination differences (see Figure 4c). The topographically corrected image is calculated by

$$ref_c = ref_o * \frac{\cos(z) + c}{IL + c} \tag{5}$$

where c is a/m from the linear regression of $ref_o = a + m * IL$ and $IL = cos(i)$ [38]. The *c-factor* approach is applied separately for each spectral band. The correction of a common hyperspectral scan usually takes less than a minute. For very dark and deeply shaded regions of the image, pixels can be heavily overcorrected. These pixels are characterised by extreme, up to infinite values, which exceed the common value range of reflectance data distinctly. The affected pixels are detected and masked using appropriate thresholds, which are set according to the spectral reflectance minimum and maximum of the topographically uncorrected image (e.g., 0 and 1).

4.7. Minimum Wavelength Mapping

The finally corrected HSI can now be used for subsequent mapping and interpretation. In the present paper, a Minimum Wavelength (MWL) mapping approach is exemplarily used to test the quality and applicability of the data for mineral mapping.

MWL mapping using the Wavelength Mapper [39,40] aims to estimate the position of the deepest absorption feature in a given wavelength range. The position of the absorption minimum is a key to link surface mineralogy to subtle variations in mineral composition (e.g., shift of the Al–OH feature depending on the coordination of the Al). First, a hull curve is calculated and divided from the spectra. Second, position and depth of the most prominent absorption are computed using a second-order polynomial function. These two parameters can be used to create MWL position maps, where the position of the investigated feature is displayed by a colour change, while the colour intensity is controlled by the absorption depth.

The success of the MWL mapping approach depends crucially on the analysis of subtle changes of position and depth of mostly small mineralogical absorption features. Therefore, it is an excellent possibility to evaluate image correction methods, which affect both the intensity ratio between single pixels of the image (topographic correction) and the shape of the spectrum itself (radiometric and atmospheric correction). In this context, the successful removal of distortions is as important as maintaining existing and real intensity relations and spectral features.

4.8. Generation of Hyperclouds

At the end of the workflow described above, each pixel of the HSI (and any HSI mapping product) has an assigned geographic position and elevation through the corresponding pixel in the projected and rasterised 2D point cloud. By deriving this information for each pixel of the spectral raster, we can create a so-called "hypercloud", which visualises the spectral data as a 3D point cloud. The displayed data can comprise any spectral data or result, such as simple reflectance data, results from decorrelation, and endmember mapping methods, or MWL mapping results as presented here. The hypercloud can be displayed and processed further with respective 3D software such as CloudCompare (open-source GPL software, retrievable from http://www.cloudcompare.org/) or SKUA-GOCAD (Emerson/Paradigm, Houston, United States). If the hyperspectral survey consisted of several scans covering different parts of the observed area, the creation of hyperclouds can be an excellent option to set the single mapping results into a spatial context by simultaneously displaying or merging multiple hyperclouds. The 3D hypercloud also allows for integration with other spatial datasets such as boreholes or structural observations.

5. Results

5.1. Nunngarut Peninsula, Maarmorillik, Greenland

Two hyperspectral scans were acquired from two different scanning locations, covering the largest part of the south and east coast of the Nunngarut Peninsula (Figure 1a). The approximate distance between sensor and observed target ranged between 2 and 5 km for the majority of all outcrop-related image pixels. Despite overall dry and sunny conditions during acquisition, numerous sharp atmospheric absorption features within the spectral data (see Figures 3 and 7) suggested a high influence of the atmospheric layer between the sensor and the target. Figure 7 displays the known major atmospheric contributions (in this case water vapour, CO_2, O_2, and O_3) to the overall observed atmospheric perturbances and the resulting calculated spectrum used for the corrections. We showcase that the radiometric correction approach presented here allows us to remove the influence of the atmosphere almost completely, whereas typical mineral-related spectral features of the Mârmorilik Formation remain. In the resulting atmospherically corrected target spectrum, the remaining absorption features are indubitably attributable to characteristic mineral features. Besides the distinct carbonate feature of the Mârmorilik marbles, the characteristic AlOH and OH/H_2O features are clearly represented. These characteristic absorptions are related either to abundant evaporitic gypsum and/or clay minerals originating from inclusions or nearby pelite horizons known to be present in this lithological unit.

Scan 1, imaging the south facing cliff of the Nunngarut Peninsula, was directly opposed to the sun during the measurements and is therefore evenly illuminated. In contrast, Scan 2, acquired in the morning and facing the eastern coast of the peninsula, featured high illumination differences, which made a topographic correction crucial for the subsequent mapping process (Figure 4c).

With atmospheric and topographic corrections successfully applied to the hyperspectral datacubes, the datacubes provide the basis for a characterisation of the mineralogical composition of the Mârmorilik Formation carbonates, with relevance for exploration mapping. The identification of different carbonates from hyperspectral data is possible using the position and depth of the carbonate-related vibrational overtone absorption band between 2310 and 2340 nm [41]. Whereas pure calcite features an absorption around 2340 nm, the absorption band of pure dolomite occurs at 2320 nm. Carbonate-related absorptions at even shorter wavelengths can indicate an occurrence of tremolite together with dolomite. This relationship is confirmed by spectroscopic analysis of representative rock samples from the Mârmorilik Formation (Figure 8a). Elemental and mineralogical composition of the samples are further validated by pXRF (portable X-ray fluorescence) and thin section analysis, respectively (see Rosa et al. [19]; pers. commun. C.A. Partin). From the pXRF results, the respective Ca/Mg ratios of four to six measurement spots on each sample were calculated and compared to the classification of limestones and dolomites of Chilingar [42]. Sample #SLA15 featured high Ca/Mg ratios between 31.2 ± 0.7 and 619.3 ± 13.7 and would be therefore classified as calcitic limestone. The ratio of sample #562032 ranged between 2.0 ± 0.5 and 5.9 ± 0.9, indicating a highly dolomitic limestone or calcareous dolomite. Sample #562048 ranges between a dolomite and magnesian dolomite with a low Ca/Mg ratio between 1.0 ± 0.1 and 2.0 ± 0.1 [42]. A simple MWL mapping approach hence provides a good means of distiguishing these different carbonate phases in the outcrop (Figure 9). Pelite horizons and noncarbonatitic rocks, which are spectrally characterised by a very weak or nonexistent carbonate features, were masked out using a threshold based on the MWL depth of the mapped carbonate feature.

The contact between the upper and lower Mârmorilik Formation is clearly visible on the east-facing slope of Nunngarut, as the lower Mârmorilik Formation is dominated by dolomite interbedded with tremolite-rich horizons [43], whereas the upper Mârmorilik Formation is calcite-dominated. Also, a dolomitisation along faults can be traced.

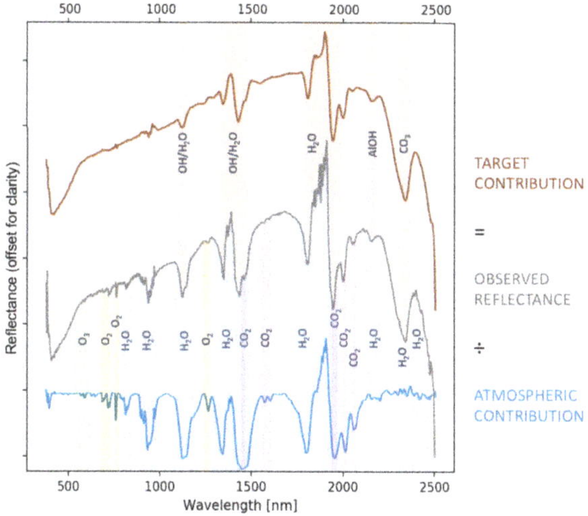

Figure 7. Contribution of geological target and atmosphere to an exemplaric observed reflectance spectrum (Nunngarut study area, Mârmorilik Formation). At this, the target contribution equals the reflectance spectrum after atmospheric correction.

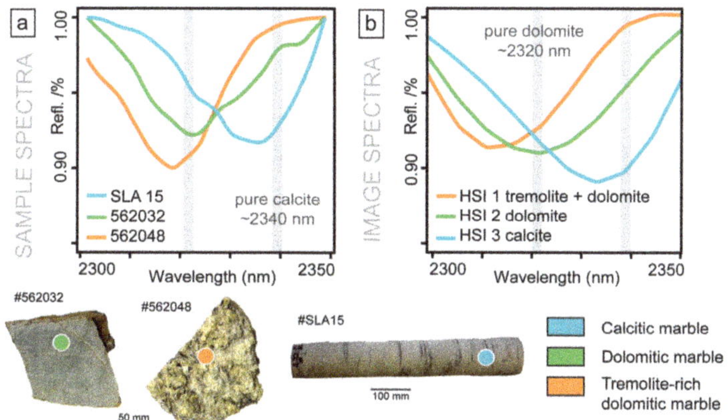

Figure 8. Spectral validation of the Minimum-Wavelength-Position-based mapping of the carbonate composition at Nunngarut test site. (**a**) Lab point spectra of three carbonate samples of the Maarmorilik formation, representing typical calcitic, dolomitic, and tremolite-rich dolomitic end members; (**b**) HSI spectral plot of the sampling positions marked in Figure 9, representing calcite-, dolomite-, and tremolite-rich dolomitic end members of the scene. A continuum removal was applied on all spectra. Elemental and mineralogical composition is further validated by portable XRF (pXRF) and thin section analysis, respectively (see Rosa et al. [19]).

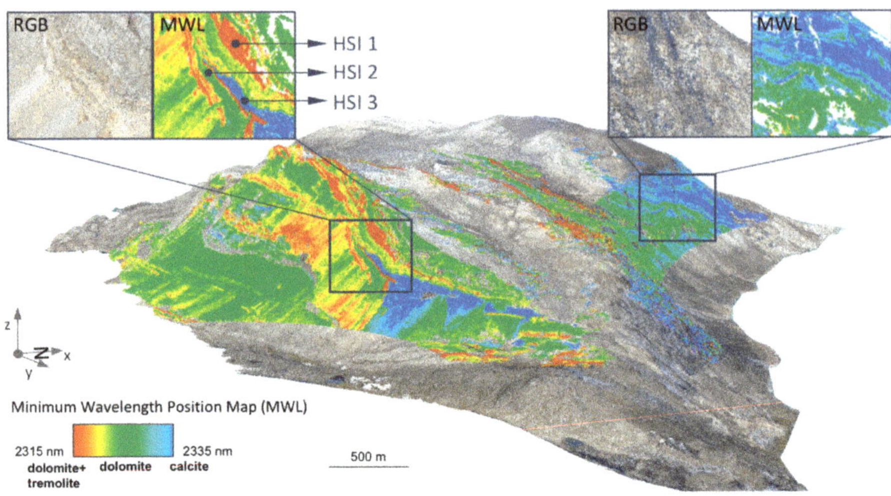

Figure 9. 3D hypercloud of two individual HSI image scenes overlain on photogrammetric RGB point cloud of the Maarmorillik marble cliffs. Minimum Wavelength Position Mapping was applied to both HSI datacubes to highlight variations in carbonate composition. HSI 1, 2, and 3 mark the sampling points of Figure 8.

5.2. Corta Atalaya, Riotinto, Spain

For the Corta Atalaya, three overlapping hyperspectral scans are used to demonstrate the described workflow (Figure 1b). The scans were acquired from the same panorama viewpoint of Corta Atalaya, but at different times: Scan 1 was acquired in March 2016, and Scans 2 and 3 were acquired in October 2016. The distance between sensor and target ranges broadly between 400 and 1100 m. The conditions on both acquisition days were dry and sunny, with a very good and constant illumination of the imaged pit wall. Despite the shorter distance to the target compared with that at Nunngarut test site and the Mediterranean climate conditions, i.e., with hot and dry summers, distinct atmospheric absorption features were observed in the image data.

All scans were atmospherically corrected and geometrically rectified using the photogrammetric pointcloud. A topographic correction was attempted but deemed unnecessary in the end, because the geologically most interesting northern and eastern part of the outcrop are evenly illuminated, and the shaded southern wall of the pit does not contain sufficient spectral information. After preprocessing and correction of the scenes, a Minimum Wavelength Position Mapping of the AlOH feature between 2190 and 2215 nm was conducted on all three scenes, to exemplarily show the capability of the corrected datasets for alteration mapping. The subsequently created hyperclouds show a great coincidence in the mapped alteration zones and could be easily merged into one final Hypercloud AlOH map (Figure 10).

The spectral validation of the mapping result was conducted using a set of field spectrometer data acquired in situ. Due to the restricted accessibility of the mine pit, the spectral readings are limited to a few pit levels. However, a wide range of lithologies could be covered and compared to the respective HSI pixel spectrum. A selection is shown in Figure 11a and proves the similarity of spectral shape and the occurrence of spectral features between image and field spectra. The given field sample density allows also us to validate the AlOH MWL position distribution. In Figure 11b, the AlOH feature position of each field spectrometer measurement within the main region of interest is displayed as coloured squares using the same colour scale as the underlying HSI mapping result.

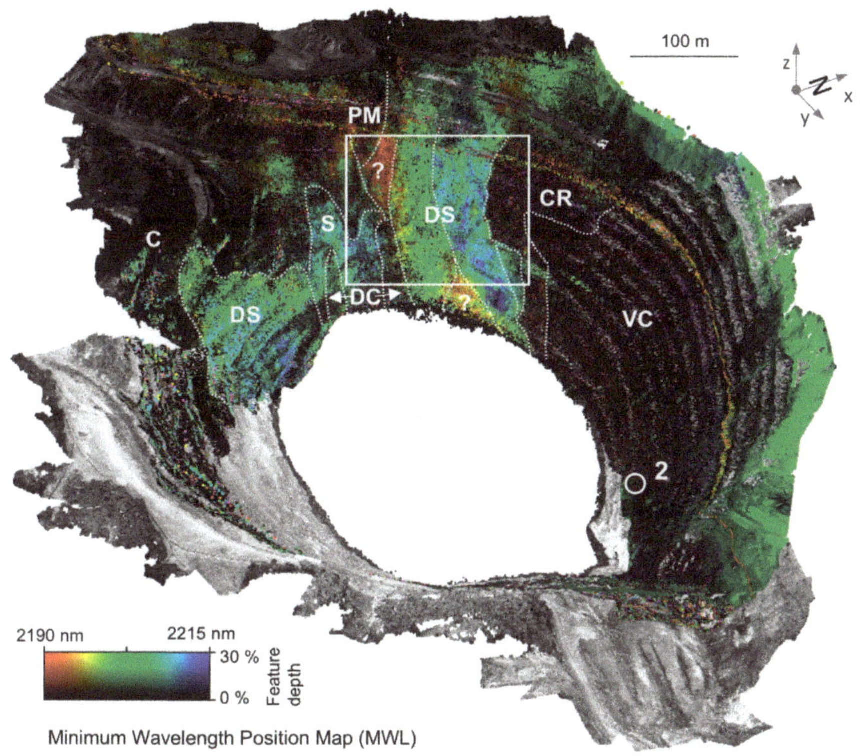

Main lithologies and alteration zones after Sáez & Donaire 2008:

DC	Dacites with chloritic alteration	CR	Conglomerates	C Culm
DS	Dacites with sericitic alteration	S	Massive sulfides	
VC	(Sub-)volcanics with chloritic alteration	PM	Pink shales/ Pizarras Moradas	

Figure 10. 3D hypercloud display of three individual HSI image scenes overlain on photogrammetric RGB point cloud of the Corta Atalaya open pit. All three scenes were used for Minimum Wavelength Position Mapping to highlight lithological variations associated with differences in the abundance of AlOH-bearing minerals. The white rectangle marks the area shown in Figure 11. The colour differences in the MWL hypercloud show excellent correlation with the known main lithologies and alteration zones [28]. Zones not described in [28] are indicated with question marks. Sample locations for Figure 11b are marked with white circles and numbers.

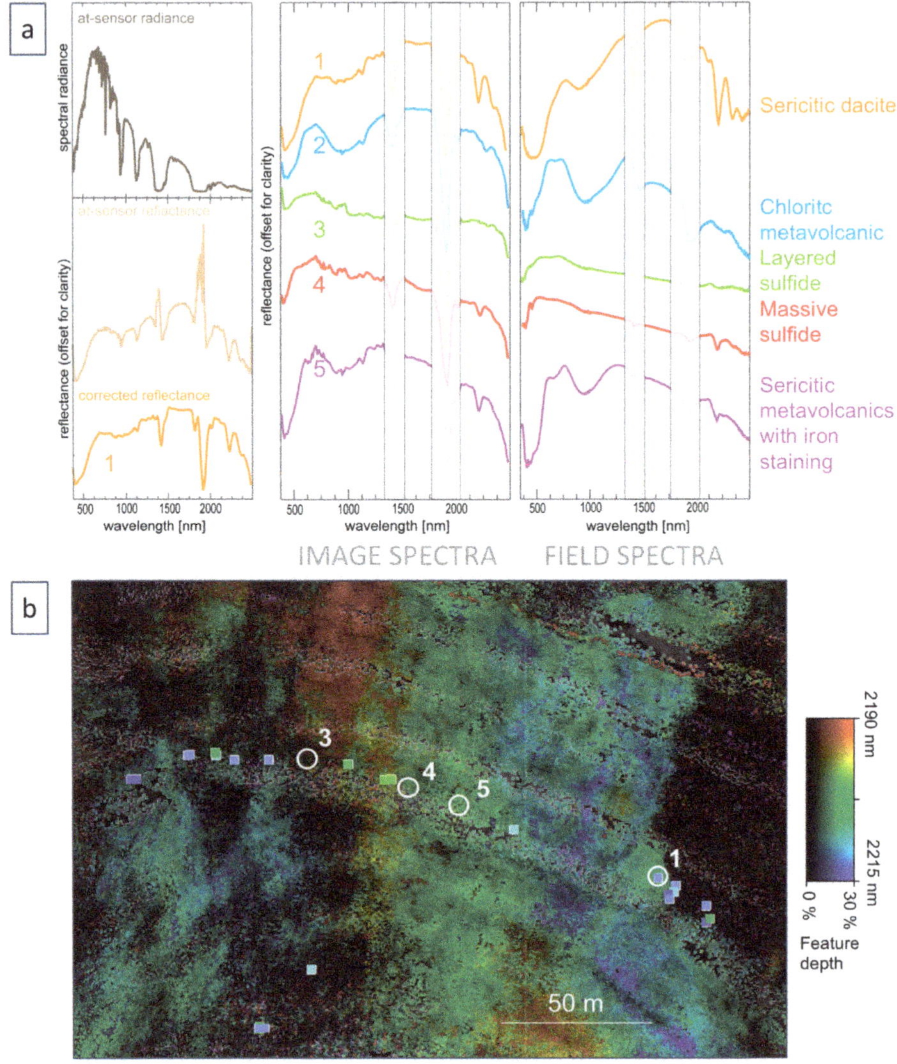

Figure 11. Validation of HSI data of the Corta Atalaya open pit. (**a**) Left: Spectral signature improvement of Sample Point 1 within different processing stages. Right: Comparison of spectral shape between field spectra and image spectra of the approximate same location. Sample locations are marked with white circles and numbers in Figures 10 and 11b; (**b**) Comparison of feature position: minimum wavelength map for AlOH (see map extent in Figure 10) and feature position of field spectra (coloured squares; same colouring scheme).

6. Discussion

6.1. Radiometric and Atmospheric Correction

Both test scenarios contain spectral distortions due to atmospheric absorption features. At Corta Atalaya/Spain, most of the observed atmospheric absorption features could originate both from upwelling water vapour of the pit lake and from dust and particles caused by the nearby mining activities in the adjacent Cerro Colorado open pit. This assumption is supported by the distribution of the atmospherically disturbed image pixels, which are not directly related to the distance of target and sensor, but mainly occur in areas where the signal needed to pass over the water surface in the mining pit. In contrast, for the Greenland site, the intensity of the atmospheric absorptions was roughly proportional to the distance between sensor and target. Here, contributions both from general air humidity and from upwelling water vapour from the fjords separating Nunngarut Peninsula and the respective observation positions on adjacent cliffs can be assumed. The overall atmospheric influence on the signal was much higher than that at Corta Atalaya, which may be related to both the distinctly increased distance to the target and the generally higher air humidity of the arctic climate. The described novel atmospheric correction workflow takes into account this variability in the composition of the atmospheric layer between sensor and target by extracting the shape of the correction curve directly from the scene and determining the correction intensity according to the pixel-specific atmospheric absorption depth and not the distance to the target.

For all five processed datasets, the atmospheric correction approach was fast and robust. Atmospheric absorptions were removed, whereas the general spectral shape and smaller mineral-related features were maintained. It was shown that the correction approach respects all abundant atmospheric components that contribute to the extracted pervasive signal and which we attribute to atmospheric perturbations. Besides water vapour, this may comprise any abundant atmospheric gases (such as CO_2 or O_3) and minor or pervasive amounts of atmospheric dust that show significant spectral absorption features in the VNIR and SWIR. Only in the rare case of an extreme amount of locally concentrated atmospheric dust or gas, e.g., due to blasting or the exhaust of waste gases within a mine, may the atmospheric correction fail for the affected image region. In this case, the local atmospheric perturbations will deviate distinctly from the used correction spectrum and cause an unsatisfactory spectral result. However, such scenarios can be avoided easily by the respective timing of the image acquisition, e.g., ahead of scheduled blasting operations.

It should be noted that for highly distorted pixels, spectral noise can remain at the former atmospheric absorption positions. The affected pixels mostly originate from extremely distant targets. Here, the proportion of the target signal on the spectral signal received at the sensor is so low that a removal of the atmospheric influence leads to an extremely low signal-to-noise ratio of the returned spectrum, which therefore appears noisy and featureless. This may suggest an upper distance limit for long-range HSI. However, this limit would be at an up to ten or more kilometre distance, depending on the atmospheric conditions of the scene. At this distance, the resulting pixel footprint on the ground would be in the range of several hundred square meters, questioning the informative value of the measurement. In conclusion, we were able to prove the successful application of the introduced atmospheric correction approach within a reasonable imaging distance.

6.2. Topographic Correction

As shown in the example of Nunngarut Peninsula in Figure 4, topographic correction is necessary under certain circumstances, as it ensures the comparability of absorption intensities between differently illuminated parts of the image. However, whereas the correction is effective for the adjustment of intensity changes, it cannot reconstruct spectral features in poorly illuminated areas of the image with associated low signal intensity, SNR, and feature detail. Therefore, we recommend a masking or at least careful interpretation of extremely poorly illuminated or deeply shadowed image parts. We further suggest evaluating the usefulness of a topographic correction for each imaged scene. From our general experience and the specific performance of the shown examples, natural targets such as mountain slopes or cliffs often have a smoother topography and therefore more consistent illumination than manmade outcrops like quarries and open pit mines. In natural targets, with the resulting smoother transitions between image parts with maximum and minimum illumination, respectively, the topographic correction usually performs well. Artificial targets often feature a terraced geometry and/or rough edges due to blasting and excavation, which generates large illumination differences. A topographic correction will not necessarily give an improvement of the image, as the applied corrections in the well-illuminated parts are minor, while the correction of the dark parts may be futile due to the mentioned reasons.

The c-factor method, despite its good performance for topographic correction, needs to be applied carefully. Due to the bandwise calculation of the correction factor using a linear regression, extreme or infinite values in one or several bands can cause an exaggeration of the correction factor for those bands and, finally, a change in the spectral shape. These peak values can be caused by bad pixels in the HSI sensor, which, due to the push-broom character of the camera, form bad pixel lines that are restricted to few adjacent bands. If a topographic correction needs to be applied, a correction or masking of those bad lines is inevitably required for a reliable image result.

6.3. Validation

The spectral validation using field spectrometer data demonstrated a great accuracy of both spectral shape and feature position of the corrected image spectra. In general, the difference between the interpolated minimum wavelength of field spectra and the corresponding library spectra for a certain absorption feature was below 5 nm in both areas of investigation. This value represents the band sampling distance of the SWIR data and lies below the achievable spectral resolution of 12 nm (FWHM). Locally, higher errors between some image and validation spectra points were observed, but these may be related to the large difference in spatial footprints of the different instruments. The field spectrometer data were retrieved from one or several 8 mm spots of a single lithologically representative sample, whereas the respective HSI pixel can easily represent a mixture of an area of some square meters of outcrop, depending on the distance to the sensor. Local variability in alteration can affect the representability of the spectrometer reading and lead to deviations from the recorded image spectrum at the same location. Additional to the spectral variations, slight mislocation of the spectrometer readings, which can be caused by the limited accuracy of the sample GPS position that can reach up to 5 m, needs to be taken into account.

6.4. 3D Integration

The potential, the spatial accuracy, and a possible application of the HSI integration with photogrammetric point clouds is discussed in more detail in Salehi et al. [20]. The current paper confirms not only the successful 3D integration for two additional examples, but further proves the capability of the workflow to integrate and merge hyperspectral datasets from different camera locations and viewing angles as well as different acquisition dates and times by eliminating the effects of topography, different illumination conditions, and atmospheric absorptions. This allows the use of hyperspectral data in a new way, as it facilitates the evaluation of spatial relationships between hyperspectral results that are not visible from one observation point or displayable in one dataset, such as opposing faces of a mountain or a mining pit.

7. Conclusions

With this paper, we present a novel approach for the atmospheric and topographic correction of long-range ground-based hyperspectral imagery. Such corrections are essential for obtaining reliable information on mineral composition in geological applications. The general workflow is partly based on the algorithms developed for drone-borne and vessel-based HSI data, which were presented and used in our previous papers [7,20], but is adapted and extended by adding radiometric and topographic correction approaches to meet the particular challenges of long-range, ground-based HSI.

The most important outcomes of this paper are the following:

1. The correction spectrum for the atmospheric correction is derived directly from the scene, and the correction intensity is determined according to the pixel-specific atmospheric absorption

depth. As a result, the workflow is independent from knowledge about the composition of the atmospheric layer or the distance to the target.

2. The incidence angles for the topographic corrections are calculated using the point normals of the photogrammetric 3D outcrop model. This allows us, for the first time, to utilise common topographic correction algorithms, such as the used c-factor method, for vertical outcrops.
3. The generation of a hypercloud, i.e., a geometrically and spectrally accurate combination of a photogrammetric point cloud and the HSI datacube, is achieved through the projective transformations of a photogrammetric 3D outcrop model. The removal of the effects of atmosphere and topography allows the integration of hyperspectral mapping results originating from different camera positions, dates, and, therefore, varying illumination conditions.
4. Two study areas with five HSI datasets in total proved the applicability and robustness of the workflow in differently challenging measuring conditions regarding climate, distance, atmospheric composition, geological diversity, and mapping objectives. A successful MWL mapping demonstrated both the geological applicability and the accuracy of spectral absorption positions and depths.
5. The accuracy and reliability of the created data and mapping results is validated by field spectra and the mineralogical analysis of geological samples.
6. The presented workflow is fast and simple and requires only a minimum of input parameters. Most of the processing steps are automatised and need no or extremely few manual actions.
7. The workflow enables (i) reliable spectral mapping of vertical and completely inaccessible outcrops; (ii) three-dimensional integration of multiple scans and other data sources; and (iii) a higher spectral resolution, range, and SNR than most drone- or air-borne HSI data.

On account of the promising quality of the presented datasets, we highly encourage the use of carefully processed and corrected long-range ground-based HSI data for geological applications and suggest a further development of highly adapted topographic and atmospheric correction algorithms. In several upcoming application-based papers, we will further present and discuss the geological interpretation of data corrected with the presented workflow and their integration with other data types such as structural data and long-wave infrared (LWIR) hyperspectral data.

Acknowledgments: The Helmholtz Institute Freiberg for Resource Technology is gratefully thanked for supporting and funding this project. The authors thank the entire group of "Exploration Technology" for their constructive feedback and extensive testing of the scripts. Further, we thank Atalaya Mining for access to Riotinto mine and IPH Ingeniería y proyectos for performing the UAS flight in Corta Atalaya. The Ministry for Mineral Resources, Government of Greenland, and the Geological Survey of Denmark and Greenland are gratefully acknowledged for funding and supporting fieldwork and data acquisition within the project "Karrat Zinc".

Author Contributions: S.L. developed the processing workflow with substantial contributions from S.S., M.K., and R.G. and implemented the workflow in Python. R.Z. and E.V.S. delivered logistic support in the field and were responsible for data acquisition and photogrammetric processing. S.L., R.Z., and G.U. processed the hyperspectral datasets and performed the geological interpretation and validation. S.L. wrote the manuscript with input from all authors. R.G. supervised the study at all stages.

Conflicts of Interest: The authors declare no conflict of interest.

References

1. Hubbard, B.E.; Crowley, C.K.; Zimbelman, D.R. Comparative alteration mineral mapping using visible to shortwave infrared (0.4–2.4 mm) Hyperion, ALI, and ASTER imagery. *IEEE Trans. Geosci. Remote Sens.* **2003**, *41*, 1401–1410. [CrossRef]
2. Kruse, F.A. Mineral mapping with AVIRIS and EO-1 Hyperion. In *Proceedings of the 12th JPL Airborne Geoscience Workshop*; Pasadena, CA, USA, 24–28 January 2003, Jet Propulsion Laboratory: Pasadena, CA, USA, 2003; Volume 41, pp. 149–156.
3. Bedini, E. Mapping lithology of the Sarfartoq carbonatite complex, southern West Greenland, using HyMap imaging spectrometer data. *Remote Sens. Environ.* **2009**, *113*, 1208–1219. [CrossRef]

4. Laukamp, C.; Cudahy, T.; Thomas, M.; Jones, M.; Cleverley, J.S.; Oliver, N.H. Hydrothermal mineral alteration patterns in the Mount Isa Inlier revealed by airborne hyperspectral data. *Aust. J. Earth Sci.* **2011**, *58*, 917–936. [CrossRef]
5. Zimmermann, R.; Brandmeier, M.; Andreani, L.; Mhopjeni, K.; Gloaguen, R. Remote Sensing Exploration of Nb-Ta-LREE-Enriched Carbonatite (Epembe/Namibia). *Remote Sens.* **2016**, *8*, 620. [CrossRef]
6. Jakob, S.; Gloaguen, R.; Laukamp, C. Remote Sensing-Based Exploration of Structurally-Related Mineralizations around Mount Isa, Queensland, Australia. *Remote Sens.* **2016**, *8*, 358. [CrossRef]
7. Jakob, S.; Zimmermann, R.; Gloaguen, R. The Need for Accurate Geometric and Radiometric Corrections of Drone-Borne Hyperspectral Data for Mineral Exploration: MEPHySTo—A Toolbox for Pre-Processing Drone-Borne Hyperspectral Data. *Remote Sens.* **2017**, *9*, 88. [CrossRef]
8. Gao, B.-C.; Heidebrecht, K.B.; Goetz, A.F.H. Derivation of scaled surface reflectances from AVIRIS data. *Remote Sens. Environ.* **1993**, *44*, 165–178. [CrossRef]
9. Adler-Golden, S.M.; Matthew, W.M.; Bernstein, L.S.; Levine, R.Y.; Berk, A.; Richtsmeier, S.C.; Acharya, P.K.; Anderson, G.P.; Felde, J.W.; Gardner, J.A.; et al. Atmospheric correction for shortwave spectral imagery based on MODTRAN4. In *Summaries of the Eighth JPL Airborne Earth Science Workshop*; Jet Propulsion Laboratory: Pasadena, CA, USA, 1999; Volume 99–17, pp. 21–29.
10. Richter, R.; Schlaepfer, D. Geo-atmospheric processing of airborne imaging spectrometry data, Part 2: Atmospheric/topographic correction. *Int. J. Remote Sens.* **2002**, *23*, 2631–2649. [CrossRef]
11. Smith, G.M.; Milton, E.J. The use of the empirical line method to calibrate remotely sensed data to reflectance. *Int. J. Remote Sens.* **1999**, *20*, 2653–2662. [CrossRef]
12. Roberts, D.A.; Yamaguchi, Y.; Lyon, R. Comparison of various techniques for calibration of AIS data. In *Proceedings of the 2nd Airborne Imaging Spectrometer Data Analysis Workshop*; Pasadena, CA, USA, 6–8 May 1986; Jet Propulsion Laboratory: Pasadena, CA, USA, 1986; Volume 86–35, pp. 21–30.
13. Chavez, P.S. An improved dark-object subtraction technique for atmospheric scattering correction of multispectral data. *Remote Sens. Environ.* **1988**, *24*, 459–479. [CrossRef]
14. Clark, R.N.; Swayze, G.A.; Livo, K.E.; Kokaly, R.F.; King, T.V.V.; Dalton, J.B.; Vance, J.S.; Rockwell, B.W.; Hoefen, T.; McDougal, R.R. Surface Reflectance Calibration of Terrestrial Imaging Spectroscopy Data: A Tutorial Using AVIRIS. In *Proceedings of the 10th Airborne Earth Science Workshop*; Jet Propulsion Laboratory: Pasadena, CA, USA, 2002; Volume 02-1.
15. Laliberte, A.S.; Goforth, M.A.; Steele, C.M.; Rango, A. Multispectral remote sensing from unmanned aircraft: Image processing workflows and applications for rangeland environments. *Remote Sens.* **2011**, *3*, 2529–2551. [CrossRef]
16. Kurz, T.H.; Buckley, S.J.; Howell, J.A. Close-range hyperspectral imaging for geological field studies: Workflow and methods. *Int. J. Remote Sens.* **2013**, *34*, 1798–1822. [CrossRef]
17. Kurz, T.H.; Buckley, S.J. A review of hyperspectral imaging in close range applications. *Int. Arch. Photogramm. Remote Sens. Spat. Inf. Sci.* **2016**, *41*, 865–870. [CrossRef]
18. Murphy, R.J.; Taylor, Z.; Schneider, S.; Nieto, J. Mapping clay minerals in an open-pit mine using hyperspectral and LiDAR data. *Eur. J. Remote Sens.* **2015**, *48*, 511–526. [CrossRef]
19. Rosa, D.; Dewolfe, M.; Guarnieri, P.; Kolb, J.; Laflamme, C.; Partin, C.A.; Salehi, S.; Sørensen, E.V.; Thaarup, S.; Thrane, K.; et al. *Architecture and Mineral Potential of the Paleoproterozoic Karrat Group, West Greenland: Results of the 2016 Season*; GEUS Rapport 2017/5; Geological Survey of Denmark and Greenland: Copenhagen, Denmark, 2017; p. 112.
20. Salehi, S.; Lorenz, S.; Sørensen, E.V.; Zimmermann, R.; Fensholt, R.; Heincke, B.H.; Kirsch, M.; Gloaguen, R. Integration of Vessel-Based Hyperspectral Scanning and 3D-Photogrammetry for Mobile Mapping of Steep Coastal Cliffs in the Arctic. *Remote Sens.* **2008**, *10*, 175. [CrossRef]
21. Kolb, J.; Keiding, J.K.; Steenfeld, A.; Secher, K.; Keulen, N.; Rosa, D.; Stensgaard, B. M Metallogeny of Greenland. *Ore Geol. Rev.* **2016**, *78*, 493–555. [CrossRef]
22. Sørensen, L.L.; Stensgaard, B.M.; Thrane, K.; Rosa, D.; Kalvig, P. *Sediment-Hosted Zinc Potential in Greenland*; GEUS Rapport 2013/56; Geological Survey of Denmark and Greenland: Copenhagen, Denmark, 2013; p. 184.
23. Grocott, J.; McCaffrey, K.J.W. Basin evolution and destruction in an Early Proterozoic continental margin: The Rinkian fold–thrust belt of central West Greenland. *J. Geol. Soc.* **2017**, *174*, 453–467. [CrossRef]
24. Pedersen, F.D. Remobilization of the massive sulfide ore of the Black Angel Mine, central West Greenland. *Econ. Geol.* **1980**, *75*, 1022–1041. [CrossRef]

25. Henderson, G.; Pulvertaft, T.C.R. *Geological Map of Greenland, 1:100 000. Mârmorilik 71 V.2 Syd, Nûgâtsiaq 71 V.2 Nord, Pangnertôq 72 V.2 Syd. Lithostratigraphy and Structure of a Lower Proterozoic Dome and Nappe Complex;* Geological Survey of Greenland: Copenhagen, Denmark, 1987; p. 72.
26. Guarnieri, P.; Partin, C.; Rosa, D. *Palaeovalleys at the Basal Unconformity of the Palaeoproterozoic Karrat Group, West Greenland;* Geological Survey of Denmark and Greenland Bulletin; Geological Survey of Denmark and Greenland: Copenhagen, Denmark, 2016; pp. 63–66.
27. Rosa, D.; Guarnieri, P.; Hollis, J.; Kolb, J.; Partin, C.A.; Petersen, J.; Sørensen, E.V.; Thomassen, B.; Thomsen, L.; Thrane, K. *Architecture and Mineral Potential of the Paleoproterozoic Karrat Group, West Greenland: Results of the 2015 Season;* Geological Survey of Denmark and Greenland: Copenhagen, Denmark, 2016; p. 98.
28. Sáez, R.; Donaire, T. Corta atalaya. In *Geología de Huelva: Lugares de Interés Geológico;* Universidad de Huelva, Facultad de Ciencias Experimentales: Huelva, Spain, 2008; pp. 106–111.
29. Soriano, C.; Casas, J. Variscan tectonics in the Iberian Pyrite Belt, South Portuguese Zone. *Int. J. Earth Sci.* **2002**, *91*, 882–896. [CrossRef]
30. Green, A.A.; Berman, M.; Switzer, P.; Craig, M.D. A transformation for ordering multispectral data in terms of image quality with implications for noise removal. *IEEE Trans. Geosci. Remote Sens.* **1988**, *26*, 65–74. [CrossRef]
31. Phillips, R.D.; Blinn, C.E.; Watson, L.T.; Wynne, R.H. An Adaptive Noise-Filtering Algorithm for AVIRIS Data with Implications for Classification Accuracy. *IEEE Trans. Geosci. Remote Sens.* **2009**, *47*, 3168–3179. [CrossRef]
32. Westoby, M.J.; Brasington, J.; Glasser, N.F.; Hambrey, M.J.; Reynolds, J.M. 'Structure-from-Motion' photogrammetry: A low-cost, effective tool for geoscience applications. *Geomorphology* **2012**, *179*, 300–314. [CrossRef]
33. Eltner, A.; Kaiser, A.; Castillo, C.; Rock, G.; Neugirg, F.; Abellán, A. Image-based surface reconstruction in geomorphometry—Merits, limits and developments. *Earth Surf. Dyn.* **2016**, *4*, 359–389. [CrossRef]
34. Carrivick, J.L.; Smith, M.W.; Quincey, D.J. *Structure from Motion in the Geosciences;* John Wiley & Sons, Ltd.: Chichester, UK, 2016; ISBN 978-1-118-89581-8.
35. James, M.R.; Robson, S.; d'Oleire-Oltmanns, S.; Niethammer, U. Optimising UAV topographic surveys processed with structure-from-motion: Ground control quality, quantity and bundle adjustment. *Geomorphology* **2017**, *280*, 51–66. [CrossRef]
36. Lowe, D.G. Object recognition from local scale-invariant features. In Proceedings of the International Conference on Computer Vision, Kerkyra, Greece, 20–27 September 1999; Volume 2, pp. 1150–1157.
37. Muja, M.; Lowe, D. Fast approximate nearest neighbors with automatic algorithm configuration. In Proceedings of the Fourth International Conference on Computer Vision Theory and Applications, Lisboa, Portugal, 5–8 February 2009; Volume 1, pp. 1–10.
38. Teillet, P.M.; Guindon, B.; Goodenough, D.G. On the Slope-Aspect Correction of Multispectral Scanner Data. *Can. J. Remote Sens.* **1982**, *8*, 84–106. [CrossRef]
39. Bakker, W.H.; van Ruitenbeek, F.J.A.; van der Werff, H.M.A. Hyperspectral image mapping by automatic color coding of absorption features. In Proceedings of the 7th EARSEL Workshop of the Special Interest Group in Imaging Spectroscopy, Edinburgh, UK, 11–13 April 2011; pp. 56–57.
40. van der Meer, F.; Kopačková, V.; Koucká, L.; van der Werff, H.M.A.; van Ruitenbeek, F.J.A.; Bakker, W.H. Wavelength feature mapping as a proxy to mineral chemistry for investigating geologic systems: An example from the Rodalquilar epithermal system. *Int. J. Appl. Earth Obs. Geoinf.* **2018**, *64*, 237–248. [CrossRef]
41. Gaffey, S.J. Reflectance spectroscopy in the visible and near-infrared (0.35–2.55 μm): Applications in carbonate petrology. *Geology* **1985**, *13*, 270–273. [CrossRef]
42. Chilingar, G.V. Classification of Limestones and Dolomites on Basis of Ca/Mg Ratio. *SEPM J. Sediment. Res.* **1957**, *27*, 187–189. [CrossRef]
43. Garde, A.A. *The Lower Proterozoic Marmorilik Formation, East of Mârmorilik, West Greenland;* Nyt Nordisk Forlag Arnold Busck: Copenhagen, Denmark, 1978; Volume 200, ISBN 978-87-17-02525-7.

Article

Quantitative Estimation of Soil Salinity Using UAV-Borne Hyperspectral and Satellite Multispectral Images

Jie Hu [1], Jie Peng [1,2], Yin Zhou [1,3], Dongyun Xu [1], Ruiying Zhao [1], Qingsong Jiang [1,4], Tingting Fu [1], Fei Wang [5] and Zhou Shi [1,*]

1 Institute of Agricultural Remote Sensing and Information Technology Application, College of Environmental and Resource Sciences, Zhejiang University, Hangzhou 310029, China; jiehu@zju.edu.cn (J.H.); pjzky@163.com (J.P.); zhouyin@zju.edu.cn (Y.Z.); xudongyun@zju.edu.cn (D.X.); ruiyingzhao@zju.edu.cn (R.Z.); qingsongjiang0827@126.com (Q.J.); ftt321@zju.edu.cn (T.F.)
2 College of Plant Science, Tarim University, Alar 843300, China
3 Institute of Land Science and Property, School of Public Affairs, Zhejiang University, Hangzhou 310029, China
4 College of Information Engineering, Tarim University, Alar 843300, China
5 Xinjiang Common University Key Lab of Smart City and Environmental Stimulation, College of Resource and Environmental Sciences, Xinjiang University, Urumqi 830046, China; Wangfei1986@xju.edu.cn
* Correspondence: shizhou@zju.edu.cn; Tel.: +86-0571-8898-2831

Received: 20 December 2018; Accepted: 23 March 2019; Published: 27 March 2019

Abstract: Soil salinization is a global issue resulting in soil degradation, arable land loss and ecological environmental deterioration. Over the decades, multispectral and hyperspectral remote sensing have enabled efficient and cost-effective monitoring of salt-affected soils. However, the potential of hyperspectral sensors installed on an unmanned aerial vehicle (UAV) to estimate and map soil salinity has not been thoroughly explored. This study quantitatively characterized and estimated field-scale soil salinity using an electromagnetic induction (EMI) equipment and a hyperspectral camera installed on a UAV platform. In addition, 30 soil samples (0~20 cm) were collected in each field for the lab measurements of electrical conductivity. First, the apparent electrical conductivity (EC_a) values measured by EMI were calibrated using the lab measured electrical conductivity derived from soil samples based on empirical line method. Second, the soil salinity was quantitatively estimated using the random forest (RF) regression method based on the reflectance factors of UAV hyperspectral images and satellite multispectral data. The performance of models was assessed by Lin's concordance coefficient (CC), ratio of performance to deviation (RPD), and root mean square error (RMSE). Finally, the soil salinity of three study fields with different land cover were mapped. The results showed that bare land (field A) exhibited the most severe salinity, followed by dense vegetation area (field C) and sparse vegetation area (field B). The predictive models using UAV data outperformed those derived from GF-2 data with lower RMSE, higher CC and RPD values, and the most accurate UAV-derived model was developed using 62 hyperspectral bands of the image of the field A with the RMSE, CC, and RPD values of 1.40 dS m^{-1}, 0.94, and 2.98, respectively. Our results indicated that UAV-borne hyperspectral imager is a useful tool for field-scale soil salinity monitoring and mapping. With the help of the EMI technique, quantitative estimation of surface soil salinity is critical to decision-making in arid land management and saline soil reclamation.

Keywords: soil salinity; unmanned aerial vehicle; hyperspectral imager; random forest regression; electromagnetic induction

1. Introduction

Salt-affected soils are widespread across the world, especially in arid and semi-arid regions [1]. Approximately 20% of irrigated agriculture land worldwide is affected by salinization [2], which results in soil degradation, arable lands loss and ecological environmental deterioration. Thus, it is of great significance to regularly monitor and map salt-affected areas to provide sufficient information for land informed management and salinized soil reclamation.

Conventional methods to quantitatively determine soil salinity were conducted through the measurement of the electrical conductivity (EC) of soil solution extracts or extracts at higher than normal water contents [3–5]. Because it was impractical to extract soil water from samples at typical field water contents, EC of the saturation extract made at 1:1, 1:2, and 1:5 soil:water ratios, noted as $EC_{1:1}$, $EC_{1:2}$, and $EC_{1:5}$, respectively, were generally used to estimate soil salinity. However, the use of such a traditional approach required a great deal of time and funding, usually leading to low efficiency and high cost for soil salinity characterization.

In the late 1970s, researchers in the U.S. first applied the theory of EMI technique to measure the apparent electrical conductivity (EC_a) for field-scale soil salinity mapping [6]. Soil properties such as soil salinity, soil moisture, clay content, and temperature are the dominant factors that influence EC_a [7]. By assuming relative homogeneity in other soil properties or having prior knowledge of them, the measurement of EC_a using EMI has been used extensively to noninvasively characterize and map soil salinity [8]. In order to develop relationship between EC_a with EC of the saturation extract, various conversion methods have been proposed [9]. Although much research has investigated and compared non-linear transformations, linear calibration methods were proved to be sufficiently accurate [10]. With the advantage of rapidly acquiring abundant EC_a data, the EMI technique was available to aid the spatial prediction of soil salinity with limited soil samples.

Remote sensing has gained popularity for delineating saline soils over the last two decades as a rapid, non-destructive and cost-effective method [11–13]. Researchers have found that saline soils present distinctive morphological features at the soil surface and spectral characteristics from non-saline soils, with an overall higher reflectance in the visible and near-infrared parts of the spectrum [14,15]. Previously, researchers used various multispectral data acquired from satellite-borne sensors in combination with field measurement to differentiate saline and non-saline soils before mapping salt-affected regions [16–18]. In 1994, Verma et al. [19] conducted an integrated approach of visual interpretation method to map salt-affected soils using Landsat TM satellite images. In 2002, Dehaan and Taylor [20] developed spectral unmixing techniques to derive indicators for characterizing and mapping soil salinity in the Murray-Darling Basin, Australia. With the occurrence of hyperspectral technique, remote sensing enabled detailed analysis of the spectral characteristics of the land surface with a large amount of narrow and contiguous wavelength bands. Soil salinity research has progressed from qualitative classification to quantitative estimation [21–23]. For example, various absorption bands have been used for quantifying salt minerals [24–27]. Farifteh et al. [28] in 2007 estimated salt concentrations in soils based on laboratory data, field measured spectral reflectance and hyperspectral images, and recommended that the useful spectral bands for salinity estimation were in the near infrared (NIR) and SWIR regions. In 2014, Pang et al. [29] improved the prediction accuracy for soil salt content based on the genetic algorithm method, using hyperspectral remote sensing data acquired in Minqin County, China.

However, the quality of satellite-borne and air-borne remote sensing images can easily be confined to bad weather and unfavorable revisit times. Also, the lack of imagery with optimum spatial and spectral resolutions was a critical limitation for real-time crop management using current satellite sensors [30]. The introduction of UAV provided an easy and cost-efficient approach for soil salinity monitoring, as UAV-borne hyperspectral sensors not only acquired images with ultra-high spatial resolution but were also convenient to operate freely in proper conditions. The sensors on board included digital camera, multispectral camera, hyperspectral imager and Light Detection and Ranging equipment (LiDAR) [31]. Although UAV has been widely used, applications were mainly focused on crops or forest mapping and vegetation feature extraction [32–34]. Studies using UAV images for soil salinity detection and mapping were still rare. Ivushkin et al. [35] have tried combining a WIRIS thermal camera, a Rikola hyperspectral camera and a Riegl VUX-SYS LiDAR scanner to measure salt stress in quinoa plants, and they found UAV-borne remote sensing to be a useful technique for salt stress measurements. Romero-Trigueros et al. [36] concluded that the red and near-infrared bands were critical to assess the saline stress Citrus suffered from. However, no existing literature has discussed the potential of synthesizing UAV-borne hyperspectral data and EMI measurements for soil salinity estimation.

Our research aimed to (i) evaluate the potential for quantitative estimation of soil salinity and its spatial distribution at field-scale, using a UAV-borne hyperspectral imager (0.50–0.89 μm) and (ii) compare these to the predictions of soil salinity from GF-2, a multispectral satellite remote sensor (0.45–0.89 μm). In both cases, random forest (RF) regression was used to relate spectral information to soil salinity contents. Meanwhile, a fairly large number of soil samples and spatially dense EMI measurements were available to provide the electrical conductivity data taken as the dependent variable of the RF models for quantitative estimation of field-scale soil salinity.

2. Materials and Methods

2.1. Study Area

The study site was located in the center of Aksu (79°39′~82°01′E, 39°30′~41°27′N), western Xinjiang, China. It included three fields with variable vegetation cover (A: bare land with no vegetation cover; B: sparse vegetation cover; C: dense vegetation cover); each covered about 1 ha (100 × 100 m) in area (Figure 1). The region was close to Taklimakan, the biggest desert in China, with a low average annual rainfall of 67 mm and a high average annual evaporation of 2110 mm. The average annual temperature varied from 9.9 °C to 11.5 °C. The soil type was Typic Aridi-Orthic Halosols in Chinese soil taxonomy. The average pH values of soil samples collected in the study areas were 8.7, 8.4 and 9.1 for fields A, B, and C, respectively. The dominant species in the study areas were halophytes, belonging to the family of Chenopodiaceae and Tamaricaceae. To be specific, the typical halophytes presented in the field B was *Tamarix ramosissima*, and the ones presented in the field C were *Halostachys belangeriana* and *Halocnemum strobilaceum* [37].

Due to the extremely arid local climate, intense evapotranspiration and relatively high ground water level, salt in the profile tends to accumulate on the surface soil, resulting in visible salt crust and salt crystals in UAV images (Figure 1a–c). The salts were mainly of sulphates in chemical composition.

2.2. EMI Measurements

The field measurement of EMI was carried out in late October of 2017. In each field, the EC_a data were measured along crisscrossed grid lines with an interval size of 20 m using an EM38-MK2 (Geonics Ltd., Mississauga, Ontario, Canada) instrument in both vertical (EC_{av}, mS m^{-1}) and horizontal (EC_{ah}, mS m^{-1}) dipole modes with measuring depths of approximately 1.5 m and 0.75 m [7]. A built-in Global Positioning System (GPS) was used to record spatial information. The EM38-MK2 had a measuring range of 0~1000 mS m^{-1}, its measurement accuracy was ±0.1%, the working frequency was 14.5 KHz, and the working temperature ranged from −30 to 50 °C. It weighed approximately 5.4 kg, containing two receiver coils spaced at 0.5 m and 1 m from the transmitter coil.

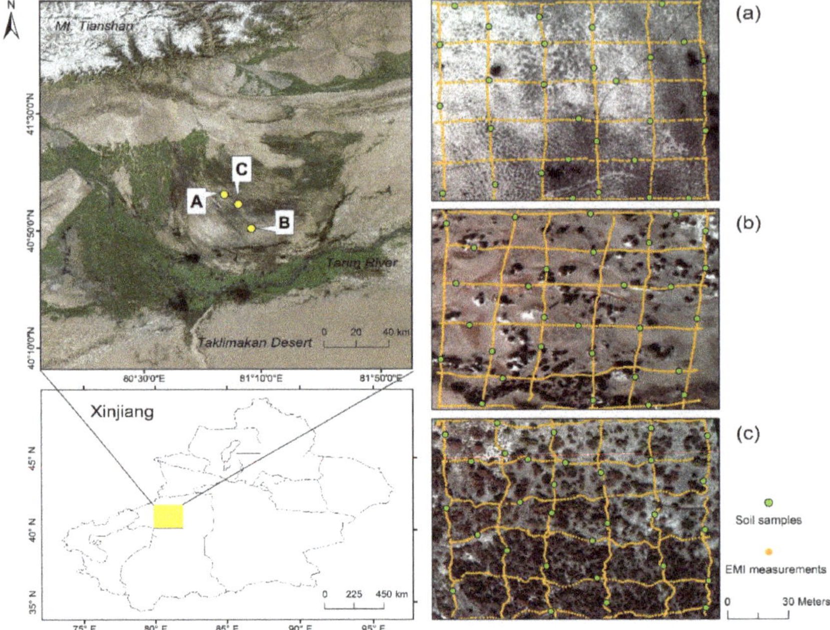

Figure 1. Location of study area and electromagnetic induction (EMI) measurements in fields A (**a**), B (**b**), and C (**c**) within the Xinjiang Autonomous Region.

The EM38-MK2 measured EC_a by first inducing an electrical current in the soil. Then, a fraction of the secondary induced electromagnetic field from each loop was intercepted by the receiver. Finally, the sum of these signals was formed into an output voltage which is linearly related to a depth-weighted soil EC_a [7]. In this case, the EM38-MK2 sensor was carried out in auto-collecting mode through the fields by an operator on foot. It took about an hour to survey a field with the EMI, there was no significant temperature change during the surveys. Compared with other EM38 devices, the EM38-MK2 we used implemented the temperature-compensation circuitry to avoid thermal drift as a consequence of internal temperature influence [38], hence temperature correction on the EMI sensor signal could be waived. For each site more than 2000 points have been collected via EMI, however, when the EMI measurements were conducted in auto-collecting mode, inevitably there were some densely clustered points within a very small region when the operator stopped to avoid the road bumps or stones. After removing those densely clustered points, there were 1500 points for each site. We later converted their EC_a values to $EC_{1:5}$ using empirical line method.

2.3. Soil Sampling and Laboratory Measurement

In the same days as the EMI data were obtained in auto-collecting mode, 30 sample points were chosen on the EMI measurement lines in each field. When selecting sample points, we tried to cover the different values of EC_a measurements, including high, medium, and low values in each field [5]. First, the EC_a of every sample point was measured via handheld EM38-MK2. Then, a total of 90 soil samples of the chosen points were collected to the depths of 0~0.20 m. Soil sampling for each site was conducted within one day. After that, soil samples were transferred to laboratory, air-dried, crushed and sieved to 1 mm size. Finally, the leachate was extracted from the suspension to measure the $EC_{1:5}$ using a LeiCi DDS-307 (ShengKe, Shanghai, China) conductivity meter [39].

The EC values of the EM38-MK2 measurement points were predicted from linear regression relationship [39]. Empirical linear regression was established between the EC_a and $EC_{1:5}$ of the

30 sampling points for each field. Such method derived the coefficients needed to fit original EMI measurements and then converted all the 1500 EC_a values to $EC_{1:5}$.

2.4. Remote Sensing Data Processing

A frame-based hyperspectral camera (Rikola Ltd., Oulu, Finland) was loaded on the UAV platform. The camera had 62 spectral bands in the visible-near infrared (Vis-NIR) region with a spectral resolution of approximately 10 nm. The narrow bands could provide sufficient data for salinity prediction, but the camera we used did not capture data in the shortwave wavelength region less than 0.50 μm. The UAV-borne hyperspectral images of fields B, C, and A were collected on 27, 29, and 30 October, 2017. The ground pixel size was 0.1 m with the flying height of approximately 154 m. The camera weighed approximately 720 g and had a maximum image size of 1010 × 1010 pixels. The image field of view (FOV) was 36.5°, which was suitable for field-scale to regional-scale investigations.

Hyperspectral Imager 2.0 software (Rikola Ltd., Oulu, Finland) could help users of the UAV-borne Rikola hyperspectral sensor carry out sensor parameter settings, real-time imaging, image quality evaluation, and image preprocessing such as dark current correction. The dark current correction was carried out using a dark current measurement taken before the flight by covering the lens, and the raw images were converted to at-sensor-radiance images after the dark current correction [35]. The radiance images were then transformed into reflectance factor images through empirical line method using the measurement of the reference panel taken before each UAV flight [40]. Due to intrinsic characteristics of the Fabry–Pérot interferometer (FPI) technology, the UAV images on different wavelengths were captured at different times, thus band-to-band alignment was performed to correct the difference between the extents of each wavelength. Thereafter, the reflectance factor images were coordinated after orthorectification and georeferencing.

In addition, the GF-2 images were acquired on 27 October, 2017. The Chinese GF-2 environmental satellite was launched on 19 August, 2014. Each image consists of 5 spectral bands, and the spatial resolution is relatively high among environmental satellite data. Radiometric calibrations were applied, and the raw GF-2 images were converted to radiance images using the absolute calibration coefficients provided by the China Centre for Resources Satellite Data and Application (CRESDA). The atmospheric correction was carried out using the Fast Lin-of-Site Atmospheric Analysis of Spectral Hypercubes (FLAASH) [41] algorithm and the GF-2 spectral response function provided by CRESDA. Details on the remote sensing sensors and platforms were given in Table 1.

Table 1. Remote sensing sensors used for detection and mapping of soil salinity. UAV: unmanned aerial vehicle; NIR: near infrared.

Sensor	No. of Bands	Spectral Range (μm)	Spatial Resolution	Platform
Hyperspectral Imager	62	Visible and NIR B1~62: 0.50–0.89	0.1 m (flight height: 154 m)	UAV
GF-2	5	Visible and NIR Band1: 0.45–0.52 (Blue) Band2: 0.52–0.59 (Green) Band3: 0.63–0.69 (Red) Band4: 0.77–0.89 (NIR) Panchromatic: 0.45–0.90	1 m (Panchromatic)/ 4 m (Multispectral)	Satellite

The ultra-high spatial resolution of UAV images may bring noises such as shadows into quantitative estimation of soil properties. Additionally, the scale differences between the EMI sampling interval and the spatial resolution of UAV data attributed to the poor prediction results of the models derived from original UAV data. In our case, the spatial distance between two adjacent EMI measurements was approximately 1 m, while the spatial resolution of the original UAV data was

0.1 m. We have tried a series of grid sizes, and the model got relatively more accurate predictions when using spatial resampled UAV data with resampling size of 1 m.

To make comparison between UAV-borne and satellite-borne data, this research used three data sets for building RF regression models; 1) the hyperspectral UAV data set which was spatially resampled to 1 m spatial resolution from the original images, 2) the multispectral GF-2 data set and 3) the multispectral UAV data set produced from spectral resampling of the hyperspectral UAV data set. The spectral resampling was undertaken by turning narrow bands into broad bands similar to that of the GF-2 data, the GF-2's spectral response function was used in the process.

Matrices of the input variables of the RF method was made by combining the $EC_{1:5}$ data (n = 1500) with the reflectance factors of spectral data. For each point of the $EC_{1:5}$ samples, the reflectance factors of hyperspectral or multispectral bands were extracted according to their spatial location. The data rows of each matrices were later split into a training set and a validation set following the ratio of 2:1 [42]. The training set was used to build the prediction model of each field by tuning model parameters (in this case, the number of trees in the forest and the number of randomly selected independent variables at each mode), and the validation set was used to evaluate the model's robustness and prediction accuracy.

2.5. Soil Salinity Prediction Using RF

Random forest (RF) was an ensemble learning method proposed by Ho in 1998, then developed by Breiman and Cutler [43–45]. Due to its high accuracy, the novel method of determining variable importance and the ability to model complex interactions among predictor variables, RF has been increasingly used for classification and regression in recent years [46–48]. In this study, the RF regression method was used to develop the soil salinity prediction models due to its proved robustness and efficiency when dealing with abundant variables.

RF regression was operated by constructing a multitude of single regression trees and outputting the mean prediction of the individual trees, it predicted the dependent variable (the soil salinity) from a set of independent variables (the reflectance factors of 62 UAV-derived hyperspectral bands or 4 satellite-derived multispectral bands). Each regression tree was independently constructed using a bootstrap sample of the training data set (the 1000 $EC_{1:5}$ samples which were used to build the model). Then, for each independent variable, the data were split at several split points. The sum of squared error (SSE) at each split point between the predicted $EC_{1:5}$ and the actual $EC_{1:5}$ was calculated, and the variable resulting in the minimum SSE was selected for the node splitting. This process was recursively continued until the entire data set was covered. In our case, RF regression was operated using the package 'randomForest' [49] within R environment software [50].

RF required no assumption of the probability distribution of the target predictors as with linear regression [51]. Moreover, the variable importance analysis of RF was a useful tool to describe the significance of any variable in the model. In carrying out the procedure, first, the mean square error (MSE) on the out-of-bag (OOB) portion of the data (the $EC_{1:5}$ samples which were left out when constructing a regression tree using the bootstrap sampling) was calculated in the whole regression model, then the values of a variable were randomly shuffled to compute the MSE again on the perturbed data, and finally the normalized difference between these two MSE was taken as the importance score for this variable [49]. The statistical definition can be found in Zhu et al. [52].

After training the models using the training datasets, the validation datasets were taken as the input of these models. Several prediction accuracy indicators, including CC [53], RPD, and RMSE were adopted to compare and evaluate the prediction results. CC quantified the agreement between the $EC_{1:5}$ samples and the predicted $EC_{1:5}$ of a RF model, it ranged from -1 to 1, also represented how well the measured versus predicted data follows the 1:1 line. RPD calculated the ratio of the standard error of prediction to the standard deviation of the samples. RMSE explained the difference between the samples and the model predictions. Generally, a model that performed well would have high CC and RPD values, and a low RMSE value [28,54].

$$CC = \frac{2rs_{\hat{y}}s_y}{s_{\hat{y}}^2 + s_y^2 + (\overline{\hat{y}} - \overline{y})^2} \quad (1)$$

$$RMSE = \sqrt{\frac{1}{n}\sum_{i=1}^{n}(y_i - \hat{y}_i)^2} \quad (2)$$

$$RPD = \frac{s_y}{RMSE} \quad (3)$$

where r is the usual Pearson product-moment correlation coefficient between the observed and predicted values, s_y and $s_{\hat{y}}$ are the standard deviation of the observed and predicted values, s_y^2 and $s_{\hat{y}}^2$ are the variances of the observed and predicted values, $\overline{y}$ and $\overline{\hat{y}}$ are the mean of the observed and predicted values, n is the number of the observation samples used, and y_i and $\hat{y}_i$ are the observed and predicted values of sample point i, respectively.

3. Results

3.1. Soil Salinity Content and Variation

The descriptive summary of the EC_a and the $EC_{1:5}$ value of each point measured by hand-hold EM38-MK2 and chemical analysis were presented in Table 2.

Table 2. Descriptive summary of electrical conductivity EC_a and $EC_{1:5}$ measured on the samples in fields A, B, and C.

Field	Conductivity	Descriptive Statistics (EC_a, mS m^{-1}; $EC_{1:5}$, dS m^{-1})						
		N	Min	Max	Mean	Median	Std.Dev.	CV
A	EC_{ah}	30	571.15	955.72	765.05	766.72	119.02	16%
	EC_{av}	30	598.15	1065.57	846.74	865.22	144.53	17%
	$EC_{1:5}$	30	20.25	54.90	37.64	35.80	9.21	25%
B	EC_{ah}	30	450.20	1092.15	830.47	903.09	200.58	24%
	EC_{av}	30	585.67	1035.90	824.51	779.02	154.27	19%
	$EC_{1:5}$	30	7.20	14.68	11.73	11.91	2.38	20%
C	EC_{ah}	30	695.86	1126.99	890.15	861.09	136.54	15%
	EC_{av}	30	560.17	955.56	778.00	782.09	118.92	15%
	$EC_{1:5}$	30	9.64	19.64	14.11	14.50	2.94	21%

As shown in Table 2, the minimum EC_{ah} value was 450.20 mS m^{-1}, which was measured in the field B. The maximum EC_{ah} value was 1126.99 mS m^{-1} and was found in the field C. The minimum and the maximum EC_{av} values were measured in the field C and the field A, with the values of 560.17 mS m^{-1} and 1065.57 mS m^{-1}, respectively. When it comes to $EC_{1:5}$, the highest and the lowest values were 54.90 mS m^{-1} and 7.20 mS m^{-1}, which could be found in the field A and the field B, respectively.

Taking the mean values of EC_a into consideration, the field A had the lowest average EC_{ah} value of 765.05 mS m^{-1}, and the highest average EC_{ah} value was 890.15 mS m^{-1}, which was measured in the field C. The field C had the lowest average EC_{av} value of 778.00 mS m^{-1}, and the average EC_{av} value of the field A was the highest among three fields, which was 846.74 mS m^{-1}. As for the mean values of $EC_{1:5}$, the field A had the biggest $EC_{1:5}$ value of 37.64 dS m^{-1}, and the smallest average $EC_{1:5}$ was found in the field B with the value of 11.73 dS m^{-1}.

For fields A, B, and C, the relationships between EM38-MK2-measured EC_{ah}, EC_{av}, and laboratory-analyzed $EC_{1:5}$ of the samples (n = 30) were given as Equations (4–6):

$$EC_{1:5}\left(\text{dS m}^{-1}\right) = 0.0278EC_{av} + 0.0234EC_{ah} - 5.52532\left(R^2 = 0.85, \text{ adjusted } R^2 = 0.84, \text{ RMSE} = 3.00 \text{ dS m}^{-1}\right) \quad (4)$$

$$EC_{1:5}\left(dS\,m^{-1}\right) = -0.0119EC_{av} + 0.0177EC_{ah} + 6.85742\left(R^2 = 0.75,\ \text{adjusted } R^2 = 0.73,\ \text{RMSE} = 1.14\ dS\,m^{-1}\right) \quad (5)$$

$$EC_{1:5}\left(dS\,m^{-1}\right) = 0.0275EC_{av} - 0.0042EC_{ah} + 0.28139\left(R^2 = 0.95,\ \text{adjusted } R^2 = 0.95,\ \text{RMSE} = 0.73\ dS\,m^{-1}\right) \quad (6)$$

Combining Table 2 and Equations 4–6, it was clear that the fitted linear relationship of field C produced the most accurate prediction of $EC_{1:5}$ using EC_{ah} and EC_{av}, and the prediction accuracies of all three fields were satisfying with R^2 and adjusted R^2 values no less than 0.7.

For each field, the corresponding equation was used to calibrate the ECa values (n = 1500) of the EMI survey and convert them to $EC_{1:5}$ values. The descriptive summary of calibrated $EC_{1:5}$ in study fields were presented in Table 3.

Table 3. Descriptive summary of $EC_{1:5}$ in fields A, B, and C.

Field	Descriptive Statistics ($EC_{1:5}$, dS m^{-1})						
	N	Min	Max	Mean	Median	Std.Dev.	CV
A	1500	18.81	47.14	31.54	31.22	4.17	13%
B	1500	5.04	15.20	9.89	10.00	1.64	17%
C	1500	11.98	25.94	18.13	18.37	2.10	12%

Ranging from 5.04 dS m^{-1} to 47.14 dS m^{-1}, the $EC_{1:5}$ measured in the study area had a broad value domain. As shown in Table 3, the average $EC_{1:5}$ value of the field A, B and C was 31.54 dS m^{-1}, 9.89 dS m^{-1} and 18.13 dS m^{-1}, respectively, showing considerable difference between fields with variable vegetation cover. The maximum $EC_{1:5}$ value was measured in the field A, which was bare land with no vegetation cover and with a large area of visible salt crust, and the minimum was measured in the field B, which had relatively moderate vegetation cover of clustered halophyte, *Tamarix ramosissima*. The coefficients of variation in the three fields were all greater than 10%, indicating moderate variation of soil salinity within the study areas. The $EC_{1:5}$ was directly taken as the proxy of soil salinity [55,56], and was denoted as EC hereafter.

3.2. Prediction Accuracy of RF Regression Models

Table 4 showed the soil salinity prediction accuracy (training and validation) of RF regression models using UAV, GF-2 and spectral resampled UAV data. Although the prediction accuracies of training and validation were quite similar, showing robustness in each of the RF prediction models, the training statistics were generally better than the validation stats as expected.

Table 4. Training and validation statistics of random forest (RF) regression models. CC: concordance coefficient; RPD: ratio of performance to deviation; RMSE: root mean square error.

Data Set	Source	A			B			C		
		CC	RPD	RMSE (dS m^{-1})	CC	RPD	RMSE (dS m^{-1})	CC	RPD	RMSE (dS m^{-1})
Training (n = 1000)	UAV	0.96	3.92	1.05	0.94	3.29	0.49	0.81	1.91	1.07
	GF-2	0.93	2.93	1.40	0.92	2.75	0.58	0.74	1.67	1.22
	Resampled UAV	0.95	3.22	1.28	0.92	2.72	0.59	0.70	1.55	1.31
Validation (n = 500)	UAV	0.94	2.98	1.40	0.86	2.15	0.74	0.56	1.29	1.59
	GF-2	0.88	2.23	1.87	0.84	2.00	0.80	0.44	1.20	1.71
	Resampled UAV	0.89	2.35	1.78	0.81	1.85	0.86	0.40	1.12	1.83

In each field, it was true both for training and validation sets that the CC and RPD values of UAV model were generally greater, and the RMSE values smaller than that of GF-2 and resampled UAV models. It indicated that the prediction performance of the UAV model was the best among three types of models. Comparing the validation results of three different fields, the models constructed from the UAV hyperspectral data of bare land (A) showed the best prediction performances with the highest CC and RPD values of 0.94 and 2.98, whereas the resampled UAV model of the area with dense vegetation cover (C) produced the worst prediction performance with the lowest CC and RPD values of 0.40 and 1.12. It suggested that dense vegetation cover might deteriorate the predicting capability of soil salinity through covering soil surface and blurring the spectral information of surface soil. In addition, the prediction accuracy was sharply higher (lower RMSE) for the field B with moderate vegetation cover.

The fitted lines of the field A (Figure 2a,d) were the closest to the 1:1 lines, showing the best prediction performance of the UAV models among all three fields, while the measured versus predicted points using the validation data set were dispersed in the scatter plot of the field C (Figure 2f). Moreover, the field B exhibited the lowest RMSE values of the prediction models.

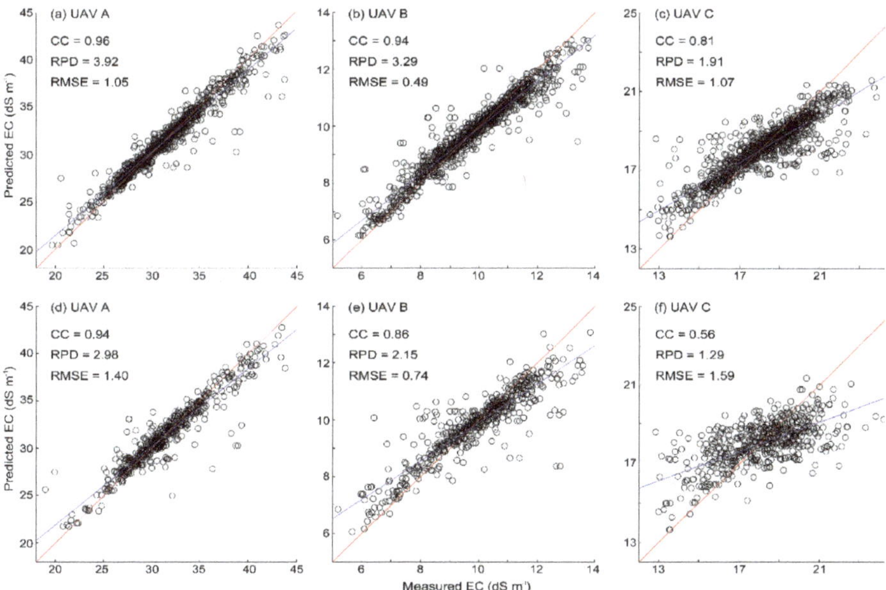

Figure 2. Scatter plots of measured versus predicted electrical conductivity (EC)-derived from RF regression models using UAV data for the field A (**a**,**d**), B (**b**,**e**), and C (**c**,**f**); the upper three (**a**–**c**) are the training results; the lower three (**d**–**f**) are the validation results; the blue lines are the fitted lines and the red lines are the 1:1 lines.

3.3. Soil Salinity Maps Derived from UAV and GF-2 Data

Figure 3 showed the soil salinity maps of the study areas developed using RF regression models. Since the resampled UAV data didn't produce better prediction accuracy than the original UAV data did (Table 4), only the salinity maps derived from the original UAV and GF-2 data were shown to make comparisons. For both the UAV and GF-2 prediction models, the predicted EC values of fields A, B and C covered the range of around 20.0~44.0 dS m^{-1}, 6.0~14.0 dS m^{-1}, and 13.0~22.0 dS m^{-1}, respectively. In general, the maps of the field A (Figure 3a,b) and B (Figure 3c,d) showed distinct spatial variation pattern of soil salinity, whereas the GF-2 map of field C (Figure 3f) was too fragmented and scattered to recognize any salinity spatial pattern due to its dense vegetation cover.

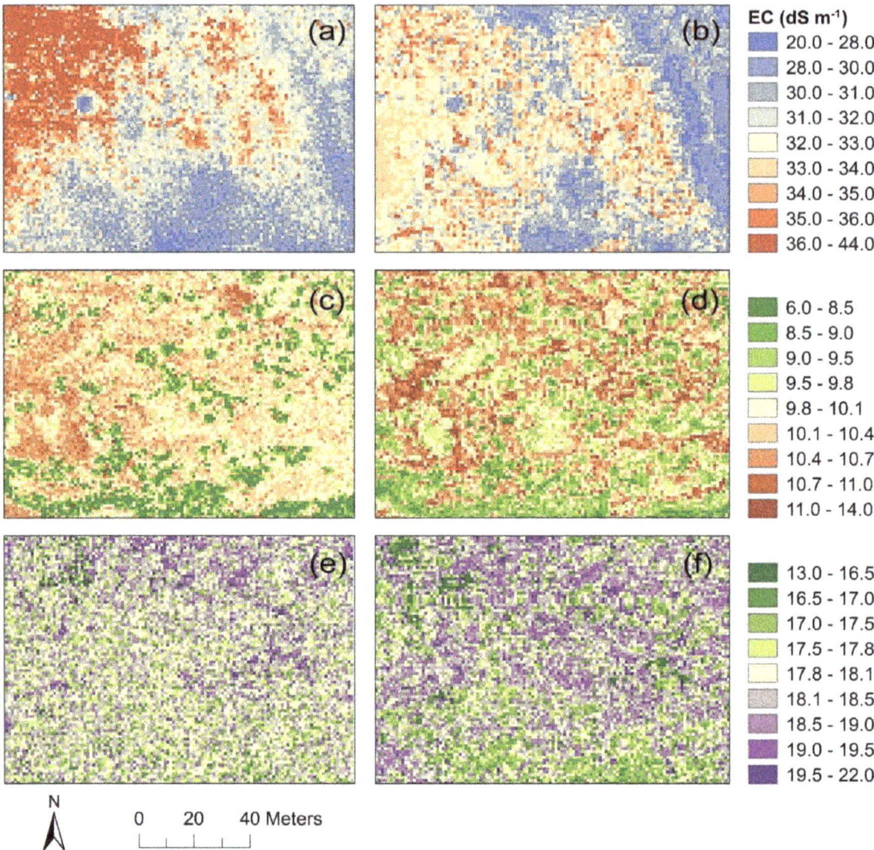

Figure 3. EC maps derived from RF regression models using UAV (the left three: **a**,**c**,**e**) and GF-2 (the right three: **b**,**d**,**f**) data for the field A (**a**,**b**), B (**c**,**d**), and C (**e**,**f**).

The field A had the most extreme soil salinity. High salts (>35.0 dS m^{-1}) were mostly located in the northwest area (Figure 3a). An obvious difference was visible between the UAV and GF-2 models of the field A. For example, a large area with high salt content (>35.0 dS m^{-1}) using the UAV model (Figure 3a) exhibited with relatively lower salt contents (32~35 dS m^{-1}) using the GF-2 model (Figure 3b). In the UAV prediction map of the field B (Figure 3c), relatively high EC values (>10.4 dS m^{-1}) were mostly found in the north and the west part of the study area, and EC values less than 9.0 dS m^{-1} were mainly distributed in the southern region and places with clustered populations of halophytes (*Tamarix ramosissima*). However, the GF-2 prediction map of the field B (Figure 3d) showed fewer areas of moderate EC values (9.8~10.4 dS m^{-1}) and more areas of relatively high EC values (>11.0 dS m^{-1}). In the soil salinity map of the field C derived from UAV data (Figure 3e), high salt content (>19.0 dS m^{-1}) soils were located in the northeast part of the area. Compared with the UAV prediction map (Figure 3e), there was a greater area with EC values higher than 18.5 dS m^{-1} in the GF-2 prediction map of the field C.

4. Discussion

4.1. Comparison of RF Regression Models Based on UAV and GF-2

Accurate atmospheric correction was critical to remote sensing-based soil properties estimation as mainly atmospheric scattering distorted the real surface reflectance especially for the blue bands. However, conventional atmospheric correction methods for satellite images were not directly applicable for UAV-borne hyperspectral images. Although the lack of atmospheric correction may lead to inaccurate retrieval of soil salinity because of atmospheric perturbations, the Rikola camera we used did not capture data in the shortwave wavelength region less than 0.50 μm. The reflectance of the UAV-borne hyperspectral data was more detailed and intense than the reflectance of the satellite-borne multispectral data. Thus, the fully empirical approach with the RF can be applied without atmospheric correction. Even so, many researchers have tried to develop different radiometric correction methods especially for UAV-borne hyperspectral data. Honkavaara et al. [57] constructed a physically-based method which includes a radiometric block adjustment utilizing radiometric tie points and utilized in situ irradiance measurements. Lorenz et al. [58] performed a radiometric correction using a single atmospheric correction spectrum for each scene.

The RF regression modelling permitted reliable estimations and mapping of soil salinity at the field-scale (Table 4). The RF regression models using the UAV data source had higher CC and RPD values and lower RMSE values than the models using the GF-2 data source. This indicated that the 62-band hyperspectral images provided better prediction results than the multispectral GF-2 data in all three fields, despite lacking spectral information in the wavelength range from 0.45 to 0.50 μm. After spectral resampling, the accuracy of the UAV prediction models reduced, revealing that narrow bands, compared with broad bands, provide more detailed spectral information which could contribute to improving model performance. The RF regression models of field A were accurate with RPD values of 2.98 and 2.23, and the predictions of RF regression models of field B were good with RPD values of 2.15 and 2.00, but both the UAV and GF-2 prediction models of field C were ineffective as the RPD values were below 1.80 [59].

Because EMI measurements were conducted densely at field-scale, OOB and validation samples were often almost identical to samples used in the training of the models. It inevitably resulted in overestimation of the model's prediction accuracy [60,61]. Even so, this study developed a novel approach of combining EMI and remote sensing techniques to map field-scale soil salinity. Our results presented relatively reliable spectral inversion of salinity in three fields with variable vegetation cover. In future research, spatial independence selection methods such as spatial blocking will be employed to conduct cross-validation in order to address the overoptimism of the prediction models.

4.2. Soil Salinity under Various Vegetation Cover Conditions

In this study, the highest surface soil EC value (47.14 dS m^{-1}) was measured in the field A where no vegetation existed. However, although the field C had the densest vegetation cover, the soil salinity was generally higher than that of the field B where vegetation cover was sparse. It suggested that soil salinity was not simply negatively correlated with vegetation cover. As given in Table 2, the average soil salinity in the field C was above 15 dS m^{-1}, which was much greater than the soil salinity in the field B. One possible reason is that halophytes, unlike other plants or crops, were adapted to moderate and even high contents of salt in soils. For the phreatophyte *Tamarix ramosissima* in the field B, their root could reach deep down in the soil, and the physiological activity and biomass accumulation majorly rely on the stable groundwater [62]. Moreover, for halophytes in the field C, their physiological characteristic enabled them to not only survive but also flourish with optimal growth in saline conditions that would kill other species [63].

With the increase of vegetation cover, the prediction performance of spectral retrieval models presented a decreasing trend with higher RMSE and lower CC and RPD values. It was reasonable because reflectance of the canopy rather than surface soil was collected via UAV. Although the canopy spectra did not directly depict the salt content in soils, it could be an indirect indicator of salinity. Under salt stress, the spectral reflectance and morphology of plant or crops on the ground would change due to insufficient water uptake and specific ion toxicity. Existing literatures have proposed methods to assess soil salinity using environmental indicators, including spectral vegetation indices such as normalized difference vegetation index (NDVI). Peng et al. [42] used a variety of environmental and ecological covariates, including NDVI, to quantitatively characterize the salinity of arid-area soils, the prediction accuracy of the cubist model was good with the R^2, RMSE, MAE and RPD values of 0.91, 5.18 dS m^{-1}, 3.76 dS m^{-1}, and 3.15, respectively. In the Yellow River Delta of China, Zhang et al. [64] assessed the applicability of monitoring soil salinization utilizing vegetation indices derived from the MODIS time series data. Additionally, Allbed et al. [65] analyzed NDVI values and salinity index properties to monitor changes in soil salinity and vegetation cover from multispectral images. Further study about delineating soil salinity using salinity indices will be carried out to overcome the low prediction accuracy of models derived from a dense vegetation area.

4.3. Evaluation of the Variable Importance for Hyperspectral Soil Salinity Modeling

To understand which variables were the most significant among the 62 hyperspectral bands, variable importance analysis of RF regression models was utilized and the result was shown in Figure 4.

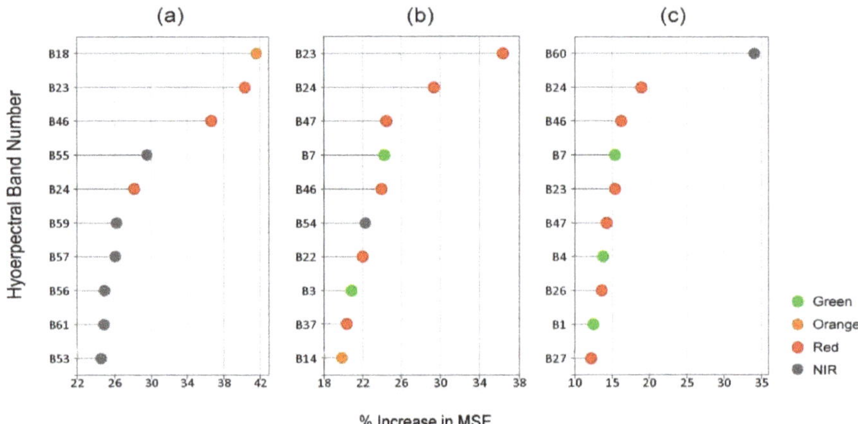

Figure 4. Variable importance measured by percentage increase in mean squared error (MSE) for the UAV-derived prediction models of field A (**a**), B (**b**), and C (**c**).

As shown in Figure 4, B18 (0.61 µm), B23 (0.65 µm), and B60 (0.87µm) were the most important bands for the UAV-derived prediction models of fields A, B, and C, respectively. They provided approximately 42%, 36%, and 34% increase in MSE for the regression models of the study area. It indicated that the red bands of fields A and B were of great significance, and the bands in the NIR spectral range were more important for field C when estimating soil salinity using UAV-derived hyperspectral data. As shown in Figure 4a, six of the top ten important bands for the prediction model of field A were NIR bands. The accumulated variable importance of NIR bands in Figure 4a reached up to 156%, suggesting that those bands are also critical to the modeling of soil salinity in the field A. The results were in accordance with the results of existing research. Sidike et al. [66] selected soil salinity sensitive bands using PLSR method, and the results indicated that the near-infrared band had the most contribution to the estimation of soil salinity. The statistical analysis of Fan et al. [67] demonstrated that soil salinity was more correlated with NIR and SWIR bands with larger negative correlation coefficients. Resulting from raw reflectance correlogram, first derivative reflectance correlogram, and

PLSR carried out by Zhang et al. [68], wavelengths at 395~410, 483~507, 632~697, 731~762, 812~868, 884~909, and 918~930 nm were found to be the most sensitive wavebands. In the spectral range of 500~890 nm, wavelengths at 632~697, 731~762, and 812~868 nm covered B22~B31, B40~B47, and B53~B60 of the hyperspectral data in this research, respectively. Regarding to Figure 4, it was worth noticing that B23, B24, and B46 were among the top five most important hyperspectral bands for RF models in all three fields. Meanwhile, some NIR bands, including B53, B55, B56, B57, and B59, were all presented as important variables for RF model of the field A, as shown in Figure 4a.

5. Conclusions

This paper examined unmanned aerial vehicle-borne hyperspectral data and Chinese GF-2 satellite data for RF modeling to quantitatively estimate soil salinity in fields with various vegetation cover conditions. The strongest linear relationships between EM38-MK2-measured EC_{ah}, EC_{av}, and laboratory-analyzed $EC_{1:5}$ of the samples was found in the field C with R^2 values 0.95. The bare land (field A) had the most saline soil, and its average $EC_{1:5}$ of the soil samples was 37.64 dS m^{-1}. The results showed that bare land with high salt content in soil had the most accurate estimation result among three fields. In addition, resampling UAV data to 1 m was necessary to get a reasonable relation to EMI measurements. For UAV-derived prediction models, the most important spectral band for salinity prediction was B18, B23, and B60 for the fields A, B, and C, respectively. Whereas B23, B24, and B46 were all significant to RF models of three fields. While the UAV platform was satisfactory for collecting spectral information to establishing regression models between EC and soil surface reflectance, soil salinity estimation achieved more accurate results for bare land and sparse vegetation area than dense vegetation area. As the acquired ultra-high-resolution images can capture details of ground objects, the UAV-borne hyperspectral imager was recommended for very accurate soil salinity mapping, monitoring and assessment in order to assist decision making in precision agriculture.

Author Contributions: Conceptualization, J.H., J.P. and Z.S.; Data curation, J.H. and R.Z.; Funding acquisition, J.P. and Z.S.; Investigation, J.P. and R.Z.; Methodology, J.H., J.P., Y.Z., and D.X.; Project administration, Z.S.; Resources, Q.J.; Software, D.X.; Supervision, Y.Z. and Z.S.; Visualization, J.H., Y.Z. and T.F.; Writing—original draft, J.H. and T.F.; Writing—review and editing, J.H., J.P., Y.Z., D.X., R.Z., Q.J., and F.W.

Funding: This research was funded by the National Key Research and Development Program of China (2016FYC0501400) and the National Natural Science Foundation of China (No. 41361048; No. 41261083).

Conflicts of Interest: The authors declare no conflict of interest. The funders had no role in the design of the study; in the collection, analyses, or interpretation of data; in the writing of the manuscript, or in the decision to publish the results.

References

1. Dehni, A.; Lounis, M. Remote sensing techniques for salt affected soil mapping: Application to the oran region of algeria. *Procedia Eng.* **2012**, *33*, 188–198. [CrossRef]
2. Ghassemi, F.; Jakeman, A.J.; Nix, H.A. *Salinisation of Land and Water Resources: Human Causes, Extent, Management and Case Studies*; CAB International: Wallingford, UK, 1995.
3. Corwin, D. *Past, Present, and Future Trends of Soil Electrical Conductivity Measurement Using Geophysical Methods*; FAO: Rome, Italy, 2008.
4. Ding, J.; Yu, D. Monitoring and evaluating spatial variability of soil salinity in dry and wet seasons in the werigan–kuqa oasis, china, using remote sensing and electromagnetic induction instruments. *Geoderma* **2014**, *235–236*, 316–322. [CrossRef]
5. Guo, Y.; Huang, J.; Shi, Z.; Li, H. Mapping spatial variability of soil salinity in a coastal paddy field based on electromagnetic sensors. *PLoS ONE* **2015**, *10*, e0127996. [CrossRef] [PubMed]
6. de Jong, E.; Ballantyne, A.K.; Cameron, D.R.; Read, D.W.L. Measurement of apparent electrical conductivity of soils by an electromagnetic induction probe to aid salinity surveys. *Soil Sci. Soc. Am. J.* **1979**, *43*, 810–812. [CrossRef]

7. McNeill, J.D. Rapid, accurate mapping of soil salinity by electromagnetic ground conductivity meters. In *Advances in Measurement of Soil Physical Properties: Bringing Theory into Practice*; Topp, G.C., Reynolds, W.D., Green, R.E., Eds.; Soil Science Society of America: Madison, WI, USA, 1992; pp. 209–229.
8. Corwin, D.L.; Lesch, S.M. Apparent soil electrical conductivity measurements in agriculture. *Comput. Electron. Agric.* **2005**, *46*, 11–43. [CrossRef]
9. Guo, Y.; Shi, Z.; Li, H.Y.; Triantafilis, J. Application of digital soil mapping methods for identifying salinity management classes based on a study on coastal central china. *Soil Use Manag.* **2013**, *29*, 445–456. [CrossRef]
10. Triantafilis, J.; Odeh, I.O.A.; McBratney, A.B. Five geostatistical models to predict soil salinity from electromagnetic induction data across irrigated cotton. *Soil Sci. Soc. Am. J.* **2001**, *65*, 869–878. [CrossRef]
11. Csillag, F.; Pasztor, L.; Biehl, L.L. Spectral band selection for the characterization of salinity status of soils. *Remote Sens. Environ.* **1993**, *43*, 231–242. [CrossRef]
12. Eldeiry, A.A.; Garcia, L.A. Detecting soil salinity in alfalfa fields using spatial modeling and remote sensing. *Soil Sci. Soc. Am. J.* **2008**, *72*, 201–211. [CrossRef]
13. Kalra, N.K.; Joshi, D.C. Potentiality of landsat, spot and irs satellite imagery, for recognition of salt affected soils in indian arid zone. *Int. J. Remote Sens.* **1996**, *17*, 3001–3014. [CrossRef]
14. Ben-Dor, E.; Metternicht, G.; Goldshleger, N.; Mor, E.; Mirlas, V.; Basson, U. *Review of Remote Sensing-Based Methods to Assess Soil Salinity*; Taylor & Francis Group: Abingdon, UK, 2009; pp. 39–60.
15. Metternicht, G.I.; Zinck, J.A. Remote sensing of soil salinity: Potentials and constraints. *Remote Sens. Environ.* **2003**, *85*, 1–20. [CrossRef]
16. Dwivedi, R.S.; Sreenivas, K. Image transforms as a tool for the study of soil salinity and alkalinity dynamics. *Int. J. Remote Sens.* **1998**, *19*, 605–619. [CrossRef]
17. Hick, P.T.; Russell, W.G.R. Some spectral considerations for remote-sensing of soil-salinity. *Aust. J. Soil Res.* **1990**, *28*, 417–431. [CrossRef]
18. Mougenot, B.; Pouget, M.; Epema, G.F. Remote sensing of salt affected soils. *Remote Sens. Rev.* **1993**, *7*, 241–259. [CrossRef]
19. Verma, K.S.; Saxena, R.K.; Barthwal, A.K.; Deshmukh, S.N. Remote-sensing technique for mapping salt-affected soils. *Int. J. Remote Sens.* **1994**, *15*, 1901–1914. [CrossRef]
20. Dehaan, R.L.; Taylor, G.R. Field-derived spectra of salinized soils and vegetation as indicators of irrigation-induced soil salinization. *Remote Sens. Environ.* **2002**, *80*, 406–417. [CrossRef]
21. Douaoui, A.E.K.; Nicolas, H.; Walter, C. Detecting salinity hazards within a semiarid context by means of combining soil and remote-sensing data. *Geoderma* **2006**, *134*, 217–230. [CrossRef]
22. Eldeiry, A.A.; Garcia, L.A. Comparison of ordinary kriging, regression kriging, and cokriging techniques to estimate soil salinity using landsat images. *J. Irrig. Drain. Eng. Asce* **2010**, *136*, 355–364. [CrossRef]
23. Fernandez-Buces, N.; Siebe, C.; Cram, S.; Palacio, J.L. Mapping soil salinity using a combined spectral response index for bare soil and vegetation: A case study in the former lake texcoco, mexico. *J. Arid Environ.* **2006**, *65*, 644–667. [CrossRef]
24. Ben-Dor, E.; Patkin, K.; Banin, A.; Karnieli, A. Mapping of several soil properties using dais-7915 hyperspectral scanner data—A case study over clayey soils in israel. *Int. J. Remote Sens.* **2002**, *23*, 1043–1062. [CrossRef]
25. Cloutis, E.A. Hyperspectral geological remote sensing: Evaluation of analytical techniques. *Int. J. Remote Sens.* **1996**, *17*, 2215–2242. [CrossRef]
26. Ghosh, G.; Kumar, S.; Saha, S.K. Hyperspectral satellite data in mapping salt-affected soils using linear spectral unmixing analysis. *J. Indian Soc. Remote Sens.* **2012**, *40*, 129–136. [CrossRef]
27. Weng, Y.; Qi, H.; Fang, H.; Zhao, F.; Lu, Y. Plsr-based hyperspectral remote sensing retrieval of soil salinity of chaka-gonghe basin in qinghai province. *Acta Pedol. Sin.* **2010**, *47*, 1255–1263.
28. Farifteh, J.; Van der Meer, F.D.; Atzberger, C.; Carranza, E.J.M. Quantitative analysis of salt-affected soil reflectance spectra: A comparison of two adaptive methods (plsr and ann). *Remote Sens. Environ.* **2007**, *110*, 59–78. [CrossRef]
29. Pang, G.; Wang, T.; Liao, J.; Li, S. Quantitative model based on field-derived spectral characteristics to estimate soil salinity in minqin county, china. *Soil Sci. Soc. Am. J.* **2014**, *78*, 546–555. [CrossRef]
30. Berni, J.A.J.; Fereres Castiel, E.; Suárez Barranco, M.D.; Zarco-Tejada, P.J. Thermal and narrow-band multispectral remote sensing for vegetation monitoring from an unmanned aerial vehicle. *Ieee Trans. Geosci. Remote Sens.* **2009**, *47*, 722–738. [CrossRef]

31. Feng, Q.; Liu, J.; Gong, J. Uav remote sensing for urban vegetation mapping using random forest and texture analysis. *Remote Sens.* **2015**, *7*, 1074–1094. [CrossRef]
32. Aasen, H.; Burkart, A.; Bolten, A.; Bareth, G. Generating 3d hyperspectral information with lightweight uav snapshot cameras for vegetation monitoring: From camera calibration to quality assurance. *Isprs J. Photogramm. Remote Sens.* **2015**, *108*, 245–259. [CrossRef]
33. de Castro, A.I.; Torres-Sanchez, J.; Pena, J.M.; Jimenez-Brenes, F.M.; Csillik, O.; Lopez-Granados, F. An automatic random forest-obia algorithm for early weed mapping between and within crop rows using uav imagery. *Remote Sens.* **2018**, *10*, 285. [CrossRef]
34. Lelong, C.; Burger, P.; Jubelin, G.; Roux, B.; Labbé, S.; Baret, F. Assessment of unmanned aerial vehicles imagery for quantitative monitoring of wheat crop in small plots. *Sensors* **2008**, *8*, 3557. [CrossRef] [PubMed]
35. Ivushkin, K.; Bartholomeus, H.; Bregt, A.K.; Pulatov, A.; Franceschini, M.H.D.; Kramer, H.; van Loo, E.N.; Jaramillo Roman, V.; Finkers, R. Uav based soil salinity assessment of cropland. *Geoderma* **2019**, *338*, 502–512. [CrossRef]
36. Romero-Trigueros, C.; Nortes, P.A.; Alarcón, J.J.; Hunink, J.E.; Parra, M.; Contreras, S.; Droogers, P.; Nicolás, E. Effects of saline reclaimed waters and deficit irrigation on citrus physiology assessed by uav remote sensing. *Agric. Water Manag.* **2017**, *183*, 60–69. [CrossRef]
37. Zhao, K.; Song, J.; Feng, G.; Zhao, M.; Liu, J. Species, types, distribution, and economic potential of halophytes in china. *Plant Soil* **2011**, *342*, 495–526. [CrossRef]
38. Heil, K.; Schmidhalter, U. Comparison of the em38 and em38-mk2 electromagnetic induction-based sensors for spatial soil analysis at field scale. *Comput. Electron. Agric.* **2015**, *110*, 267–280. [CrossRef]
39. Peng, J.; Ji, W.; Ma, Z.; Li, S.; Chen, S.; Zhou, L.; Shi, Z. Predicting total dissolved salts and soluble ion concentrations in agricultural soils using portable visible near-infrared and mid-infrared spectrometers. *Biosyst. Eng.* **2016**, *152*, 94–103. [CrossRef]
40. Roosjen, P.; Suomalainen, J.; Bartholomeus, H.; Kooistra, L.; Clevers, J. Mapping reflectance anisotropy of a potato canopy using aerial images acquired with an unmanned aerial vehicle. *Remote Sens.* **2017**, *9*, 417. [CrossRef]
41. Adler-Golden, S.M.; Matthew, M.W.; Bernstein, L.S.; Levine, R.Y.; Berk, A.; Richtsmeier, S.C.; Acharya, P.K.; Anderson, G.P.; Felde, J.W.; Gardner, J.A.; et al. Atmospheric correction for shortwave spectral imagery based on modtran4. *SPIE* **1999**, *3753*, 61–69.
42. Peng, J.; Biswas, A.; Jiang, Q.S.; Zhao, R.Y.; Hu, J.; Hu, B.F.; Shi, Z. Estimating soil salinity from remote sensing and terrain data in southern xinjiang province, china. *Geoderma* **2019**, *337*, 1309–1319. [CrossRef]
43. Ho, T.K. The random subspace method for constructing decision forests. *Ieee Trans. Pattern Anal. Mach. Intell.* **1998**, *20*, 832–844.
44. Breiman, L. Random forests. *Mach. Learn.* **2001**, *45*, 5–32. [CrossRef]
45. Cutler, A.; Cutler, D.R.; Stevens, J.R. Random forests. In *Ensemble Machine Learning: Methods and Applications*; Zhang, C., Ma, Y., Eds.; Springer US: Boston, MA, USA, 2012; pp. 157–175.
46. Cutler, D.R.; Edwards, T.C.; Beard, K.H.; Cutler, A.; Hess, K.T.; Gibson, J.; Lawler, J.J. Random forests for classification in ecology. *Ecology* **2007**, *88*, 2783–2792. [CrossRef] [PubMed]
47. Mutanga, O.; Adam, E.; Cho, M.A. High density biomass estimation for wetland vegetation using worldview-2 imagery and random forest regression algorithm. *Int. J. Appl. Earth Obs. Geoinf.* **2012**, *18*, 399–406. [CrossRef]
48. Were, K.; Bui, D.T.; Dick, Ø.B.; Singh, B.R. A comparative assessment of support vector regression, artificial neural networks, and random forests for predicting and mapping soil organic carbon stocks across an afromontane landscape. *Ecol. Indic.* **2015**, *52*, 394–403. [CrossRef]
49. Liaw, A.; Wiener, M. Classification and regression by randomforest. *R News* **2002**, *2*, 18–22.
50. R Core Team. *R: A Language and Environment for Statistical Computing*; R Foundation for Statistical Computing: Vienna, Austria, 2018.
51. Forkuor, G.; Hounkpatin, O.K.L.; Welp, G.; Thiel, M. High resolution mapping of soil properties using remote sensing variables in south-western burkina faso: A comparison of machine learning and multiple linear regression models. *PLoS ONE* **2017**, *12*, e0170478. [CrossRef]
52. Zhu, R.; Zeng, D.; Kosorok, M.R. Reinforcement learning trees. *J. Am. Stat. Assoc.* **2015**, *110*, 1770–1784. [CrossRef]

53. Lawrence, I.K.L. A concordance correlation coefficient to evaluate reproducibility. *Biometrics* **1989**, *45*, 255–268.
54. Zhou, Y.; Hartemink, A.E.; Shi, Z.; Liang, Z.; Lu, Y. Land use and climate change effects on soil organic carbon in north and northeast china. *Sci. Total Environ.* **2019**, *647*, 1230–1238. [CrossRef] [PubMed]
55. Farifteh, J.; van der Meer, F.; van der Meijde, M.; Atzberger, C. Spectral characteristics of salt-affected soils: A laboratory experiment. *Geoderma* **2008**, *145*, 196–206. [CrossRef]
56. Mashimbye, Z.E.; Cho, M.A.; Nell, J.P.; De Clercq, W.P.; Van Niekerk, A.; Turner, D.P. Model-based integrated methods for quantitative estimation of soil salinity from hyperspectral remote sensing data: A case study of selected south african soils. *Pedosphere* **2012**, *22*, 640–649. [CrossRef]
57. Ji, W.J.; Li, S.; Chen, S.C.; Shi, Z.; Rossel, R.A.V.; Mouazen, A.M. Prediction of soil attributes using the chinese soil spectral library and standardized spectra recorded at field conditions. *Soil Tillage Res.* **2016**, *155*, 492–500. [CrossRef]
58. Roberts, D.R.; Bahn, V.; Ciuti, S.; Boyce, M.S.; Elith, J.; Guillera-Arroita, G.; Hauenstein, S.; Lahoz-Monfort, J.J.; Schröder, B.; Thuiller, W.; et al. Cross-validation strategies for data with temporal, spatial, hierarchical, or phylogenetic structure. *Ecography* **2017**, *40*, 913–929. [CrossRef]
59. Honkavaara, E.; Saari, H.; Kaivosoja, J.; Pölönen, I.; Hakala, T.; Litkey, P.; Mäkynen, J.; Pesonen, L. Processing and assessment of spectrometric, stereoscopic imagery collected using a lightweight uav spectral camera for precision agriculture. *Remote Sens.* **2013**, *5*, 5006–5039. [CrossRef]
60. Lorenz, S.; Salehi, S.; Kirsch, M.; Zimmermann, R.; Unger, G.; Vest Sørensen, E.; Gloaguen, R. Radiometric correction and 3d integration of long-range ground-based hyperspectral imagery for mineral exploration of vertical outcrops. *Remote Sens.* **2018**, *10*, 176. [CrossRef]
61. Le Rest, K.; Pinaud, D.; Monestiez, P.; Chadoeuf, J.; Bretagnolle, V. Spatial leave-one-out cross-validation for variable selection in the presence of spatial autocorrelation. *Glob. Ecol. Biogeogr.* **2014**, *23*, 811–820. [CrossRef]
62. Xu, H.A.O.; Li, Y.A.N.; Xu, G.; Zou, T. Ecophysiological response and morphological adjustment of two central asian desert shrubs towards variation in summer precipitation. *Plantcell Environ.* **2007**, *30*, 399–409. [CrossRef] [PubMed]
63. Flowers, T.J.; Colmer, T.D. Salinity tolerance in halophytes*. *New Phytol.* **2008**, *179*, 945–963. [CrossRef]
64. Zhang, T.T.; Qi, J.G.; Gao, Y.; Ouyang, Z.T.; Zeng, S.L.; Zhao, B. Detecting soil salinity with modis time series vi data. *Ecol. Indic.* **2015**, *52*, 480–489. [CrossRef]
65. Allbed, A.; Kumar, L.; Aldakheel, Y.Y. Assessing soil salinity using soil salinity and vegetation indices derived from ikonos high-spatial resolution imageries: Applications in a date palm dominated region. *Geoderma* **2014**, *230*, 1–8. [CrossRef]
66. Sidike, A.; Zhao, S.; Wen, Y. Estimating soil salinity in pingluo county of china using quickbird data and soil reflectance spectra. *Int. J. Appl. Earth Obs. Geoinf.* **2014**, *26*, 156–175. [CrossRef]
67. Fan, X.; Liu, Y.; Tao, J.; Weng, Y. Soil salinity retrieval from advanced multi-spectral sensor with partial least square regression. *Remote Sens.* **2015**, *7*, 488–511. [CrossRef]
68. Zhang, T.T.; Zeng, S.L.; Gao, Y.; Ouyang, Z.T.; Li, B.; Fang, C.M.; Zhao, B. Using hyperspectral vegetation indices as a proxy to monitor soil salinity. *Ecol. Indic.* **2011**, *11*, 1552–1562. [CrossRef]

Technical Note

Multi-Channel Optical Receiver for Ground-Based Topographic Hyperspectral Remote Sensing

Sean E. Salazar * and Richard A. Coffman

Department of Civil Engineering, University of Arkansas, 4190 Bell Engineering Center, Fayetteville, AR 72701, USA; rick@uark.edu

* Correspondence: ssalazar@uark.edu

Received: 21 January 2019; Accepted: 5 March 2019; Published: 9 March 2019

Abstract: Receiver design is integral to the development of a new remote sensor. An effective receiver delivers backscattered light to the detector while optimizing the signal-to-noise ratio at the desired wavelengths. Towards the goal of effective receiver design, a multi-channel optical receiver was developed to collect range-resolved, backscattered energy for simultaneous hyperspectral and differential absorption spectrometry (LAS) measurements. The receiver is part of a new, ground-based, multi-mode lidar instrument for remote characterization of soil properties. The instrument, referred to as the soil observation laser absorption spectrometer (SOLAS), was described previously in the literature. A detailed description of the multi-channel receiver of the SOLAS is presented herein. The hyperspectral channel receives light across the visible near-infrared (VNIR) to shortwave infrared (SWIR) spectrum (350–2500 nm), while the LAS channel was optimized for detection in a narrower portion of the near-infrared range (820–850 nm). The range-dependent field of view for each channel is presented and compared with the beam evolution of the SOLAS instrument transmitter. Laboratory-based testing of each of the receiver channels was performed to determine the effectiveness of the receiver. Based on reflectance spectra collected for four soil types, at distances of 20, 35, and 60 m from the receiver, reliable hyperspectral measurements were gathered, independent of the range to the target. Increased levels of noise were observed at the edges of the VNIR and SWIR detector ranges, which were attributed to the lack of sensitivity of the instrument in these regions. The suitability of the receiver design, for the collection of both hyperspectral and LAS measurements at close-ranges, is documented herein. Future development of the instrument will enable the combination of long-range, ground-based hyperspectral measurements with the LAS measurements to correct for absorption, due to atmospheric water vapor. The envisioned application for the instrument includes the rapid characterization of bare or vegetated soils and minerals, such as are present in mine faces and tailings, or unstable slopes.

Keywords: instrument development; hyperspectral; spectroradiometry; telescope; receiver; soil

---

1. Introduction

All remote sensors, including various types of lidar instruments, employ receivers to collect backscattered energy. The receiver design is commonly dependent on the sensor type and the instrument application. While some lidar receivers use one or more lenses to focus and collimate incoming light, others utilize custom, large-aperture optical arrays to maximize, split, or otherwise manipulate the received energy. Ground-based, atmosphere-focused laser absorption spectrometry (LAS) instruments, commonly identified as differential absorption lidars (DIAL), have often employed a telescope as the primary aperture of the receiver [1–4]. Compact, large-diameter telescopes have been favored because the relative light grasp of a telescope is directly proportional to the square of the aperture area, aiding in long-range atmospheric measurements.

While most examples in the literature utilize simple, single-channel, configurations to receive light, some researchers have designed multi-channel optical receivers, placed between the primary aperture (telescope) and the data acquisition system. For example, Moore et al. [5] split light into separate channels to allow for simultaneous low-gain/high-gain detection and laser-to-telescope alignment. Likewise, Repasky [6] and Moen [7] split light into near-field and far-field receiver channels to provide atmospheric measurements over short (1 km) and long (up to 12 km) ranges, respectively. In another iteration of the Moen [7] two-channel DIAL receiver, a shared telescope for transmission and receiving enabled stable alignment and eye-safe beam expansion [4].

As DIAL instruments have historically been developed to collect atmospheric backscatter from water vapor and aerosols in the troposphere, there are limited examples of DIAL instruments operating in horizontal orientations to collect backscatter from a topographic (hard) target [1,8–10]. In the aforementioned instances, the topographic targets served as a test for bias, due to differential spectral reflectance [8,10], or as a measurement of spectral purity [9]. Typical DIAL configurations provide information for two wavelengths (one wavelength centered on a molecular species absorption line, λ_{on}, while the second, nearby wavelength, λ_{off}, serves as a reference).

In this paper, a multi-channel optical receiver is described. The receiver was developed to enable simultaneous range-resolved hyperspectral measurements of hard targets and differential laser absorption measurements for atmospheric corrections of the hyperspectral measurements. The receiver is part of a new ground-based remote sensing instrument, called the soil observation laser absorption spectrometer (SOLAS), previously described in the literature by Salazar et al. [11]. The instrument was developed for rapid characterization of bare soil, rock surfaces, and/or vegetation. There is also potential for cross-platform calibration and validation (ground-truth) of airborne or upcoming spaceborne hyperspectral missions, such as PRISMA, EnMAP, HISUI, and HyspIRI [12–15]. The SOLAS instrument transmits two amplitude-modulated continuous-wave (AM-CW) near-infrared (NIR) lasers with wavelengths of 823.20 nm and 847.00 nm. The SOLAS receives backscattered light with a hyperspectral sensor and a pair of near-infrared photodetectors. The hyperspectral receiver detects light continuously across the visible to shortwave infrared (SWIR) range (350–2500 nm). A balanced photodetector is used to determine the range to the target using a frequency-modulated continuous-wave (FMCW) lidar, while an avalanche photodetector is used to determine the horizontal concentration of atmospheric water vapor en route to the target via a differential laser absorption measurement technique. The atmospheric measurements will be used in the future to correct the hyperspectral reflectance from long-range targets. Although the SOLAS instrument was described previously [11], a more detailed discussion of the development and testing of the multi-channel receiver portion of the instrument, as used to collect the backscattered energy, is discussed in the following sections.

2. Materials and Methods

The primary aperture of the SOLAS instrument receiver consists of a Meade LX200-ACF Schmidt-Cassegrain catadioptric telescope (Meade Instruments; Irvine, California, USA). The surfaces of the telescope optics are coated with a proprietary Ultra High Transmission Coating (UHTC). The UHTC is designed to reduce reflections while maximizing light transmission. Various compounds are used in the coating (aluminum and titanium oxides on the front and back of the corrector lens; titanium and silicon dioxides on the reflecting surface of the primary and secondary mirrors). The telescope has a diameter of 203 mm and an effective focal length of 2032 mm that focuses light into a multi-channel, polarization insensitive, optical relay mounted to the rear port of the telescope. An uncoated Thorlabs LB1471 field lens (Thorlabs Inc.; Newton, NJ, USA), positioned at the focal plane of the telescope, gathers the received light from the rear port. Positioned behind the field lens is a 0.8–25.0 mm diameter adjustable Thorlabs SM1D25 iris and an uncoated Thorlabs LBF254-050 spherical singlet collimator lens. A Thorlabs BPD254-G Polka-Dot 50:50 beamsplitter positioned at 45° splits the collimated light evenly into two separate channels; one hyperspectral channel and one LAS channel.

The hyperspectral channel, referred to in this paper as Channel 1, is reserved for hyperspectral backscatter measurements. For this channel, light is focused with two uncoated aspheric lenses (Thorlabs AL1512 and AL108) and coupled into a high radiometric-resolution spectroradiometer (Analytical Spectral Devices (ASD) FieldSpec 4 Hi-Res; Malvern Panalytical, Longmont, CO, USA) via a multimode fiber bundle. The ASD FieldSpec 4 instrument detects light continuously over the visible to SWIR wavelengths using 2151 bands. The visible near-infrared (VNIR) bands, ranging in wavelength from 350 to 1000 nm, use a silicon detector to provide a spectral resolution of 3 nm and a sampling interval of 1.4 nm. Two sets of SWIR bands, ranging in wavelength from 1001 to 1800 nm and 1801 to 2500 nm, each using a thermoelectric-cooled indium gallium arsenide (InGaAs) detector, provide a spectral resolution of 8 nm and a sampling interval of 1.1 nm. The wavelength reproducibility is 0.1 nm and the wavelength accuracy is 0.5 nm.

The LAS channel, referred to in this paper as Channel 2, focuses light via two, coated, positive achromatic doublet lenses (Thorlabs AC127-050-B and AC080-10-B) and optionally filters the light using one of two interchangeable narrow bandpass filters, centered at 820 nm or 850 nm (Thorlabs FB820-10 and FB850-10, respectively), each with full-width at half-maximum (FWHM) filtering of 10 ± 2 nm. After focusing and filtering, the light in Channel 2 is collimated into a 50-μm core diameter, anti-reflective-coated, step-index multimode, fiber optic cable (Thorlabs M50L02S-B) via a Thorlabs PAF-SMA-5-B aspheric lens fiber-coupling stage. The aforementioned light on Channel 2 is delivered to a pair of near-infrared photodetectors as part of a topographic LAS measurement system. The LAS measurement system is described in further detail in Salazar et al. [11]. A labeled photograph of the receiver is presented in Figure 1 and a schematic of the receiver is presented in Figure 2.

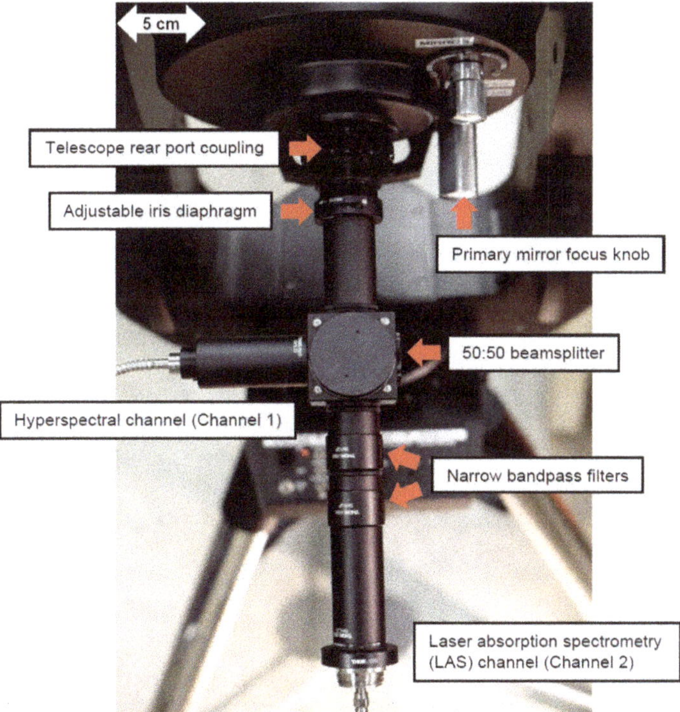

Figure 1. Labeled photograph of the multi-channel optical receiver for the soil observation laser absorption spectrometer (SOLAS).

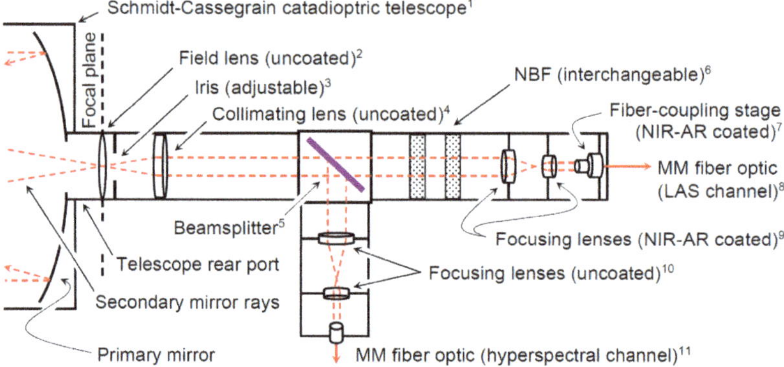

Figure 2. Schematic of the multi-channel optical receiver for the soil observation laser absorption spectrometer (SOLAS) instrument (not to scale). Key: [1] Primary aperture (Meade Instruments LX200-ACF telescope), Ø 203 mm, f_{eff} = 2032 mm, f/10; [2] Uncoated biconvex lens (Thorlabs (TL) LB1471), f = 50 mm; [3] Adjustable iris diaphragm (TL SM1D25), Ø 0.8–25 mm; [4] Uncoated spherical singlet lens (TL LBF254-050), f = 50 mm; [5] Uncoated broad transmission 50:50 polka-dot beamsplitter (TL BPD254-G); [6] Narrow bandpass filters (NBF): 820 nm (TL FB820-10) or 850 nm (TL FB850-10); [7] Near-infrared anti-reflective (NIR-AR) coated aspheric lens fiber-coupling stage (TL PAF-SMA-5-B), 4.9 mm clear aperture, f = 4.6 mm; [8] AR coated multi-mode (MM) fiber optic cable (TL M50L02S-B), Ø 50 μm, numerical aperture = 0.22; [9] NIR-AR coated achromatic doublet lenses, f = 25 mm (TL AC127-050-B), f = 10 mm (TL AC080-10-B); [10] Uncoated aspheric lenses, f = 12 mm (TL AL1512), f = 8 mm (TL AL108); [11] MM fiber optic bundle to ASD FieldSpec 4 Hi-Res spectroradiometer.

The field of view (FOV) for each of the receiver channels was determined using Equation (1) [16]. The diameter of the fiber core, D_f, and the focal length, f, of the primary mirror of the telescope were used to determine the FOV.

$$FOV = \frac{D_f}{f} \tag{1}$$

The placement of the optical components (focusing and collimating lenses) between the telescope and the fiber for each channel of the receiver magnifies the image onto the core of the fiber, thereby increasing the FOV of the channel [9]. Thus, the image is magnified by factors of 50/12 and 12/8 for Channel 1 (hyperspectral channel), where light is focused onto the bare end of the fiber bundle. The 105 μm core diameter for the VNIR bands and 200 μm core for the SWIR bands resulted in a FOV of 0.321 mrad and 0.612 mrad for the VNIR and SWIR bands, respectively. For Channel 2 (LAS channel), where light is focused into 50μm fiber using a fiber-coupling stage, the image is magnified by factors of 50/25, 25/10, and 10/4.6, resulting in a FOV of 0.267 mrad. As part of the LAS functionality of the SOLAS instrument, the actively transmitted laser has a variable beam diameter of 2.0 mm up to a maximum of 8.0 mm and a beam divergence of 0.285 mrad. A plot of the FOV diameter as a function of range, for each of the receiver channels, is presented in Figure 3. For comparison, the laser beam evolution is included, though the relationship between the FOV and the laser beam diameter is only important for the LAS measurements, which are not presented in this paper. For completeness, the specifications for each of the receiver channels are summarized in Table 1.

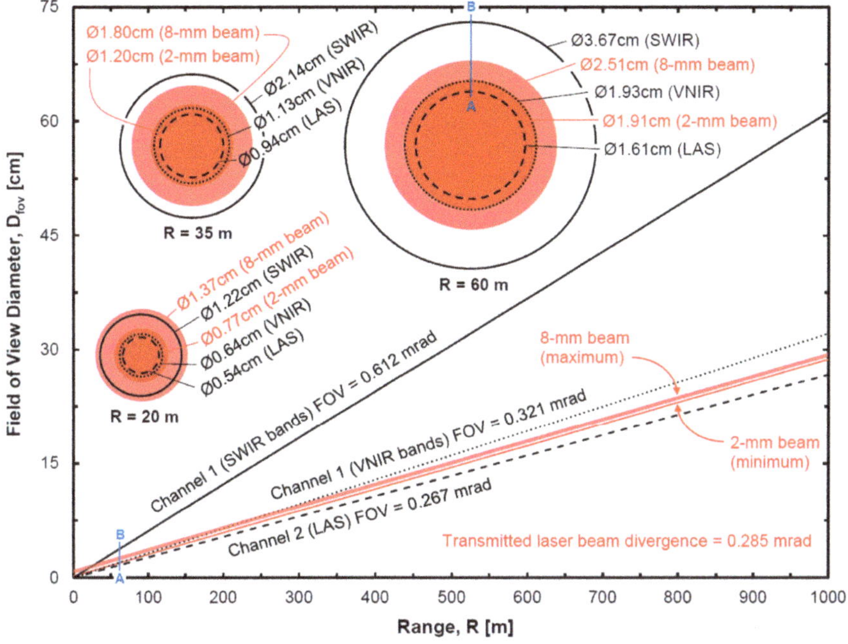

Figure 3. Diameter of the field of view as a function of range for each of the soil observation laser absorption spectrometer (SOLAS) receiver channels including graphical representation of the field of view cross-sections for the three range distances (20, 35, and 60 m) tested in this paper (transmitted laser beam evolution as a function of range shown for reference). Key: SWIR = Shortwave Infrared (1001–2500 nm); VNIR = Visible Near-Infrared (350–1000 nm); LAS = Laser Absorption Spectrometry; R = Range; FOV = Field of View.

Table 1. Specifications for the SOLAS instrument multi-channel receiver.

Primary aperture	Unit	Specification
Telescope	-	Schmidt-Cassegrain
Diameter (D)	(mm)	203
Focal length (f)	(mm)	2032
N (f/D)	-	10
Common channel	**Unit**	**Specification**
Field lens (uncoated)	-	Thorlabs LB1471
Iris (adjustable)	-	Thorlabs SM1D25
Diaphraghm diameter	(mm)	0.8–25.0
Collimating lens (uncoated)		Thorlabs LBF254-050
f (at λ = 835 nm)	(mm)	50.4
Beamsplitter	-	Thorlabs BPD254-G
Type	-	50:50 Polka-Dot, B270 glass
Hyperspectral channel (Channel 1)	**Unit**	**Specification**
Field of view (FOV)	(mrad)	0.32 (VNIR); 0.61 (SWIR)
Focusing lenses (uncoated)	-	Thorlabs AL1512 and AL108
f (at λ = 1425 nm)	(mm)	12.2 and 8.2
Fiber optic cable	-	Multimode bundle (57 fibers)
Core diameter (D_f)	(µm)	105 (VNIR); 200 (SWIR)
Acceptance angle (θ_a)	(rad)	0.22

Table 1. *Cont.*

LAS channel (Channel 2)	Unit	Specification
FOV	(mrad)	0.27
Focusing lenses (NIR-AR coated)	-	Thorlabs AC127-025-B and AC080-010-B
f (at λ = 835 nm)	(mm)	25.0 and 10.0
Narrow bandpass filters	-	Thorlabs FB820-10 and FB850-10
CWL	(nm)	820 and 852 (tested)
FWHM	(nm)	11.0 and 10.7 (tested)
Fiber-coupling stage (NIR-AR coated)	-	Thorlabs PAF-SMA-5-B
f (at λ = 835 nm)	(mm)	4.6
Fiber optic cable	-	Thorlabs M50L02S-B
Type	-	Step-index multimode (AR-coated)
Core diameter (D_f)	(µm)	50
Acceptance angle (θ_a)	(rad)	0.22

Key: N = F-number; VNIR = Visible Near-Infrared (350–1000 nm); SWIR = Shortwave Infrared (1001–2500 nm); ASD = Analytical Spectral Devices; InGaAs = Indium Gallium Arsenide; LAS = Laser Absorption Spectrometry; NIR = Near-Infrared; AR = Anti-Reflective; CWL = Center Wavelength; FWHM = Full-Width at Half-Maximum.

Receiver Testing

The receiver was tested, in a laboratory setting, to verify the transmission of the wavelengths of interest through each channel. A 25 by 25 cm, calibrated Spectralon® (Labsphere Inc., North Sutton, NH, USA) diffuse reflectance reference panel was positioned with an incidence angle of 32° relative to the receiver and the receiver was focused on the center of the panel at a range of 5 m. To achieve focus, the primary mirror of the telescope was adjusted until the focal plane aligned with the receiver optics. The correct alignment was verified by observing the maximum amplitude response, as measured with the ASD FieldSpec 4 instrument. An ASD "Illuminator" direct-current powered tungsten quartz halogen lamp provided full-spectrum illumination across the reference panel. The ASD FieldSpec 4 instrument collected 10 reflectance spectra of the panel through each of the receiver channels. The reflectance measurement from the panel, as observed through Channel 1, provided a reference (baseline) for the measurements observed through Channel 2.

Four specimens, consisting of different types of soil, were prepared for observation with the receiver. The soil types included: (i) KaoWhite-S, a commercial kaolinite soil (Thiele Kaolin Co., Sandersville, Georgia, USA); (ii) Ottawa sand, a pure silica (O2Si) sand (Humboldt Mfg. Co., Elgin, Illinois, USA); (iii) coarse, quartzitic, Arkansas River sand (Arkhola, Van Buren, Arkansas, USA); and (iv) Donna Fill, a synthetic nepheline synetite material (Donna Fill Co., Little Rock, Arkansas, USA). Each specimen was 25 cm in diameter and 0.5 cm thick. The aforementioned Spectralon® reference panel was placed in view of the receiver at a distance of 20 m, with an effective incidence angle of 32°, and the panel was illuminated with the full-spectrum halogen lamp shining perpendicular to the surface of the panel. Baseline reflectance values were recorded for the panel, followed by the collection of reflectance spectra for each of the soil specimens placed in view of the receiver at the same range and incidence angle as the reference panel. Ten spectra were gathered for each specimen via Channel 1. This procedure was repeated for distances of 35 and 60 m (maximum distance available within the laboratory).

For the data that were collected for the Spectralon® panel and the soil specimens, each set of spectra were averaged, normalized with respect to the reference panel, and plotted as a function of wavelength. A splice correction procedure [17] was applied to the reflectance values for $\lambda > 1000$ nm to eliminate offsets that occurred at the transition wavelengths (1000 nm, 1800 nm) between the VNIR and two SWIR channel bands. A Savitzky-Golay [18] filter was also applied to smooth the spectra.

3. Results and Discussion

The spectral reflectance of the reference panel, as acquired via each of the receiver channels, is presented as a function of wavelength in Figure 4. The reflectance spectrum collected via Channel 1 was characteristic of a Lambertian reflector across the range of wavelengths (reflectance values close to 1.0). Although Channel 2 was designed to deliver light to a pair of near-infrared photodetectors used for the LAS measurements, as discussed previously in this paper and in [11], the specifications of the ASD FieldSpec 4 instrument were well suited for also assessing the functionality of the Channel 2 optical design across the near-infrared wavelength range. This also enabled direct comparison between receiver channels. Analysis of the spectrum collected via Channel 2 revealed that transmission was significantly reduced outside of the VNIR range. These findings were explained by the inclusion of the broadband NIR-AR coatings, optimized for the 650–1050 nm range, that exist on the optical elements within Channel 2; Channel 1 delivers light without any additional optical coatings. The spectra collected via Channel 2, with the addition of each of the interchangeable narrow bandpass filter (820 or 850 nm), indicated the effectiveness of the filters, allowing only collection around the wavelengths of interest (λ_{on} = 823.20 nm or λ_{off} = 847.00 nm) for the LAS measurements. The filters may be employed to isolate the λ_{on} or λ_{off} backscatter in cases where sunlight saturates the returns.

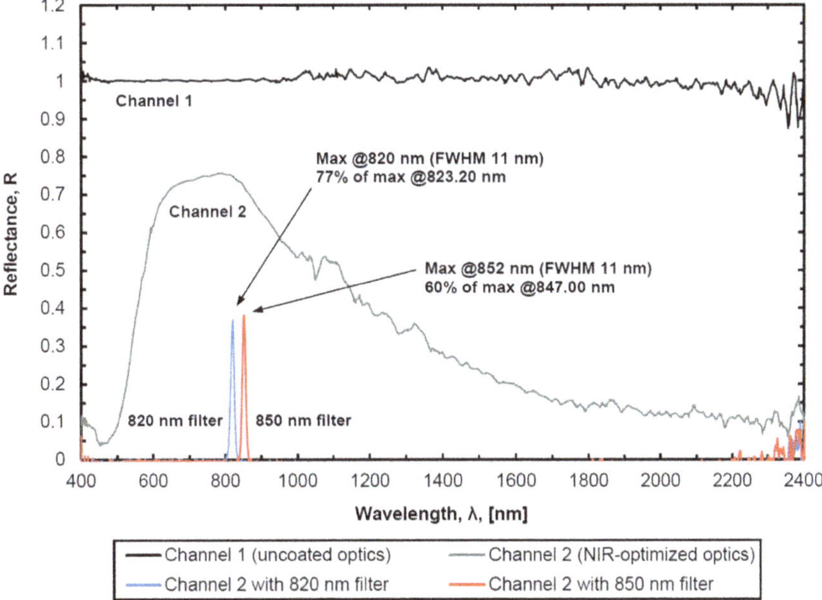

Figure 4. Spectral reflectance as a function of wavelength for Spectralon® white reference panel, as acquired with the ASD FieldSpec 4 spectroradiometer through (1) receiver Channel 1 (uncoated, full-spectrum optics), and (2) receiver Channel 2 (NIR-optimized optics) without additional filtering, and (3) receiver Channel 2 with interchangeable narrowband filters (measured transmission peaks of 820 nm and 852 nm and full-width at half-maximum (FWHM) of 11 nm).

For each of the spectra, increased levels of noise were observed for the wavelengths near the edges of each detector range. The noise was primarily attributed to the lack of sensitivity of the silicon and InGaAs detectors at the edges of the ranges [19,20]. The statistical metrics for each of the three detector ranges of a typical baseline spectrum, as observed via Channel 1 (presented previously in Figure 4), are summarized in Table 2. The SWIR 1 range (1001–1800 nm) was the most stable, followed by the VNIR range (350–1000 nm), and then the SWIR 2 range (1801–2500 nm). The measured signal-to-noise ratio

(SNR) was greatest for the VNIR range. These findings matched other findings in the literature [19,20]. Furthermore, it is hypothesized that the mismatch in the FOV between the VNIR and SWIR bands, as illustrated previously in Figure 3, may be a factor in the spectral noise, due to inconsistent specimen uniformity (surface roughness) between different FOV. Although the maximum range tested was 60 m (Figure 5), the effect that the difference in the FOV between the VNIR and SWIR bands has on the SNR is hypothesized to increase at longer distances. This hypothesis will continue to be tested in future work, especially when performing field measurements at long ranges.

Table 2. Statistical metrics for the baseline spectrum (Spectralon® panel) observed via Channel 1.

Statistical Metric (Reflectance Units)	VNIR Range * (350–1000 nm)	SWIR 1 Range (1001–1800 nm)	SWIR 2 Range * (1801–2500 nm)
Mean	1.00	1.01	0.976
Variance	1.11×10^{-3}	2.00×10^{-4}	4.10×10^{-3}
Sum of Squares of Deviations	7.07×10^{-1}	1.60×10^{-1}	2.83
Standard Deviation	3.34×10^{-2}	1.41×10^{-2}	6.40×10^{-2}
Noise Equivalent Radiance ($W \cdot cm^{-2} \cdot nm^{-1} \cdot sr^{-1}$) †	9.2×10^{-10}	1.7×10^{-9}	7.5×10^{-10}
Signal-to-Noise Ratio (Radiance Units) †	42	25	26

* Erroneous reflectance values greater than 1.2 at the near (350 nm) and far (2500 nm) edges of the wavelength range were excluded from the statistical summary (approximately 1% of the 2151 individual wavelength bands). † Typical values for the midpoint of each wavelength range (measured at 700, 1400, and 2100 nm).

Both receiver channels shared common optical elements ("coated" and "uncoated"), namely the UHTC-coated telescope, and the uncoated field lens, collimating lens, and beamsplitter (see Table 1 for specifications). Although the UHTC was optimized by the telescope manufacturer for wavelengths in the visible range (450–700 nm) for astronomic observations, there was no evidence that the UHTC adversely affected transmission of light outside of this range. To maximize the transmission of full-spectrum light through the hyperspectral channel (Channel 1), the remaining optical elements (common field lens, common collimating lens, common beamsplitter, and the focusing lenses within Channel 1) were uncoated. However, the lens substrates reduced transmission efficiency at longer wavelengths. For example, according to data provided by Thorlabs, transmission of light at 2200 nm was reduced by 10.7% and 9.7% from maximum for the common lenses and the Channel 1 lenses, respectively. Furthermore, due to the wavelength-dependent focal length of the lenses, defocusing of the light most likely occurred at the shortest and longest wavelengths in the spectrum. To optimize detection of the λ_{on} and λ_{off} backscattered signals for the LAS measurements, the design wavelength of the common lenses, after the light was collected by the telescope, was 835 nm (mean wavelength between absorption lines). Similarly, the focal lengths of the lens pair within Channel 2 were optimized for 835 nm. However, the design wavelength of the hyperspectral channel was 1425 nm (mean wavelength of receiver bandwidth). According to data provided by Thorlabs, the sum of the focal length shifts for the pair of uncoated lenses in the hyperspectral channel was +0.79 mm at 2200 nm and −0.52 mm at 500 nm. The effects of transmission losses and defocusing were noted for completeness, but were considered to have an insignificant impact on the measurements, based on the observed SNR.

The relative reflectance spectra of the four tested soil specimens, as acquired via the uncoated optical elements on Channel 1 (hyperspectral channel), are presented as a function of wavelength in Figure 5. The kaolinite soil was the most reflective, followed by the Ottawa sand, while the coarse river sand was less reflective than the Donna Fill at wavelengths below 1000 nm and more reflective than the Donna Fill at wavelengths above 1000 nm. The kaolinite soil spectra exhibited water absorption features around the 970 nm, 1400 nm and 1900 nm wavelength bands with characteristic doublets in the 1400 nm and 2200 nm regions. The Ottawa sand, coarse river sand, and the Donna Fill spectra exhibited absorption features around the 1900 nm wavelength band, with otherwise milder or non-distinguishable features. Although the specimens tested in this study were dry, the hygroscopic moisture content likely affected the fine-grained kaolinite soil more than the other specimens. Typical hygroscopic moisture contents (gravimetric) were determined to be ~1% for the kaolinite soil, <0.2% for the Donna Fill, ~0.1% for the Ottawa sand, and <0.1% for the coarse river sand.

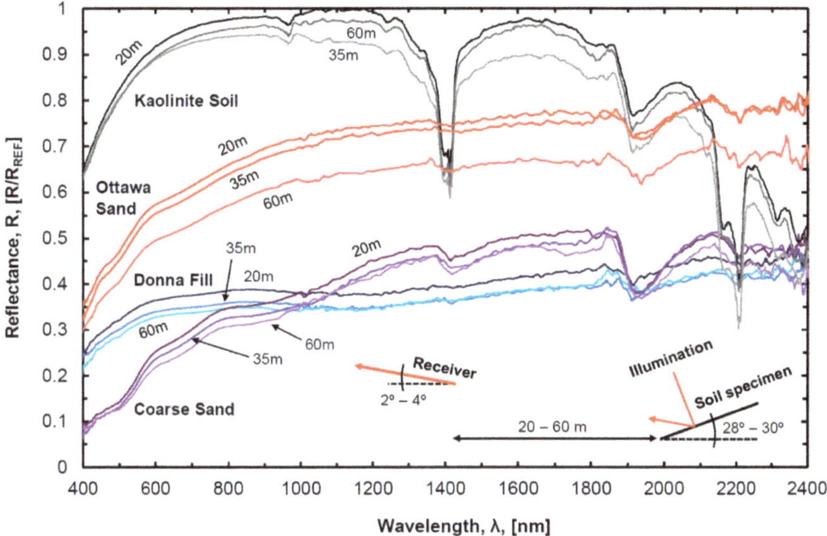

Figure 5. Relative spectral reflectance as a function of wavelength for four soil types (kaolinite, Ottawa sand, Donna Fill, and coarse river sand), as acquired with the ASD FieldSpec 4 spectrometer through receiver Channel 1, in a laboratory setting, for distances of 20, 35, and 60 m and an incidence angle of 32°.

As the observation distance increased, the magnitude of the reflectance for each of the tested specimens generally decreased across the range of wavelengths. However, the shape of each of the spectra was consistent, regardless of distance from the receiver, indicating collection of reliable measurements, independent of the range to target, was possible. The relatively large drop in reflectance, observed for the Ottawa sand specimen at a distance of 60 m, was attributed to the specimen sliding gently due to gravity (resulting in a slightly shallower incidence angle for this measurement). As the specimens were tested in an indoor laboratory environment and under direct illumination of an artificial full-spectrum lamp, no long-path atmospheric absorption or solar absorption features were observed [19]. Thus, the presence of absorption features indicated that even under laboratory conditions (low relative humidity), the measurements were sensitive to absorption and scattering en route to the receiver. The general decrease in reflectance with an increase in range is believed to be attributed to the absorption and scattering, while the increase in the FOV diameter may also be a factor. Future experimental verification is required to verify these hypotheses.

The stable environmental conditions of the laboratory setting minimized the temperature-induced radiometric errors [20] that are typical of the spectroradiometer instrument. A 1-hour warm-up period, before measurements were collected, further minimized these errors. Although frequent referencing of the Spectralon® standard to establish a baseline for subsequent measurements is recommended by the manufacturer, only one reference was collected for each range (20, 35, and 60 m). Future measurements performed in an outdoor field setting will be more sensitive to changes in temperature and illumination conditions (solar irradiation) and may require more frequent referencing of the Spectralon® panel or a companion spectrometer to measure a reference simultaneously. Atmospheric attenuation, due to absorption and scattering by water vapor and other aerosols along the receiver path, will necessitate corrections to derive exact reflectance measurements. These corrections will be achieved using the LAS measurement system of the SOLAS instrument, as described previously [11], and will be addressed in future work.

4. Conclusions

A multi-channel optical receiver was designed and tested for inclusion within a new ground-based, topographic, hyperspectral lidar instrument, called the soil observation laser absorption spectrometer (SOLAS). The primary aperture of the receiver is a 203-mm diameter telescope that focuses backscattered light into an optical beamsplitting array to enable simultaneous data collection via two channels. One of the channels collects hyperspectral radiometric measurements across the visible near-infrared (VNIR) and shortwave infrared (SWIR) ranges (350–2500 nm), while the other channel directs light into a pair of near-infrared photodetectors for range-resolved, laser absorption spectrometry (LAS) measurements in the 820–850 nm region. Testing of each of the channels, in a laboratory setting, demonstrated the suitability of the receiver design for measurements of the wavelengths of interest. Specifically, the hyperspectral channel was optimized to collect light from 350 nm to 2500 nm, while the LAS channel was optimized to detect backscattered energy from transmitted laser absorption lines of 823.20 nm and 847.00 nm.

Testing of four different soil specimens (kaolinite, Ottawa sand, Donna Fill, and coarse river sand), at various distances from the receiver (20, 35, and 60 m), indicated that reliable hyperspectral measurements could be collected, independent of the range to target. Increased noise was observed in the VNIR and SWIR bands, particularly for the wavelengths near the edges of each detector measurement range (350, 1000, 1800, and 2500 nm), which was attributed to lack of instrument sensitivity in these bands. Some of the observed noise was also attributed to diverging fields of view for the VNIR and SWIR bands and wavelength-dependent transmission losses and defocusing of the received light. Future development of the LAS channel will enable atmospheric corrections for long-range hyperspectral measurements (up to 1 km or greater) and has the potential to improve ground-based optical remote sensing practices. Envisioned applications for the receiver, as part of the SOLAS instrument, include rapid classification of soils, rocks and minerals, and vegetation for ecological or agronomic research, forensic investigations of natural hazards (e.g., wildfire-induced erosion and debris flows), or monitoring of earth construction sites (e.g., mine tailings). Future measurements from the terrestrial platform of the SOLAS may provide ground-truth data for airborne or forthcoming spaceborne missions, such as PRISMA, EnMAP, HISUI, and HyspIRI [12–15]. More information on the complete SOLAS instrument is available in Salazar et al. [11].

Author Contributions: Conceptualization: S.E.S. and R.A.C.; Methodology: S.E.S.; Software: S.E.S.; Validation: S.E.S. and R.A.C.; Formal Analysis: S.E.S.; Investigation: S.E.S. and R.A.C.; Resources: R.A.C.; Data Curation: S.E.S.; Writing—Original Draft Preparation: S.E.S.; Writing—Review and Editing: S.E.S. and R.A.C.; Visualization: S.E.S.; Supervision: R.A.C.; Project Administration: R.A.C.; Funding Acquisition: S.E.S. and R.A.C.

Funding: This project was funded by the U.S. Department of Transportation (USDOT) through the Office of the Assistant Secretary for Research and Technology (OST-R) under USDOT Cooperative Agreement No. OASRTRS-14-H-UARK. The views, opinions, findings and conclusions reflected in this publication are solely those of the authors and do not represent the official policy or position of the USDOT/OST-R, or any State or other entity. USDOT/OST-R does not endorse any third party products or services that may be included in this publication. This material is also based upon work supported by the National Science Foundation Graduate Research Fellowship Program under Grant No. DGE-1450079. Any opinions, findings, and conclusions or recommendations expressed in this material are those of the authors and do not necessarily reflect the views of the National Science Foundation.

Conflicts of Interest: The authors declare no conflict of interest. The funders had no role in the design of the study; in the collection, analyses, or interpretation of data; in the writing of the manuscript, or in the decision to publish the results.

Abbreviations

The following abbreviations are used in this manuscript:

AM	Amplitude Modulation
AR	Anti-Reflective
ASD	Analytical Spectral Devices Inc. (a Malvern Panalytical Company, Longmont, CO, USA)
CW	Continuous-Wave
CWL	Center Wavelength
D	Diameter of Primary Mirror (Telescope)
D_f	Diameter of the Fiber (Core)
D_{fov}	Diameter of the Field of View
DIAL	Differential Absorption Lidar
EnMAP	Environmental Mapping and Analysis Program (Germany)
f	Focal Length
f_{eff}	Effective Focal Length
FMCW	Frequency-Modulated Continuous-Wave
FWHM	Full-Width at Half-Maximum
FOV	Field of View (Angular)
HISUI	Hyperspectral Imager SUIte (Japan)
HyspIRI	Hyperspectral Infrared Imager (USA)
θ_a	Acceptance Angle
InGaAs	Indium Gallium Arsenide
λ	Wavelength (Light)
λ_{on}	On-Line Wavelength
λ_{off}	Off-Line Wavelength
LAS	Laser Absorption Spectrometry
LIDAR	Light Detection and Ranging (commonly Lidar)
MM	Multimode (Fiber)
N	F-Number
NIR	Near-Infrared
PRISMA	PRecursore IperSpettrale della Missione Applicativa (Italy)
R	Range to Target
R	Reflectance
R_{REF}	Reference (Reflectance)
SNR	Signal-to-Noise Ratio
SOLAS	Soil Observation Laser Absorption Spectrometer
SWIR	Shortwave Infrared
TL	Thorlabs Inc. (Newton, NJ, USA)
UHTC	Ultra-High Transmission Coating (Meade Instruments Corporation, Irvine, California, USA)
VNIR	Visible Near-Infrared

References

1. Hardesty, R.M. Coherent DIAL Measurement of Range-Resolved Water Vapor Concentration. *Appl. Opt.* **1984**, *23*, 2545–2553. [CrossRef] [PubMed]
2. Little, L.M.; Papen, G.C. Fiber-Based Lidar for Atmospheric Water-Vapor Measurements. *Appl. Opt.* **2001**, *40*, 3417–3427. [CrossRef] [PubMed]
3. Machol, J.; Ayers, T.; Schwenz, K.; Koenig, K.; Hardesty, R.; Senff, C.; Krainak, M.; Abshire, J.; Bravo, H.; Sandberg, S. Preliminary Measurements with an Automated Compact Differential Absorption LIDAR for Profiling Water Vapor. *Appl. Opt.* **2004**, *43*, 3110–3121. [CrossRef] [PubMed]
4. Spuler, S.M.; Repasky, K.S.; Morley, B.; Moen, D.; Hayman, M.; Nehrir, A.R. Field-Deployable Diode-Laser-Based Differential Absorption Lidar (DIAL) for Profiling Water Vapor. *Atmos. Meas. Tech.* **2015**, *8*, 1073–1087. [CrossRef]
5. Moore, A.S.; Brown, K.E.; Hall, W.M.; Barnes, J.C.; Edwards, W.C.; Petway, L.B.; Little, A.D.; Luck, W.S.; Jones, I.W.; Antill, C.W.; et al. Development of the Lidar Atmospheric Sensing Experiment (LASE)—An

Advanced Airborne DIAL Instrument. In *Advances in Atmospheric Remote Sensing with Lidar: Selected Papers of the 18th International Laser Radar Conference (ILRC), Berlin, Germany, 22–26 July 1996*; Ansmann, A., Neuber, R., Rairoux, P., Wandinger, U., Eds.; Springer: Berlin/Heidelberg, Germany, 1996; pp. 281–288., ISBN 9783540618874.

6. Repasky, K.S.; (Montana State University, Bozeman, MT, USA). Personal Correspondence, 19 July 2016.
7. Moen, D.R. Two Channel Receiver Design and Implementation for a Ground Based Micro-Pulse Differential Absorption Lidar (DIAL) Instrument. Master's Thesis, Montana State University, Bozeman, MT, USA, 2016.
8. Grant, W.B. Effect of Differential Spectral Reflectance on DIAL Measurements Using Topographic Targets. *Appl. Opt.* **1982**, *21*, 2390–2394. [CrossRef] [PubMed]
9. Nehrir, A.R. Development of an Eye-Safe Diode-Laser-Based Micro-Pulse Differential Absorption Lidar (MP-DIAL) for Atmospheric Water-Vapor and Aerosol Studies. Ph.D. Thesis, Montana State University, Bozeman, MT, USA, 2011.
10. Ishii, S.; Koyama, M.; Baron, P.; Iwai, H.; Mizutani, K.; Itabe, T.; Sato, A.; Asai, K. Ground-Based Integrated Path Coherent Differential Absorption Lidar Measurement of CO_2: Foothill Target Return. *Atmos. Meas. Tech.* **2013**, *6*, 1359–1369. [CrossRef]
11. Salazar, S.E.; Garner, C.D.; Coffman, R.A. Development of a Multimode Field Deployable Lidar Instrument for Topographic Measurements of Unsaturated Soil Properties: Instrument Description. *Remote Sens.* **2019**, *11*, 289. [CrossRef]
12. Loizzo, R.; Ananasso, C.; Guarini, R.; Lopinto, E.; Candela, L.; Pisani, A.R. The PRISMA Hyperspectral Mission. In Proceedings of the Living Planet Symposium 2016, Prague, Czech Republic, 9–13 May 2016.
13. Guanter, L.; Kaufmann, H.; Segl, K.; Foerster, S.; Rogass, C.; Chabrillat, S.; Kuester, T.; Hollstein, A.; Rossner, G.; Chlebek, C.; et al. The EnMAP Spaceborne Imaging Spectroscopy Mission for Earth Observation. *Remote Sens.* **2015**, *7*, 8830–8857. [CrossRef]
14. Tanii, J.; Kashimura, O.; Ito, Y.; Iwasaki, A. Flight Model of HISUI Hyperspectral Sensor Onboard ISS (International Space Station). In *Sensors, Systems, and Next-Generation Satellites XXI, Proceedings of SPIE Remote Sensing, Warsaw, Poland, 11–14 September 2017*; Neeck, S.P., Bézy, J.-L., Kimura, T., Eds.; SPIE Proceedings: Bellingham, WA, USA, 2017. [CrossRef]
15. HyspIRI Mission Concept Team. *HyspIRI Final Report*; Jet Propulsion Laboratory, California Institute of Technology: Pasadena, CA, USA, 2018.
16. Chourdakis, G.; Papayannis, A.; Porteneuve, J. Analysis of the Receiver Response for a Noncoaxial Lidar System with Fiber-Optic Output. *Appl. Opt.* **2002**, *41*, 2715–2723. [CrossRef] [PubMed]
17. Danner, M.; Locherer, M.; Hank, T.; Richter, K. *Spectral Sampling with the ASD FieldSpec 4—Theory, Measurement, Problems, Interpretation*; EnMAP Field Guides Technical Report; GFZ Data Services: Potsdam, Germany, 2015.
18. Savitzky, A.; Golay, M.J.E. Smoothing and Differentiation of Data by Simplified Least Squares Procedures. *Anal. Chem.* **1964**, *36*, 1627–1639. [CrossRef]
19. Analytical Spectral Devices. *ASD Technical Guide*, 3rd ed.; Hatchell, D., Ed.; Analytical Spectral Devices, Inc.: Boulder, CO, USA, 1999.
20. Hueni, A.; Bialek, A. Cause, Effect, and Correction of Field Spectroradiometer Interchannel Radiometric Steps. *IEEE J. Sel. Top. Appl. Earth Obs. Remote Sens.* **2017**, *10*, 1542–1551. [CrossRef]

Article

Tree Species Classification Based on Hybrid Ensembles of a Convolutional Neural Network (CNN) and Random Forest Classifiers

Uwe Knauer [1,*], Cornelius Styp von Rekowski [2], Marianne Stecklina [2], Tilman Krokotsch [2], Tuan Pham Minh [2], Viola Hauffe [2], David Kilias [1], Ina Ehrhardt [1], Herbert Sagischewski [3], Sergej Chmara [3] and Udo Seiffert [1]

1 Fraunhofer Institute for Factory Operation and Automation IFF, 39106 Magdeburg, Germany; david.kilias@iff.fraunhofer.de (D.K.); ina.ehrhardt@iff.fraunhofer.de (I.E.); udo.seiffert@iff.fraunhofer.de (U.S.)

2 Otto von Guericke University, Faculty of Computer Science, 39106 Magdeburg, Germany; cornelius.vonrekowski@20bn.com (C.S.v.R.); marianne@omnius.com (M.S.); tilman.krokotsch@tu-berlin.de (T.K.); tuan.pham@uni-kassel.de (T.P.M.); viola.hauffe@st.ovgu.de (V.H.)

3 Forstliches Forschungs- und Kompetenzzentrum, ThüringenForst AöR, 99867 Gotha, Germany; Herbert.Sagischewski@forst.thueringen.de (H.S.); Sergej.Chmara@forst.thueringen.de (S.C.)

* Correspondence: uwe.knauer@iff.fraunhofer.de

Received: 29 October 2019; Accepted: 22 November 2019; Published: 26 November 2019

Abstract: In this paper, we evaluate different popular voting strategies for fusion of classifier results. A convolutional neural network (CNN) and different variants of random forest (RF) classifiers were trained to discriminate between 15 tree species based on airborne hyperspectral imaging data. The spectral data was preprocessed with a multi-class linear discriminant analysis (MCLDA) as a means to reduce dimensionality and to obtain spatial–spectral features. The best individual classifier was a CNN with a classification accuracy of 0.73 +/− 0.086. The classification performance increased to an accuracy of 0.78 +/− 0.053 by using precision weighted voting for a hybrid ensemble of the CNN and two RF classifiers. This voting strategy clearly outperformed majority voting (0.74), accuracy weighted voting (0.75), and presidential voting (0.75).

Keywords: hyperspectral imaging; tree species; multiple classifier fusion; convolutional neural network; random forest; rotation forest

1. Introduction

Tree species classification is a challenging and important task for large-area monitoring and managing of forests. For many applications, it is an important step to first retrieve the tree species in order to enable the detection of specific traits such as nutrition state, water content, various stresses, diseases, and other relevant parameters in a species-specific context.

In 2015, Fraunhofer IFF conducted measurement flights over forests in Saxony Anhalt and Thuringia, Germany for two reasons: (1) to develop new methods for detection of biotic stresses related to oak feeding society (green oak-leaf roller, *Tortrix viridana*), mottled umber (*Erannis defoliaria*), and winter moth (*Operophthera brumata*); and (2) to support ground-based forest inventory by airborne assessment of tree species and tree vitality. Both require reliable detection of oak trees as well as other tree species in mixed forests. Hyperspectral imaging was chosen as a means to address these challenging tasks. It is an emerging measurement technique frequently used for optical non-invasive characterization of surfaces and materials. In contrast to the analysis of conventional digital images, multiple image channels that correspond to reflected or transmitted light of a certain small wavelength range, must be taken into account. A typical hyperspectral image consists of hundreds of image bands. A good introduction to hyperspectral image classification was provided by Bioucas-Dias et al. [1]. Challenges mentioned are the limited number of labeled samples and the need to combine spectral and spatial information to obtain good results. Typically, the high-dimensional nature of the data and strong correlations among spectral bands require the extraction of suitable features prior to training of a classifier. The authors also highlight the importance of high spatial image resolution, as most classification techniques assume a single predominant spectral signature per pixel. In contrast, a low image resolution leads to mixed pixels, which requires spectral unmixing as an additional preprocessing step [2].

Wang et al. [3] dedicated a chapter of their book "Remote Sensing of Natural Resources" to the classification of tree species. They listed principal component analysis (PCA), minimum noisef fraction (MNF), canonical discriminant analysis, partial least squares regression (PLS), and wavelet transform as common feature extraction methods. They also provide references where these techniques have been applied. Moreover, an overview on the application of classifiers containing support vector machines (SVM), classification and regression trees (CART), and artificial neural networks (ANN) is given. Fassnacht et al. [4] published a recent survey on tree species classification. The authors considered different types of discriminant analysis (linear, quadratic, canonical, stepwise, regularized, and penalized), SVM, random forest, and maximum likelihood classifiers. With reference to experiences from previous work (e.g., [5]), the authors draw the conclusion that the choice of the classifier is less important than adequate data preprocessing to obtain good results.

Féret et al. [6] investigated the discrimination of 17 tree species in tropical forests. A comparison of different parametric and non-parametric methods showed that SVM with linear or radial basis function kernels outperformed other classifiers given a large number of training samples. For smaller training sets, the authors recommend regularized discriminant analysis. Richter et al. [7] introduced a modified version of discriminant analysis based on PLS. The authors compared their approach to random forests and SVMs on an airborne hyperspectral dataset recorded over a German forest area and obtained better results compared to these classifiers. In this study, random forests performed notably worse than SVMs. To a smaller extent, this is also the result of studies in the Southern Alps conducted by Dalponte et al. [8]. The authors show that the overall accuracy of tree species identification increased by the integration of LiDAR data in addition to hyperspectral measurements.

Xia et al. [9] have demonstrated successful usage of an ensemble classifier on hyperspectral benchmark datasets. In addition, the concept of rotation forests, originally introduced by Rodriguez et al. [10], was successfully applied using CARTs as base classifiers. Compared to MNF, independent component analysis (ICA), and local Fisher discriminant analysis (LFDA) as transformation approaches, the choice of PCA yields the highest accuracy in their studies.

Instead of searching for a favorable transformation of the feature space, certain spectral bands can be selected based on close correspondence to biochemical traits and physical properties of the observed plants. This concept led to a variety of general or trait-specific vegetation indices. Lausch et al. [11] list spectral indices related to greenness and other vitality parameters, mainly using bands in the spectral range of 450–890 nm. A similar overview can be found in a disease detection report published by Sankaran et al. [12]. Mutanga et al. [13] investigated the relationship between water content and spectral variables. Besides known water absorption features located at three different bands, related spectral indices (NDWI [14], WI [15]) and continuum-removed features were included in their correlation analysis.

Airborne hyperspectral imaging generates a large amount of data of which usually only a small fraction is labeled with reference data. Therefore, semi-supervised approaches for classifier training aim to make use of unlabeled data for hyperspectral image classification as well. Here, the term semi-supervised is used for all approaches that includes both labeled and unlabeled data in the training of classifiers to enhance the classification performance. It emphasizes the fact that reference data is still required, but a much larger dataset contributes to the final classifier.

Ayerdi et al. [16] describe a data augmentation method as a means to use the information of unlabeled data. After performing a clustering in the spectral domain, labeled pixels propagate their label to neighboring unlabeled pixels in the spatial domain if they belong to the same cluster.

Wenzhi et al. [17] introduce a semi-supervised feature extraction algorithm named semi-supervised local discriminant analysis (SELD). The combination of an unsupervised local linear feature extraction method with linear discriminant analysis (LDA) results in a projection that separates the different classes while preserving local neighborhoods. In their evaluation, SELD is compared to different feature extraction methods such as PCA and LDA and obtained the best results on several datasets with different classifiers.

Since SVMs are reported to perform well on hyperspectral datasets, an extension to the semi-supervised case is desirable. Vapnik et al. [18] proposed the idea of transductive SVMs, which was applied to hyperspectral data later on (e.g., by Bruzzone et al. [19]).

It is common practice to use both spectral and spatial information for classification. For the problem of tree species classification, a popular approach is to identify individual tree crowns (ITCs), before starting the classification process. Féret et al. [6] used mean shift clustering to segment ITCs and evaluated two different classification methods: object-based classification of the mean spectrum per ITC and majority class assignment based on the decisions per pixel within the ITC. The authors report an increased accuracy for both approaches compared to pixelwise classification without prior tree crown segmentation. Dalponte et al. [20] incorporated the assumption that all pixels belonging to an ITC should have the same label within the training process of a semi-supervised SVM. An improvement over conventional SVMs was observed, especially if the number of training samples is small. The problems related to small numbers of training samples can also be effectively addressed by tensor-based linear and non-linear models as proposed by Makantasis et al. in [21].

Ghamisi et al. [22] proposed a spectral-spatial classifier based on hidden Markov random fields and SVMs. This fully automatic approach performed better than SVMs on widely used datasets. Li et al. [23] integrate spectral and spatial information in a robust Bayesian framework, addressing problems related to noise, the presence of mixed pixels, and small number of training samples. The authors evaluated their algorithm along with several state-of-the-art methods and achieved notably better results in terms of accuracy. Zhang et al. [24] introduced a sparse ensemble learning method, which allows for information sharing between neighboring pixels during the optimization process. While any classifier can potentially be used, the experiments in this study used CARTs. The authors report not only better accuracy, but also a lower runtime of the trained ensemble compared to traditional ensemble methods.

Since the extraction of handcrafted features is time consuming and complex, Makantasis et al. [25] suggest a convolutional neural network (CNN) to automatically construct high-level features. The dimensionality of the input data is reduced by a PCA first and then three convolutional layers hierarchically detect features, which are classified by two fully connected layers in the end. The CNN outperformed different types of SVMs on benchmark datasets.

Ayerdi et al. [16] suggested a regularization step after classification: each pixel adapts its label to the majority class in its neighborhood. If rather homogeneous tree species distributions in the area under consideration can be assumed, this technique leads to more plausible results.

The review of the literature shows a number of promising approaches to classify tree species from hyperspectral imaging data. It was especially shown that recent advances in image classification with CNNs are transferable to hyperspectral image classification. However, for any new application and dataset, several state-of-the-art classifiers and different feature spaces in question must be tested to prove the suitability and superiority of an approach [26]. Given this requirement, the question arises: what benefits the combination of the best available classification techniques and turns them into an ensemble classifier capable of providing for the problem of tree species classification? To answer this question, we focus on popular voting strategies, which are easy to apply to any group of classifiers.

2. Materials and Methods

2.1. Datasets

Figure 1 illustrates the acquisition of the datasets. Hyperspectral imaging data has been recorded during two measurement flights over forests in Saxony Anhalt and Thuringia (Germany) to determine tree species and investigate oak tree specific biotic stresses. Used cameras were NEO Hyspex VNIR 1600 and NEO Hyspex SWIR 320m-e. Cameras and inertial measurement unit (IMU) Novatel SPAN CPT were fixed on a stabilized mount (see Figure 1). The recording of the flight path and orientation of the line scanning camera systems by the IMU is crucial for later alignment of the scanlines into an image. Experiments reported here are based on Hyspex VNIR 1600 images recorded during the second flight. The reasons are to minimize potential errors from alignment and interpolation of the low resolution Hyspex SWIR 320m-e camera to match the grid of the VNIR camera, better weather conditions during the second flight (no cloud cover), and the recovery of diseased oak trees after secondary 'lammas' shoots [27], which is expected to ease tree species classification.

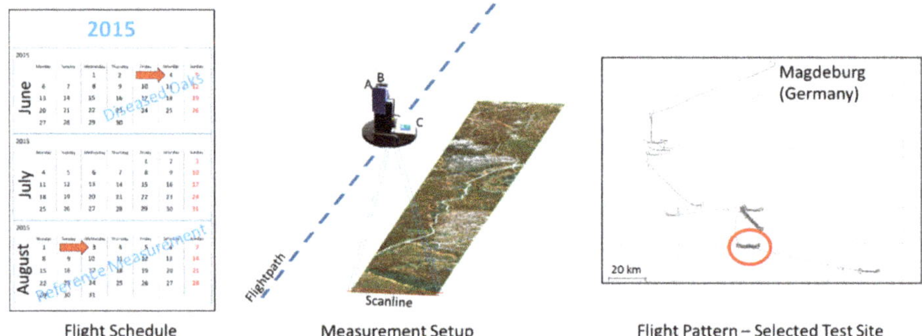

Figure 1. Left: Calendar depicts the dates and scopes of the measurement flights. Middle: Measurement setup for hyperspectral imaging data: (A) Hyspex VNIR 1600, (B) Hyspex SWIR 320m-e, and (C) Novatel SPAN CPT inertial measurement unit (IMU) mounted on stabilized platform (not depicted). Images have been recorded at an altitude of ~1000 m above ground level with FOV of 17° (VNIR) and 14° (SWIR). Right: Trajectory of the flight on 3 August.

Orthorectification of the line scanning data was done with parametric geocoding using the software PARGE [28]. Radiometric corrections were performed with the software ATCOR-4 [29].

One of the major challenges while classifying tree species is the availability of training data that represents differences between individual trees of one species, differences between development stages of individual trees as well as the differences between tree species. In order to account for the variability and to minimize the necessary fieldwork, existing databases and high-resolution aerial images were used to determine areas with a single dominant tree species. Figure 2 illustrates the results of the selection process. As shown, potential error sources such as boundary regions with a mix of tree species, individual trees and clearings were excluded to compensate for potential minor misalignment of hyperspectral images as well as discrepancies due to changes between the available historical images and the current state of the forest cover. Given the locations and shapes of the reference sites, the training data can be derived directly from the hyperspectral images. From 23 known tree species at the study sites, a subset of 15 species is represented by sufficiently large reference sites and was therefore selected to develop and evaluate classifiers. In Table 1, these tree species are listed.

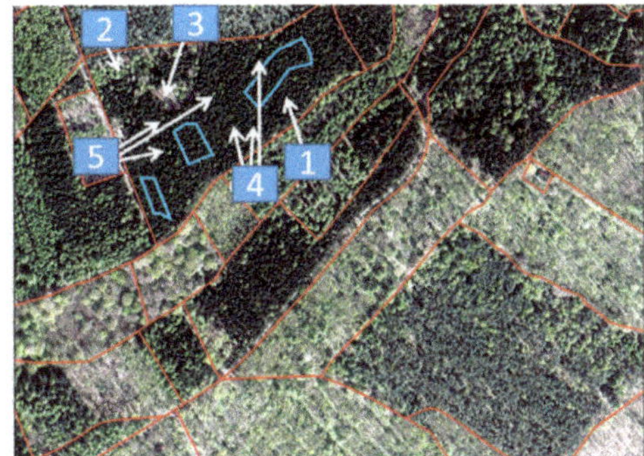

Figure 2. Identification of reference sites with a dominant tree species using existing information systems (forest inventory) and validation with existing aerial images. Left image shows a spruce reference site (1) surrounded by a mix of tree species (2). The right image shows another spruce reference site (1) with a group of birch and maple (2), large (3) as well as small clearings (4), and other isolated trees (5). The blue-bordered regions are the finally derived reference sites.

Table 1. List of 23 tree species from existing databases (forest inventory). For 15 tree species (excluded species marked with *) references sites for the classifier training have been provided forest authorities. The number of available reference sites and the total area are listed for each species.

Tree Species			Pedunculate oak	9	26,046 m^2	Littleleaf lime	1	2295 m^2
European aspen *			Norway spruce	10	20,340 m^2	*Black pine* *		
Sycamore maple *			Scotch pine	13	37,637 m^2	Sessile oak	1	1203 m^2
Birch *			European white birch	1	1117 m^2	Weymouth pine	1	969 m^2
Douglas fir	1	1102 m^2	*Japanese larch* *			Serbian spruce	1	1630 m^2
Oak	10	32,808 m^2	European beech	4	15,720 m^2	Poplar	2	17,649 m^2
European larch	4	8128 m^2	Red oak	3	5732 m^2	Robinia	1	2439 m^2
Ash *			*Red alder* *			*Sitka spruce**		

2.2. Features

As emphasized in the introduction, the selection of an appropriate feature space is crucial for successful classification. By operating in different feature spaces, diversity among the classifiers should increase in many cases. This is a crucial concept in the design of ensemble classifiers [30]. In this study, 15 different feature spaces have been selected to investigate these dependencies between feature spaces and classifiers in terms of improvements in the classification performance. Our choice of feature spaces is explained below.

The measured and preprocessed spectral data itself is a powerful data source for separation of image pixels into classes. Hence, the first feature space is the raw 160-dimensional reflectance data vector per image pixel.

A vector containing a number of spectral indices (calculated based on the reflectance spectra) is used as a second feature space. As presented in the introduction, these indices have been developed to correlate with selected physiological parameters of vegetation, vegetation health, and vegetation nutrition. The following indices have been chosen and are used as features: DI1, GNDVI, MCARI, NDVI, PRI, and WI (see Table 2 for definitions and references).

Classification based on raw spectral data, where individual image bands are simply treated as features as well as using spectral indices as features, represent the most common approaches to the analysis of hyperspectral imaging data. Here, both approaches serve just as a baseline reference for the performance of tree species classification.

Table 2. Selected spectral indices. The values of the indices are combined into a six-dimensional feature vector and used for classification of tree species. The abbreviation R800 denotes reflectance value in band with central wavelength of 800 nm.

Spectral Indices	Formulas	References
DI1	R800 − R550	Lausch et al. (2013)
GNDVI	(R780 − R550)/(R780 + R550)	Gitelson et al. (1996)
MCARI	[(R700 − R670) − 0.2*(R700 − R550)]*R700/R670	Haboudane et al. (2004)
NDVI	(R800 − R670)/(R800 + R670)	Rouse et al. (1974)
PRI	(R531 − R570)/(R531 + R570)	Gamon et al. (1992)
WI	R900/R970	Penuelas et al. (1997)

The high dimensionality of hyperspectral data imposes a number of problems, which are summarized as the curse of dimensionality in the literature [31]. Therefore, we included approaches for reduction of dimensionality prior to classifier training. Principal component analysis (PCA) is performed to calculate a projection into an orthogonal low dimensional subspace, which covers most of the variation of the original high dimensional spectral dataset. We chose 5 and 14 as the numbers of dimensions in the low dimensional representation. This choice is motivated by the number of most prominent tree species in the observed area. Using reference sites with 15 different tree species, 14 is the number of required dimensions for LDA. On the other hand, the choice of five dimensions is motivated by PCA, where the five largest Eigenvalues explain 99.5% of the variation within the spectral dataset.

Multi-class linear discriminant analysis (MCLDA) aims to generate a low dimensional representation, where a single dimension is a projection allowing a good discrimination between elements of one class and all the others. As mentioned above, 5 and 14 dimensions are used again.

SELD was included as a semi-supervised feature extraction technique. SELD combines the supervised method of LDA with the unsupervised method of local linear embedding (LLE). LLE represents the instances in a graph structure and computes a projection that preserves the local neighborhood of instances in the feature space. The combination of supervised and unsupervised feature extraction enables SELD to find features that maximize the class separation and preserve local neighborhoods in the feature space. Therefore, SELD depends on two parameters: the number of instances to consider for local neighborhoods and the number of extracted features.

Table 3 gives an overview of the above listed feature spaces. In addition to the pixelwise transformation of the spectral data into these representations, so-called spatial-spectral features are derived. Within the predefined square image blocks with a side length of $s = 2r + 1$, the statistical measures mean, standard deviation and homogeneity are calculated for each feature and used as new features, instead. This approach combines the information from the low dimensional representations of the spectral data with its local spatial distribution. We set $r = 5$ which corresponds to a side length 4.4 m given the campaigns ground sampling distance of 0.4 m per pixel. This size matches the area of typical tree crowns of boreal trees within the recorded hyperspectral images.

Table 3. Overview of the 15 selected feature spaces. First row: pixel-based feature spaces, Second row: spatial-spectral feature spaces. Due to the high number of channels of the hyperspectral images, the calculation of spatial features was performed for images of reduced dimensionality. PCA: principal component analysis; MCLDA: multi-class linear discriminant analysis; SELD: supervised local discriminant analysis.

	Reflectance Spectra	Indices	PCA 5-dim	PCA 14-dim	MCLDA 5-dim	MCLDA 14-dim	SELD 5-dim	SELD 14-dim
Pixel	·	·	·	·	·	·	·	·
Spatial		·	·	·	·	·	·	·

2.3. Classifiers

In addition to the different feature spaces, a number of classifiers were selected to find the best classifier/feature space combination for the task of tree species classification.

Differently parameterized random forests, rotation forests, SVMs, and a CNN provide state-of-the-art classification results as well as a potentially diverse pool of classifiers for integration into an ensemble classifier.

Random forests are a realization of the concept of bootstrap-aggregation (bagging) with decision trees [32]. As single decision trees tend to overfit, the bagging method creates several bootstrapped sample sets from the original data and trains one tree on each sample set. The resulting ensemble of trees can classify a new instance via majority voting. An important parameter here is the number of trees to be trained. This classifier was chosen because it is a standard approach in literature, when dealing with the classification of tree species from hyperspectral data.

Rotation forests are an advanced form of random forests and were proposed by Rodriguez et al. in 2006 [10]. We decided to test this classifier because of the good results for hyperspectral data as previously reported [9]. As in random forests, the trees are trained on bootstrapped sample sets of the original data. The difference is that the bootstrapping is not performed on a subset of the original data, but on the whole data set with only a subset of features. The feature space is divided into k subsets. For each subset, all instances of randomly drawn classes are deleted from the training data and a bootstrapped sample set with a size of 75% of the original data is created. PCA is performed on each sample set and the resulting coefficient matrices are merged into a single one. This is equivalent to k axis rotations. A decision tree is then trained on the whole PCA-transformed data set. This procedure is repeated for each tree in the ensemble. The two parameters this classifier provides are the number of trees in the ensemble and the number of subsets the feature space is divided into.

SVMs are another standard classification and regression approach. We therefore chose to include them in our experiments. An SVM tries to fit a hyperplane that separates two classes while maximizing the margin between the instances and the plane. The plane can then be described by the instances nearest to it. These instances are called support vectors. To classify more than two classes, one can either train an SVM for each pair of classes (1 versus 1) or each class against the rest (1 versus All). The final decision is then found by majority voting (1 versus 1) or by taking the result of the classifier with the highest confidence (1 versus All). In order to classify not linearly separable classes, the data can be transformed into a higher dimensional space where it is linearly separable. To make this method computable in appropriate time one can use the so-called kernel trick, where the kernel is a type of nonlinear transformation. As the number of adjustable parameters for an SVM is very large, we only chose to vary the kernel (linear or polynomial) and whether to use 1 versus 1 or 1 versus All. We tested several SVM implementations ranging from standard MATLAB classes to native C libraries connected to MATLAB via the MEX interface. We settled for the LibLinear library [33] from the National University of Taiwan for 1 versus All and SVMlin [34] by Vikas Sindhwani for 1 versus 1, both specializing in linear SVM solving. Preliminary tests revealed that polynomial SVMs did not converge in appropriate time and we therefore abandoned this kernel from further investigation.

Deep neural networks are a recent hot topic in machine learning. As Makantasis et al. [25] were able to produce good results on hyperspectral data with convolutional deep neural networks, we added this classification method to our experimental setup. Neural networks are inspired by the structure of the human brain to learn classification and regression tasks. A neural network consists of several layers of perceptron units that are interconnected. Via backpropagation, this network can learn from examples. A deep neural network has significantly more layers than a conventional network. Deep convolutional networks use filters that are shifted over the input image to generate new and more high-level features in each layer. It is not possible to test or even enumerate all possible network structures and parameter configurations. We therefore employed the network structure shown in Figure 3, which is based on the one used in the original paper [25]. The network starts with a convolutional layer containing $C_1 = 3F$ trainable filters (block size 5×5 pixels), where F denotes the number of features. The second convolutional layer then contains $C_2 = 3C_1$ trainable filters (block size 5×5). The output of this layer is then fed to a fully connected network with a hidden layer of size $C_3 = 6F$. Additionally, we added the option to use a max-pooling layer in between the convolution layers and an option to apply a specified dropout rate to the fully connected network part. A pooling layer applies a filter mask for each pixel incorporating its surroundings. Common filters are mean or maximum filter. A dropout layer randomly omits a perceptron for the next training instance with a certain probability to avoid early convergence and to improve generalization. The deep learning library MatConvNet [35] provided all functions needed to build our networks.

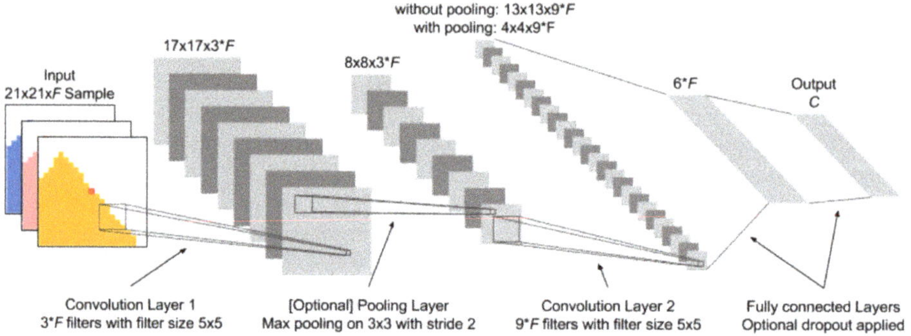

Figure 3. Structure of convolutional neural network (CNN) for classification of tree species. The variable F denotes the dimension of the input feature space.

2.4. Fusion Algorithms

Equations (1) and (2) describe the applied framework for fusion of different base classifier results. Given a feature vector x_i, its label $l = \widetilde{y}(x_i)$ is calculated as follows.

Using Equation (1), a score s_l is calculated for each class label from the results of the N base classifiers. For each label in question, the score is a sum of weights ω for the corresponding base classifiers. The indicator function I equals one, if its argument is true and zero, if not. Hence, only the weights of base classifiers with output label L are included.

$$s_l(x_i) = \sum_{j=1}^{N} \omega_j I(y_j(x_i) = l) \tag{1}$$

$$\widetilde{y}(x_i) = \underset{l}{\mathrm{argmax}} s_l(x_i) \tag{2}$$

Equation (2) simply denotes the selection of the label with the highest score. Different voting strategies are implemented by variation of the weights ω.

Majority voting is defined by setting $\forall_{i,j} : \omega_i = \omega_j$. If a majority of the N base classifiers assigns the same label, the corresponding score s_L is maximized. In our study, majority voting serves as a baseline for the performance of classifier fusion as it is a commonly used method.

Presidential voting is defined by setting $\omega_i = N - 1.5$ and $\forall_{j \neq i} : \omega_j = 1$, where the index i denotes a prior chosen classifier, e.g., the one with the best overall accuracy. Due to this constraint, the label assigned by the chosen classifier determines the ensemble result in most cases. Only if all other classifiers agree on the same divergent result, a different label is assigned.

Accuracy weighted voting uses a classification accuracy estimate to weight the different classifiers. According to the definitions in [35], the average accuracy is calculated from the elements of the confusion matrix C of each classifier. A subset of the reference data is used to determine C and to estimate the weights ω.

Hence, for the j-th classifier the weight ω_j is defined with Equation (3) as the average accuracy [36] which measures the average per-class effectiveness of a classifier.

$$\omega_j = \frac{\sum_{i=1}^{L} \frac{tp_i + tn_i}{tp_i + tn_i + fp_i + fn_i}}{L} \tag{3}$$

The $L \times L$ confusion matrix C is transformed into L 2×2 confusion matrices to obtain the required true positive tp_i, true negative tn_i, false positive fp_i, and false negative fn_i values.

For precision weighted voting, a similar approach to calculate the weights is used. The precision measure is calculated with Equation (4) as follows

$$\omega_j = \frac{\sum_{i=1}^{L} tp_i}{\sum_{i=1}^{L} (tp_i + fp_i)} \tag{4}$$

This fusion framework is a means to investigate how different popular approaches to weight a classifier influence the quality of the joint decision-making of an ensemble classifier.

3. Results

Tables 4–7 summarize the results of SVM, random forest, rotation forest, and CNN classifiers. Each table contains the mean accuracy values and standard deviations obtained with multiple runs of holdout testing. In the tables, for each feature space the best results per table row are emphasized as bold text. In addition, the feature space with the best overall classification accuracy is emphasized and the corresponding accuracy has been underlined. In each run, a complete reference site with a single dominant tree species was only used for testing the classifier, which was trained with data from the remaining reference sites. See Table 1 for the total number of reference sites and areas. If only a single reference site is available it is divided into non-overlapping sites for training and validation of equal size. Each row corresponds to one of the selected feature spaces. We discriminate between the results of pixelwise and spatial classification, where statistical measures (mean, standard deviation, homogeneity) of the original features are used as features instead.

Table 4. Overall accuracies and standard deviations for Multi-Class-SVM classifier with linear kernel using a 1-vs.-all strategy depending on the choice of feature space (1st column), pixelwise (2nd column) or with inclusion of spatial information from surrounding pixels (3rd column). The best performance within each row and the best method are emphasized with bold font to highlight the better performance of the spatial approaches.

Feature Space	Pixelwise	Spatial
Reflectances (160 dim)	0.227 ± 0.075	-
Indices (7 dim)	0.097 ± 0.061	**0.098 ± 0.04**
PCA (5 dim)	**0.231 ± 0.062**	0.195 ± 0.082
PCA (14 dim)	0.26 ± 0.114	**0.273 ± 0.113**
MCLDA (5 dim)	0.428 ± 0.049	**0.616 ± 0.064**
MCLDA (14 dim)	0.459 ± 0.033	**<u>0.651 ± 0.08</u>**
SELD (5 dim)	0.398 ± 0.1	**0.421 ± 0.071**
SELD (14 dim)	0.415 ± 0.097	**0.571 ± 0.053**

We report results for different ensemble sizes of random forest (Table 5) and rotation forest (Table 6) classifiers separately to assess the impact of this parameter and the trade-off for using a compact ensemble of 20 decision trees instead of a large ensemble of 100 decision trees.

Table 5. Overall accuracies and standard deviations for the random forest classifier with different choices of feature space (1st column) and varying complexity of models (# trees) for pixelwise classification and with consideration of spatial information from surrounding pixels. The best performance within each row and the best method are emphasized with bold font to highlight the better performance of the spatial approaches.

	Pixelwise	Spatial	Pixelwise	Spatial
# trees	20	20	100	**100**
Reflectances	0.435 ± 0.055	-	**0.447 ± 0.057**	-
Indices	0.352 ± 0.032	0.622 ± 0.039	0.361 ± 0.034	**0.633 ± 0.040**
PCA (5 dim)	0.357 ± 0.029	0.582 ± 0.056	0.366 ± 0.031	**0.594 ± 0.085**
PCA (14 dim)	0.441 ± 0.05	0.553 ± 0.088	0.452 ± 0.05	**0.572 ± 0.053**
MCLDA (5 dim)	0.487 ± 0.04	0.642 ± 0.055	0.495 ± 0.042	**0.653 ± 0.055**
MCLDA (14 dim)	0.522 ± 0.038	0.663 ± 0.052	0.533 ± 0.039	**0.684 ± 0.049**
SELD (5 dim)	0.367 ± 0.027	0.621 ± 0.03	0.377 ± 0.029	**0.63 ± 0.03**
SELD (14 dim)	0.486 ± 0.041	0.622 ± 0.06	0.499 ± 0.042	**0.639 ± 0.059**

Table 6. Overall accuracies and standard deviations for the Rotation Forest classifier with different choices of feature space (first column) and varying complexity of models (number of trees) for pixelwise classification and with consideration of spatial information from surrounding pixels (neighborhood). The best performance within each row and the best method are emphasized with bold font to highlight the better performance of the spatial approaches.

	Pixelwise	Spatial	Pixelwise	Spatial
# trees	20	20	100	**100**
Reflectances	0.501 ± 0.043	-	**0.517 ± 0.043**	-
Indices	0.33 ± 0.025	0.609 ± 0.039	0.331 ± 0.028	**0.619 ± 0.038**
PCA (5 dim)	0.342 ± 0.031	0.561 ± 0.053	0.347 ± 0.032	**0.575 ± 0.055**
PCA (14 dim)	0.442 ± 0.05	0.525 ± 0.078	0.45 ± 0.052	**0.548 ± 0.081**
MCLDA (5 dim)	0.467 ± 0.042	0.658 ± 0.048	0.472 ± 0.042	**0.665 ± 0.047**
MCLDA (14 dim)	0.522 ± 0.038	0.683 ± 0.044	0.539 ± 0.04	**0.705 ± 0.044**
SELD (5 dim)	0.345 ± 0.031	0.615 ± 0.033	0.351 ± 0.03	**0.624 ± 0.033**
SELD (14 dim)	0.487 ± 0.041	0.615 ± 0.059	0.498 ± 0.042	**0.634 ± 0.058**

Transformation of the reflectance data with MCLDA into a low dimensional representation yields the best results for all tested classifiers in separating between the 15 tree species. Moreover, inclusion of the spectral-spatial features significantly improves classification accuracy compared to pixelwise classification of spectral features in most cases. The best individual classifier was the CNN (see Table 7) with an overall accuracy of 0.732 ± 0.086, followed by a rotation forest of 100 decision trees with an overall accuracy of 0.705 ± 0.044. The CNN intrinsically combines spectral and spatial information. Hence, the CNN operating on raw or preprocessed hyperspectral image data achieved a similar performance to fandom rorests and SVMs, which make use of handcrafted spectral and spatial-spectral features. However, even the CNN the initial transformation with MCLDA was crucial to achieve the gain in overall accuracy.

The results of combining a small number of base classifiers into a hybrid ensemble are summarized in Figure 4. The diagram shows the accuracies and standard deviations of individual base classifiers together with results of different voting strategies. An accuracy gain of 5.1% was achieved by precision weighted voting compared to the best individual classifier. It significantly outperformed the three other tested voting methods.

To better understand these improvements, Figure 5 shows net diagrams of class-wise performance measures precision and recall. The subplots A–C show the performances of the included base classifiers, while subplot D represents the performance of the fusion framework with precision weighted voting using Equation (4) to calculate the weights.

Table 7. Overall accuracies and standard deviations for the CNN classifier. The gaps in the table reflect the stepwise approach to identify the best CNN configuration. First, we identified a best performing image transformation, second we evaluated different promising modifications of the chosen CNN. Bold setting and underlining highlight the parameter combination with best overall accuracy.

	Pixelwise Dropout = 0 no Pooling	Pixelwise Dropout = 0 with Pooling	Pixelwise Dropout = 0.5 no Pooling	Pixelwise Dropout = 0.5 with Pooling	Spatial Dropout = 0 no Pooling
Indices	0.485 ± 0.054	-	-	-	-
PCA (5 dim)	0.443 ± 0.072	-	-	-	-
PCA (14 dim)	0.439 ± 0.086	-	-	-	-
MCLDA (5 dim)	0.589 ± 0.074	-	-	-	-
MCLDA (14 dim)	0.665 ± 0.072	0.656 ± 0.077	0.675 ± 0.08	0.664 ± 0.087	**<u>0.732 ± 0.086</u>**
SELD (5 dim)	0.47 ± 0.058	-	-	-	-
SELD (14 dim)	0.412 ± 0.058	-	-	-	-

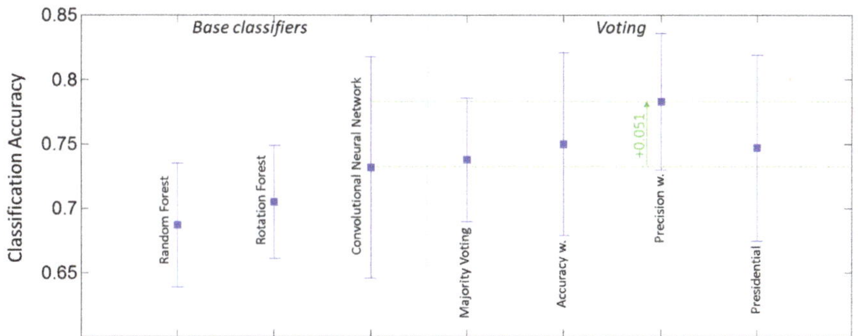

Figure 4. Classification accuracies for best base classifiers and ensembles of classifiers.

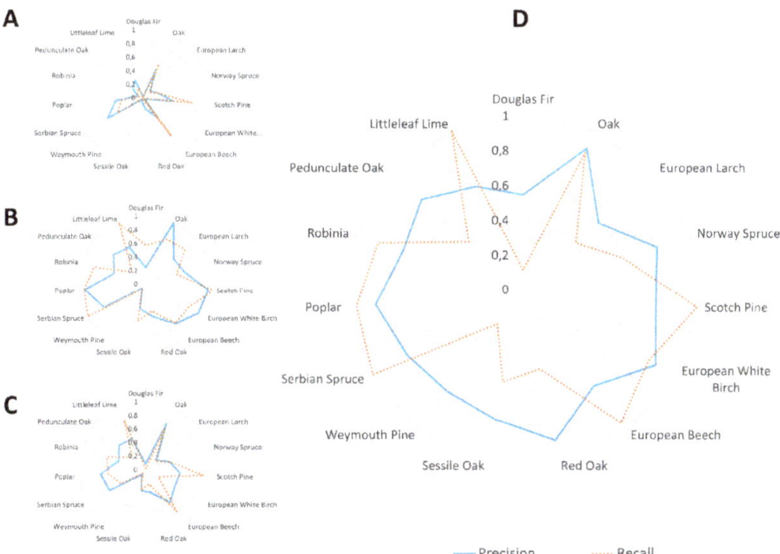

Figure 5. Performance measures of base classifiers (**A**–**C**) and hybrid ensemble based upon precision weighted fusion (**D**) per class. Net diagrams show precision and recall values of the following base classifiers: Random forest classifier for spectral indices (**A**), CNN classifier for MCLDA-transformed images (**B**), and random forest classifier for MCLDA-transformed images (**C**).

The expected general improvement of class-wise precision values is shown by the more convex shape and the much larger area within the precision curve in subplot D. The recall value of the hybrid ensemble is still dominated by the best base classifier, the CNN (compare subplots B and D, dotted curves). However, a class0wise direct comparison reveals a few minor differences. For some tree species the recall values increase (e.g., poplar, robinia, oak), while for a few others (e.g., larch, Douglas dir) they decrease slightly.

4. Discussion

The need for a reliable tree species classification motivated our investigation of several state-of-the-art classifiers as well as their combination into an ensemble classifier. This demand has led to the development of a processing pipeline, where the positions of suitable training sites with one dominant tree species are obtained from existing databases of the forest authorities. Hence, it was possible to create a large training dataset covering 15 tree species without additional groundwork. The alternative, the assessment of tree species of individual trees in mixed forests by experts to create a reference database, either from ground or from aerial stereo imagery is expensive, time-consuming, and error-prone.

Although, we developed a promising framework for classifier fusion and did careful validation, the method was not validated against ground truth data from real mixed forests. This is due to the fact that a large-scale validation would require the same efforts to obtain the reference data as mentioned above for the classifier training. However, the method was successfully applied to classify the complete forest in our hyperspectral dataset. This tree species map can then be used to select a number of individual trees and to validate the assigned class label in the field.

The performance of classification models based on this kind of training data was extensively tested with a combination of cross-validation and hold-out-testing. Compared to standard cross-validation not only a subset of randomly chosen pixels or patches were excluded from training for testing, but complete areas. This allowed us to study the performances on independent data. However, the spectral data belongs to a single measurement flight and has undergone the same preprocessing (e.g., radiometric and atmospheric corrections). Hence, the trained classifiers are adapted to the development stages of the trees and the conditions on the day of flight. To our best knowledge, the proposed method to acquire training data directly from the hyperspectral images at predefined locations with dominant tree species is the best way to learn classification models for future flights. Otherwise, many test flights and expenses are required to acquire training data that cover all possible appearances of leaves and needles beforehand. Locations with a single dominant tree species can be determined by an expert using existing databases and aerial images. For any region, this could be done once stored in a database and then be used for future measurement flights.

The per-class assessment of the classification performance of the fusion approach shows differences depending on the tree species. The net diagram in Figure 5 reveals that the precision of the class assignment was improved to large extends, but for some tree species the fraction of successfully detected trees remains low. However, compared to the state-of-the-art CNN a significant loss is only observed for Douglas firs. On the other hand, the rate of true Douglas firs among the reported ones increased significantly. There are different reasons, which possibly explain this behavior. First, the dataset contains a number of tree species with only subtle differences like oaks, sessile oaks, red oaks, and pedunculate oaks. While this information is of interest for the forest authorities and motivates the use of hyperspectral imaging to detect subtle differences, it might be better to take a hierarchical approach, which first detects all oaks and then tries to discriminate between oak species. Second, some of the tree species are more common than others. We account for this by balancing the dataset to have an equal number of samples for all tree species. However, the number of reference sites for rare species is also low and the natural variation between trees is better covered for frequent tree species. Here, our strict validation with holding back complete reference sites for testing may penalize the rare species.

With Scotch pines, Norway spruces, oaks, and beeches being the most frequent tree species in Saxony-Anhalt as well as Thuringia the results show, that our approach for analysis of airborne hyperspectral images already provides a useful tool to support forest inventory and to detect oak trees for subsequent analysis of vitality parameters. Moreover, the proposed fusion framework allows to easily add any other classifier.

5. Conclusions

In this paper, we applied a general and easy-to-use fusion framework based on voting to the problem of tree species classification from hyperspectral aerial images. The proposed hybrid multiple classifier system enhances the results of a state-of-the-art CNN with two random forest classifiers of different size and operating in different feature spaces. It was shown that this approach yields a significant gain in overall classification accuracy. This improvement results from a gain in precision of the class assignments by weighted fusion of the CNN and random forest results by an estimate of their individual precisions. The comparison to other popular voting techniques showed the superiority of the approach. The results provide evidence that even the best available classifiers for image data analysis can be further improved by incorporating their decisions into a multiple classifier system. MCLDA performs best among the different dimensionality reduction methods for hyperspectral imaging data. Even the CNN performance is enhanced by using MCLDA transformed images instead of the hyperspectral images as input data.

Author Contributions: conceptualization, S.C., U.K., and U.S.; methodology, H.S. and U.K.; software, C.S.v.R., M.S., T.K.,T.P.M.,V.H., and U.K.; validation, U.K., C.S.v.R., M.S., T.K., T.P.M., and V.H.; formal analysis, U.K., C.S.v.R., M.S., T.K., T.P.M., and V.H.; investigation, U.K., D.K., C.S.v.R., M.S., T.K., T.P.M, and V.H.; resources, H.S. and U.K.; data curation, H.S. and U.K.; writing—original draft preparation, U.K.; writing—review and editing, U.K., C.S.v.R., U.S., H.S., and S.C.; visualization, U.K. and H.S.; supervision, U.S. and U.K.; project administration, I.E.; funding acquisition, I.E., U.K., and U.S.

Funding: This research was partially funded by Ministry of Environment, Agriculture and Food of Saxony-Anhalt under grants A02/2014 and A02/2016.

Acknowledgments: The authors would like to thank Landesforstbetrieb Sachsen-Anhalt, Thüringen Forst, and Landeszentrum Wald Sachsen-Anhalt for supporting the project.

Conflicts of Interest: The authors declare no conflict of interest.

References

1. Bioucas-Dias, J.M.; Plaza, A.; Camps-Valls, G.; Scheunders, P.; Nasrabadi, N.M.; Chanussot, J. Hyperspectral Remote Sensing Data Analysis and Future Challenges. *IEEE Geosci. Remote Sens. Mag.* **2013**, *1*, 6–36. [CrossRef]
2. Bioucas-Dias, J.M.; Plaza, A.; Dobigeon, N.; Parente, M.; Du, Q.; Gader, P.; Chanussot, J. Hyperspectral Unmixing Overview: Geometrical, Statistical, and Sparse Regression-Based Approaches. *IEEE J. Sel. Top. Appl. Earth Obs. Remote Sens.* **2012**, *5*, 354–379. [CrossRef]
3. Wang, G.; Weng, Q. *Remote Sensing of Natural Resources*; Taylor & Francis: Boca Raton, FA, USA, 2013.
4. Fassnacht, F.E.; Latifi, H.; Sterenczak, K.; Modzelewska, A.; Lefsky, M.; Waser, L.T.; Straub, C.; Ghosh, A. Review of studies on tree species classification from remotely sensed data. *Remote Sens. Environ.* **2016**, *186*, 64–87. [CrossRef]
5. Fassnacht, F.E.; Neumann, C.; Förster, M.; Buddenbaum, H.; Ghosh, A.; Clasen, A.; Joshi, P.K.; Koch, B. Comparison of Feature Reduction Algorithms for Classifying Tree Species with Hyperspectral Data on Three Central European Test Sites. *IEEE J. Sel. Top. Appl. Earth Obs. Remote Sens.* **2014**, *7*, 2547–2561. [CrossRef]
6. Feret, J.B.; Asner, P. Tree Species Discrimination in Tropical Forests Using Airborne Imaging Spectroscopy. *IEEE Trans. Geosci. Remote Sens.* **2013**, *51*, 73–84. [CrossRef]
7. Richter, R.; Reu, B.; Wirth, C.; Doktor, D. The use of airborne hyperspectral data for tree species classification in a species-rich Central European forest area. *Int. J. Appl. Earth Obs. Geoinf.* **2016**, *52*, 464–474. [CrossRef]

8. Dalponte, M.; Bruzzone, L.; Gianelle, D. Tree species classification in the Southern Alps based on the fusion of very high geometrical resolution multispectral/hyperspectral images and LiDAR data. *Remote Sens. Environ.* **2012**, *123*, 258–270. [CrossRef]
9. Xia, J.; Du, P.; He, X.; Chanussot, J. Hyperspectral remote sensing image classification based on rotation forest. *IEEE Geosci. Remote Sens. Lett.* **2014**, *11*, 239–243. [CrossRef]
10. Rodríguez, J.J.; Kuncheva, L.I.; Alonso, C.J. Rotation Forest: A new classifier ensemble method. *IEEE Trans. Pattern Anal. Mach. Intell.* **2006**, *28*, 1619–1630. [CrossRef]
11. Lausch, A.; Heurich, M.; Gordalla, D.; Dobner, H.J.; Gwillym-Margianto, S.; Salbach, C. Forecasting potential bark beetle outbreaks based on spruce forest vitality using hyperspectral remote sensing techniques at different scales. *For. Ecol. Manag.* **2013**, *308*, 76–89. [CrossRef]
12. Sankaran, S.; Mishra, A.; Ehsani, R.; Davis, C. A review of advanced techniques for detecting plant diseases. *Comput. Electron. Agric.* **2010**, *72*, 1–13. [CrossRef]
13. Mutanga, O.; Ismail, R. Variation in foliar water content and hyperspectral reflectance of Pinus patula trees infested by Sirex noctilio. *South. For.* **2010**, *72*, 1–7. [CrossRef]
14. Hardinsky, M.A.; Klemas, V.; Smart, M. The influence of soil salinity, growth form, and leaf moisture on the spectral radiance of Spartina alterniflora canopies. *Photogramm. Eng. Remote Sens.* **1983**, *49*, 77–83.
15. Penuelas, J.; Pinol, J.; Ogaya, R.; Filella, I. Estimation of plant water concentration by the reflectance water index (R900/R970). *Int. J. Remote Sens.* **1997**, *18*, 2869–2875. [CrossRef]
16. Ayerdi, B.; Marqués, I.; Grana, M. Spatially regularized semi-supervised Ensembles of Extreme Learning Machines for hyperspectral image segmentation. *Neurocomputing* **2015**, *149*, 373–386. [CrossRef]
17. Wenzhi, L.; Pizurica, A.; Scheunders, P.; Philips, W.; Youguo, P. Semi-supervised Local Discriminant Analysis for Feature Extraction in Hyperspectral Images. *IEEE Trans. Geosci. Remote Sens.* **2013**, *51*, 184–198. [CrossRef]
18. Vapnik, V.N. *Statistical Learning Theory*; John Wiley&Sons: New York, NY, USA, 2013.
19. Bruzzone, L.; Chi, M.; Marconcini, M. A novel transductive SVM for semi-supervised classification of remote-sensing images. *IEEE Trans. Geosci. Remote Sens.* **2006**, *44*, 3363–3372. [CrossRef]
20. Dalponte, M.; Ene, L.T.; Marconcini, M.; Gobakken, T.; Naesset, E. Semi-supervised SVM for individual tree crown species classification. *ISPRS J. Photogramm. Remote Sens.* **2015**, *110*, 77–87. [CrossRef]
21. Makantasis, K.; Doulamis, A.D.; Doulamis, N.D.; Nikitakis, A. Tensor-Based Classification Models for Hyperspectral Data Analysis. *IEEE Trans. Geosci. Remote Sens.* **2018**, *56*, 6884–6898. [CrossRef]
22. Ghamisi, P.; Benediktsson, J.A.; Ulfarsson, M.O. Spectral-Spatial Classification of Hyperspectral Images Based on Hidden Markov Random Fields. *IEEE Trans. Geosci. Remote Sens.* **2014**, *52*, 2565–2574. [CrossRef]
23. Li, J.; Bioucas-Dias, J.M.; Plaza, A. Spectral-spatial hyperspectral image segmentation using subspace multinomial logistic regression and Markov random fields. *IEEE Trans. Geosci. Remote Sens.* **2012**, *50*, 809–823. [CrossRef]
24. Zhang, E.; Zhang, X.; Jiao, L.; Liu, H.; Wang, S.; Hou, B. Spectral-spatial hyperspectral image ensemble classification via joint sparse representation. *Pattern Recognit.* **2016**, *59*, 42–54. [CrossRef]
25. Makantasis, K.; Karantzalos, K.; Doulamis, A.; Doulamis, N. Deep Supervised Learning for Hyperspectral Data Classification through Convolutional Neural Networks. In Proceedings of the 2015 IEEE International Geoscience and Remote Sensing Symposium (IGARSS), Milan, Italy, 26–31 July 2015; pp. 4959–4962.
26. Wolpert, D.H.; Macready, W.G. Coevolutionary free lunches. *IEEE Trans. Evol. Comput.* **2015**, *9*, 721–735. [CrossRef]
27. Crawley, M.J.; Akhteruzzaman, M. Individual Variation in the Phenology of Oak Trees and Its Consequences for Herbivorous Insects. *Funct. Ecol.* **1988**, *2*, 409–415. [CrossRef]
28. Schläpfer, D.; Schaepman, M.E.; Itten, K.I. PARGE: Parametric geocoding based on GCP-calibrated auxiliary data. *Proc. SPIE* **1998**, *3438*, 334–344.
29. Schläpfer, D.; Richter, R.; Hueni, A. Recent developments in operational atmospheric and radiometric correction of hyperspectral imagery. In Proceedings of the Earsel SIG-IS Workshop, Tel-Aviv, Israel, 16–19 March 2009.
30. Didaci, L.; Fumera, G.; Roli, F. Diversity in Classifier Ensembles: Fertile Concept or Dead End? *Mult. Cl. Syst. Lect. Notes Comput. Sci.* **2013**, *7872*, 37–48.
31. Trunk, G.V. A Problem of Dimensionality: A Simple Example. *IEEE Trans. Pattern Anal. Mach. Intell.* **1979**, *1*, 306–307. [CrossRef]
32. Breiman, L. Random Forests. *Mach. Learn.* **2001**, *45*, 5–32. [CrossRef]

33. Fan, R.E.; Chang, K.W.; Hsieh, C.J.; Wang, X.R.; Lin, C.J. LIBLINEAR: A library for large linear classification. *J. Mach. Learn. Res.* **2008**, *9*, 1871–1874.
34. Sindhwani, V.; Keerthi, S. *Newton Methods for Fast Solution of Semi-Supervised Linear SVMs*; Book Chapter in Large Scale Kernel Machines; MIT Press: Cambridge, MA, USA, 2006.
35. Vedaldi, A.; Lenc, K.; Gupta, A. MatConvNet Convolutional Networks for Matlab. Available online: http://www.vlfeat.org/matconvnet/matconvnet-manual.pdf (accessed on 29 October 2019).
36. Sokolova, M.; Lapalme, G. A systematic analysis of performance measures for classification tasks. *Inf. Process. Manag.* **2009**, *45*, 427–437. [CrossRef]

Article

Identifying Mangrove Species Using Field Close-Range Snapshot Hyperspectral Imaging and Machine-Learning Techniques

Jingjing Cao [1], Kai Liu [1,*], Lin Liu [2,3,*], Yuanhui Zhu [2], Jun Li [1] and Zhi He [1]

1 Guangdong Provincial Engineering Research Center for Public Security and Disaster, Guangdong Key Laboratory for Urbanization and Geo-simulation, School of Geography and Planning, Sun Yat-sen University, Guangzhou 510275, China; caojj5@mail.sysu.edu.cn (J.C.); lijun48@mail.sysu.edu.cn (J.L.); hezh8@mail.sysu.edu.cn (Z.H.)

2 Center of GeoInformatics for Public Security, School of Geographic Sciences, Guangzhou University, Guangzhou 510006, China; zhuyhui2@gzhu.edu.cn

3 Department of Geography, University of Cincinnati, Cincinnati, OH 45221-0131, USA

* Correspondence: liuk6@mail.sysu.edu.cn (K.L.); lin.liu@uc.edu (L.L.); Tel.: +86-020-8411-3044 (K.L.)

Received: 25 October 2018; Accepted: 14 December 2018; Published: 16 December 2018

Abstract: Investigating mangrove species composition is a basic and important topic in wetland management and conservation. This study aims to explore the potential of close-range hyperspectral imaging with a snapshot hyperspectral sensor for identifying mangrove species under field conditions. Specifically, we assessed the data pre-processing and transformation, waveband selection and machine-learning techniques to develop an optimal classification scheme for eight mangrove species in Qi'ao Island of Zhuhai, Guangdong, China. After data pre-processing and transformation, five spectral datasets, which included the reflectance spectra R and its first-order derivative d(R), the logarithm of the reflectance spectra log(R) and its first-order derivative d[log(R)], and hyperspectral vegetation indices (VIs), were used as the input data for each classifier. Consequently, three waveband selection methods, including the stepwise discriminant analysis (SDA), correlation-based feature selection (CFS), and successive projections algorithm (SPA) were used to reduce dimensionality and select the effective wavebands for identifying mangrove species. Furthermore, we evaluated the performance of mangrove species classification using four classifiers, including linear discriminant analysis (LDA), k-nearest neighbor (KNN), random forest (RF), and support vector machine (SVM). Application of the four considered classifiers on the reflectance spectra of all wavebands yielded overall classification accuracies of the eight mangrove species higher than 80%, with SVM having the highest accuracy of 93.54% (Kappa = 0.9256). Using the selected wavebands derived from SPA, the accuracy of SVM reached 93.13% (Kappa = 0.9208). The addition of hyperspectral VIs and d[log(R)] spectral datasets further improves the accuracies to 93.54% (Kappa = 0.9253) and 96.46% (Kappa = 0.9591), respectively. These results suggest that it is highly effective to apply field close-range snapshot hyperspectral images and machine-learning classifiers to classify mangrove species.

Keywords: mangrove species classification; close-range hyperspectral imaging; field hyperspectral measurement; waveband selection; machine learning

1. Introduction

Mangroves are salt-tolerant evergreen woody trees and shrubs that are distributed in intertidal regions along tropical and subtropical coastlines [1,2]. As an important part of the wetland ecosystem, mangroves provide plenty of economic benefits and ecological value. They not only play a key role in filtering polluted seawater, providing wave prevention and embankment protection, maintaining biodiversity, and contributing to the global carbon balance, but also provide important forest products and socio-economic services [3]. Over the past 50 years, global mangrove resources have rapidly decreased due to human interference and natural causes [4,5]. Research on mangrove species composition is of great significance for mangrove ecosystem conservation, which provides basic information on wetland inventory and vegetation community changes.

Remote sensing techniques, including multispectral and hyperspectral, synthetic aperture radar (SAR) remote sensing [6–9], unmanned aerial vehicle (UAV)-based remote sensing [10,11], and light detection and ranging (LiDAR) [12], have been widely using in mangrove monitoring and management. Hyperspectral imaging can provide plenty of continuous narrow wavebands which increase the chance of distinguishing between different ground objects via their detailed spectral information [13]. It has proven to be effective for the classification of forest species and vegetation [14]. Various hyperspectral sensors, both imaging and non-imaging, have been applied to investigations related to the spectral analysis, classification and mapping of mangroves in the past 20 years, which can be divided into four categories: (1) space-borne hyperspectral sensors (e.g., Earth Observing-1 (EO-1) Hyperion sensor [9,15,16]); (2) airborne hyperspectral sensors (e.g., Airborne Visible/Infrared Imaging Spectrometer (AVIRIS) [17], Compact Airborne Spectrographic Imager (CASI) [18] and Airborne Imaging Spectrometer for Applications (AISA) [19]); (3) unmanned aircraft-mounted hyperspectral sensors (e.g., the Cubert UHD 185 hyperspectral imaging sensor [10]); (4) and hand-held hyperspectral sensors (e.g., the Analytical Spectral Device (ASD) spectrometer [20,21]). These studies mainly focused on a regional scale or large-scale mangrove classification and mapping based on onboard hyperspectral sensors, and the spectral characteristics analysis of mangrove plants based on fiber optic spectrometer under laboratory and field conditions. Generally, prior to the application of onboard hyperspectral sensors in mapping and classifying mangroves, it is necessary to conduct researches on the laboratory and ground-based measurements, which is one of the most important prerequisites for the future application of onboard hyperspectral sensors [22].

Recently, several studies on spectral analysis and species discrimination of mangroves have been carried out using laboratory and field hyperspectral data. Under laboratory conditions, the spectral reflectance of mangrove leaves measured with the ASD spectrometer has been used to classify 16 mangrove species, and identify healthy and stressed mangrove plants with 90% and 80% accuracy [22,23]. With ground-based measurements, the hyperspectral data of mangrove canopies obtained by the ASD spectroradiometer, VF921B (Anhui Institute of Optics and Fine Mechanics, Chinese Academy of Sciences (CAS), China) portable spectrometer, and Spectra Vista Corporation (SVC) GER-1500 portable transient spectrometer, has been employed to analyze mangrove reflectance spectral characteristics [24–26]. Most of these studies have reported using non-imaging spectrometers to measure laboratory or ground-based leaf spectral reflectance. For non-imaging spectrometers, they are generally difficult to control the range of ground objects covered by the sensor, and the obtained spectra are often mixed spectra of various ground objects. In comparison to the non-imaging spectrometer, the hyperspectral imaging sensor has the unique characteristic of acquiring image and spectral information of target simultaneously, it can extract pure pixel spectra of each ground object at the time of hyperspectral image capture [27]. Recent studies based on the hyperspectral imaging sensors, including line-scanning and snapshot, were mostly concerned with precision agriculture, such as crop classification and weed recognition. Shang et al. [28] used China's first field imaging spectrometer system (FISS) to obtain hyperspectral images for sophisticated classification of crop and weed. Xiao et al. [29] used the acousto-optic tunable filter (AOTF) hyperspectral imaging device to collect hyperspectral images of the leaves of Kentucky bluegrass, and realized the rapid identification of the different Kentucky bluegrass varieties. Gao et al. [30] presented the application of near-infrared hyperspectral images acquired with a snapshot mosaic hyperspectral camera in the laboratory for weed species recognition in a maize crop using a random forest machine-learning algorithm. However, the use of close-range snapshot hyperspectral imaging sensor for the classification of mangrove plants under field conditions has not been reported.

For the hyperspectral remote sensing of vegetation, selecting hyperspectral metrics, such as derivative spectra, hyperspectral vegetation indices (VIs) and the effective wavebands, is a key issue for specific analysis [13]. Derivative analysis can reflect the waveform changes and reveal the peak characteristics of spectra, which can improve the capability of using the spectral data to identify tree species [31]. Hyperspectral VIs are based on the mathematical transformations of spectral reflectance, which can be used to enhance spectral differences [32]. Furthermore, choosing the most useful wavebands is also necessary for dimensionality reduction and accurate species separation. Currently, the most widely used waveband selection techniques, such as the stepwise discriminant analysis (SDA) [20,23] and correlation-based feature selection (CFS) [10], have been employed to select informative wavebands for mangrove species classification. Moreover, scholars have explored different parametric and non-parametric methods in classifying tree species including mangroves [17,33]. Among them, several machine-learning techniques, such as random forest (RF) [17], support vector machine (SVM), rotation forest (RoF) [6], and logistic model tree (LMT) [7], can be used to construct effective classification models, which are non-parametric and do not rely on any assumption about the data distribution [34].

The aim of this study is to evaluate the use of close-range snapshot hyperspectral imaging for mangrove species identification under field conditions. The specific objectives were: (1) to investigate the applicability of field snapshot hyperspectral imaging sensor in identifying mangrove species; and (2) to determine the optimal spectral modes, relevant spectral wavebands, and effective classifiers for mangrove species identification. First, we collected the field hyperspectral data of eight mangrove species with a snapshot hyperspectral imaging sensor. Second, we performed data pre-processing and spectral transformations, and selected hyperspectral datasets in five spectral modes: (a) the reflectance spectra R, (b) the first-order derivative of the reflectance spectra d(R), (c) the logarithm of the reflectance spectra log(R), and (d) its first-order derivative d[log(R)], and (e) hyperspectral VIs. Third, we employed the SDA, CFS, and successive projections algorithm (SPA) to identify the optimal wavebands for mangrove species classification. Finally, we constructed classification models and compared the results obtained from four machine-learning classifiers, including the linear discriminant analysis (LDA), k-nearest neighbor (KNN), RF, and SVM. A detailed flowchart of this study is illustrated in Figure 1.

2. Materials and Methods

2.1. Study Area Description

Field hyperspectral measurements of this study were conducted at the Qi'ao Island Mangrove Nature Reserve, which has an area of 700 ha and is located on Qi'ao Island (22°23′40″–22°27′38″N, 113°36′40″–113°39′15″E), Zhuhai City, Guangdong Province, China [35,36] (Figure 2). Qi'ao Island is situated in Lingding Bay of the Pearl River Estuary, which has a typical tropical-subtropical transitional coastal-inland wetland ecosystem. Qi'ao Island belongs to the southern subtropical maritime monsoon climate zone, with sufficient sunshine and abundant rainfall [37,38]. The tidal pattern of Qi'ao Island is an irregular semidiurnal tide [8], which is characterized by the tidal height inequality of two adjacent high or low tides, the tide duration inequality between flood and ebb tides, and that the average tidal range varies with the flood and dry season, spring and neap tides [39]. The Mangrove Nature Reserve is the largest mangrove forest in the Pearl River Delta, it is the largest area of artificially planted mangrove forests in China and it has a rich variety of mangrove plants [8,36].

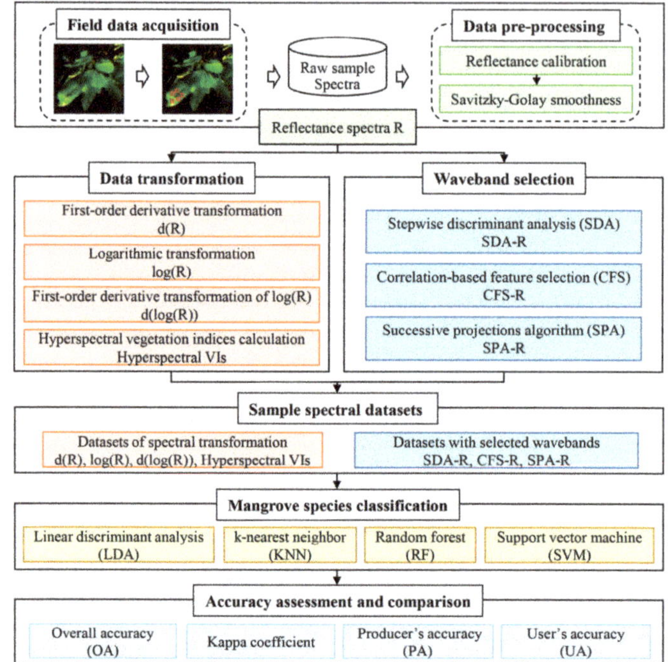

Figure 1. Flowchart of mangrove species identification using field close-range snapshot hyperspectral imaging and machine-learning classifiers.

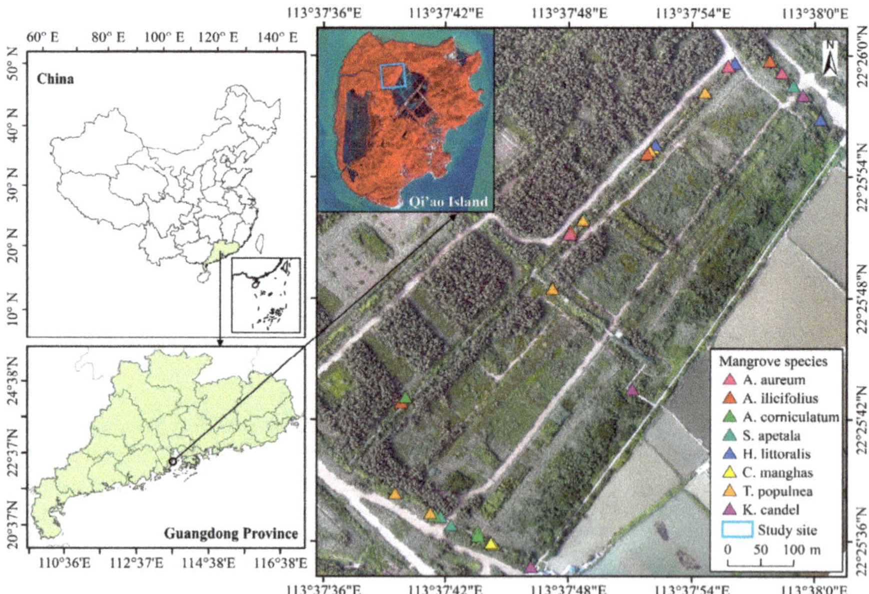

Figure 2. Location of Qi'ao Island, showing the WorldView-2 image (false color composite composed of R, band 7; G, band 5; B, band 3). The map of field survey site was an UAV image (true color composition) acquired on 11 September 2015. The right panel shows a distribution map of 33 ground survey points, where the symbols in the legend indicate the sample locations along the boardwalks.

As shown in Figure 2, the field sampling route was along the boardwalks, in which the artificially planted mangrove forests were the dominant plant types. According to an existing research [37] and previous field surveys, there are eight common mangrove species in this study site, including *Kandelia candel* (*K. candel*), *Acrostichum aureum* (*A. aureum*), *Acanthus ilicifolius* (*A. ilicifolius*), *Aegiceras corniculatum* (*A. corniculatum*), *Sonneratia apetala* (*S. apetala*), *Heritiera littoralis* (*H. littoralis*), *Cerbera manghas* (*C. manghas*) and *Therspesia populnea* (*T. populnea*), as shown in Table 1. Among them, the *K. candel* stands were arbor or frutex, the *A. aureum* stands were herbage, the *A. ilicifolius* stands were frutex, and the other five stands were arbor. The *C. manghas* and *T. populnea* stands were semi-mangroves, and the other six stands were true mangroves.

Table 1. List of mangrove species used in this study.

Mangrove Species Name	Species Code	Functional Group	Ground Survey Points	Samples
Acrostichum aureum (A. aureum)	AA	Herbage	4	60
Acanthus ilicifolius (A. ilicifolius)	AI	Frutex	4	60
Aegiceras corniculatum (A. corniculatum)	AC	Arbor	4	60
Sonneratia apetala (S. apetala)	SA	Arbor	4	60
Heritiera littoralis (H. littoralis)	HL	Arbor	4	60
Cerbera manghas (C. manghas)	CM	Arbor	4	60
Therspesia populnea (T. populnea)	TP	Arbor	6	60
Kandelia candel (K. candel)	KC	Arbor/Frutex	3	60

2.2. Data Acquisition and Sample Collection

2.2.1. Field Hyperspectral Measurement

For this study, the hyperspectral imaging system (Figure 3) was set up to acquire hyperspectral images of mangrove leaves. The main component of the experimental set-up was a commercial UHD 185 hyperspectral snapshot sensor (Figure 3a) manufactured by Cubert GmbH (http://cubert-gmbh.de/), Germany. The UHD 185 hyperspectral image consisted of a hyperspectral cube of 50 × 50 pixels and a panchromatic image with a resolution of 1000 × 1000 pixels. This sensor could capture 138 spectral bands within the spectral range of 450–998 nm with a 4-nm interval. According to previous studies [10,40], the spectral bands between 454 and 950 nm were used for analysis and classification. The UHD 185 sensor is currently applied onboard the UAV platform and can also be used for laboratory and ground-based spectrometry. This sensor was compact and lightweight with a total mass of about 470 g, which makes it highly portable and suitable for field applications, and it had a fast imaging speed of 5 cubes per second.

The main field hyperspectral measurements were performed on 3 January 2017. Due to the limitations on the accessibility of data collection site, data acquisition time, and illumination conditions on the day, several mangrove plants did not collect enough hyperspectral images. Therefore, part of the hyperspectral images of *T. populnea* and *K. candel* stands was collected as the supplementary data on 26 May 2018. Considering the influence of external illumination, the leaf spectra were collected on a cloud-free day at around solar noontime between 10:30 and 14:00. The UHD 185 hyperspectral imaging sensor could be hand-held and controlled by a notebook computer (Figure 3b). During the experiment, the spectral measurements were conducted above the mangrove canopy at approximately 20 cm height with the sensor facing and being at a near-vertical viewing direction to the canopy.

Figure 3. Field hyperspectral measurements above the mangrove canopies and the set-up used to acquire the hyperspectral images. (**a**) The UHD 185 hyperspectral imaging sensor, and (**b**) a notebook computer.

2.2.2. Data Pre-Processing

The field-collected spectra were radiometrically corrected with a standard white reference and dark measurements according to the pre-processing procedure used in a previous study [10]. Field-collected spectra are susceptible to variable illumination, contamination from the background environment and instrumental noise. To avoid noise associated with specific bands, the Savitzky–Golay algorithm [41,42] was introduced to smooth the raw hyperspectral data by eliminating glitch noise existing in the spectral curves. Based on least-squares fitting, this algorithm could remove high frequency noise and smooth the original data sequence by replacing the original values with fitted values [43], thereby preserving the original features of the spectrum. In this study, the Savitzky–Golay algorithm was implemented using the MATLAB R2014b software (MathWorks Inc., Natick, MA, USA).

2.2.3. Sample Spectra Preparation

Hyperspectral images of the eight mangrove plants of ground survey points were acquired with the UHD 185 hyperspectral imaging sensor. Representative images with true color composition of each mangrove species were shown in Figure 4. The sample collection strategy is as follows: (a) for each mangrove species, there are several ground survey points (Figure 1 and Table 1); (b) for each ground survey point, a preliminary manual screening was performed to select the hyperspectral images (Figure 4); and (c) for each hyperspectral image, there are five sample spectra were chosen. Considering that the illumination and background environment were the main factors related to the high spectral variability of the plant leaves [44], we selected spectral samples from the hyperspectral images of healthy and sun-lit mangrove leaves for each species with the Cube-Pilot software (Cubert GmbH, Ulm, Germany; http://cubert-gmbh.de/). For each type of mangrove species, 60 sample patches on the corresponding hyperspectral image were randomly selected (Table 1). A sample patch corresponded to 20 × 20 pixels in the selected cube which can be considered as a region of interest, and the spectral reflectance of the sample patch was calculated as the average value of pixels within the selected cube. A total of 480 sample spectra were selected. With these sample spectra, the 10-fold cross-validation was employed for the classification training and validation.

(a) *A. aureum* (b) *A. ilicifolius* (c) *A. corniculatum* (d) *S. apetala*

(e) *H. littoralis* (f) *C. manghas* (g) *T. populnea* (h) *K. candel*

Figure 4. UHD 185 hyperspectral images of the eight mangrove species at the study site.

2.3. Hyperspectral Metrics Extraction

2.3.1. Data Transformation

Reflectance spectra from leaves in close-range imaging are often influenced by the illumination and background environment. Derivative spectra are commonly employed in hyperspectral investigations of vegetation [45], which can effectively reduce the influence of illumination variations, and eliminate the background signal and systematic errors. Previous studies have reported that the derivative analysis can further enhance the ability of the spectral data to identify tree species [46], and has been developed for mangrove species classification [47]. First-order derivative spectra can reflect the waveform changes caused by the absorption of the light by chlorophyll and other substances in plants, and reveal the peak characteristics of the spectrum [48]. Furthermore, because logarithmic transformation can enhance the spectral differences in the visible region and reduce the influence of multiplicative factors caused by changes in illumination conditions, we also performed the logarithmic transformation on spectral data. Following Pu and Gong [31], three transformations of R, including the first-order derivative of the reflectance spectra d(R), the logarithm of the reflectance spectra log(R) and its first-order derivative d[log(R)], were computed as

$$d(R) = \left(\frac{r_3 - r_1}{\Delta\lambda}, \frac{r_4 - r_2}{\Delta\lambda}, \cdots, \frac{r_n - r_{n-2}}{\Delta\lambda}\right) \tag{1}$$

$$\log(R) = [\log(r_1), \log(r_2), \cdots, \log(r_n)] \tag{2}$$

where r_i denotes the i-th wavelength, n denotes the number of wavebands, and $\Delta\lambda$ denotes the double waveband intervals (nm).

2.3.2. Vegetation Index Calculation

Vegetation indices (VIs) are generally defined as mathematical transformations of the spectral reflectance of the original wavebands. One of the advantages of VIs is their ease of use. According to the definitions of VIs, they can enhance the differences between plant species and reveal the hidden vegetation information using various combinations of ratios, differences, and linear combinations [49,50]. Moreover, VIs can eliminate the influence of the multiplicative factor associated with illumination variations and background environments [31,51]. As shown in Table 2, we calculated 25 hyperspectral VIs, which were commonly used in previous studies [14,32,50] to identify and map plant species composition. These hyperspectral VIs were selected to represent spectral variations associated with pigments including chlorophylls, the leaf area index (LAI), biomass and red edge optical parameters, and so on.

Table 2. Hyperspectral vegetation indices derived from the UHD 185 hyperspectral wavebands selected in this study.

Vegetation Indices	Definition	Commonly Related to	References
Blue Green Pigment Index 2 (BGI 2)	$\frac{R_{454}}{R_{550}}$	Chlorophylls, Carotenoids	[52]
Normalized Difference Vegetation Index (NDVI)	$\frac{R_{798}-R_{670}}{R_{798}+R_{670}}$	LAI, biomass, vegetation cover	[53]
Reformed Difference Vegetation Index (RDVI)	$\frac{R_{798}-R_{670}}{\sqrt{R_{798}+R_{670}}}$	LAI	[54,55]
Soil-Adjusted Vegetation Index(SAVI)	$\frac{(R_{798}-R_{670})\times(1+0.5)}{R_{798}+R_{670}+0.5}$	Biomass	[56]
Adjusted Transformed Soil-Adjusted VI (ATSAVI)	$\frac{a(R_{798}-aR_{670}-b)}{aR_{798}+R_{670}-ab+X(1+a^2)}$ where X = 0.08, a = 1.22, b = 0.03	LAI, biomass, soil variation	[57]
Modified SAVI (MSAVI)	$R_{798}+0.5-\sqrt{(R_{798}+0.5)^2-2(R_{798}-R_{670})}$	Biomass, soil variation	[58]
Transformed Chlorophyll Absorption in Reflectance Index (TCARI)	$3\left[(R_{702}-R_{670})-0.2(R_{702}-R_{550})\times\left(\frac{R_{702}}{R_{670}}\right)\right]$	Chlorophylls	[59]
Optimized Soil-Adjusted Vegetation Index (OSAVI)	$(1+0.16)\times\frac{R_{798}-R_{670}}{R_{798}+R_{670}+0.16}$	Biomass, soil variation	[60]
TCARI/OSAVI	$\frac{TCARI}{OSAVI}$	Chlorophylls	[59]
Modified Chlorophyll Absorption in Reflectance Index (MCARI)	$[(R_{702}-R_{670})-0.2(R_{702}-R_{550})]\times\left(\frac{R_{702}}{R_{670}}\right)$	Chlorophylls	[61]
Modified Chlorophyll Absorption Ratio Index 1 (MCARI1)	$1.2[2.5(R_{798}-R_{670})-1.3(R_{798}-R_{550})]$	LAI	[55]
Modified Chlorophyll Absorption Ratio Index 2 (MCARI 2)	$\frac{1.5[2.5(R_{798}-R_{670})-1.3(R_{798}-R_{550})]}{\sqrt{(2R_{798}+1)^2-(6R_{798}-5\sqrt{R_{670}})-0.5}}$	Chlorophylls	[55]
Photochemical Reflectance Index (PRI)	$\frac{R_{514}-R_{530}}{R_{514}+R_{530}}$	Water content	[40,62]
Triangular Vegetation Index (TVI)	$0.5[120(R_{750}-R_{550})-200(R_{670}-R_{550})]$	LAI	[63]
Modified Triangular VI 1 (MTVI 1)	$1.2[1.2(R_{798}-R_{550})-2.5(R_{670}-R_{550})]$	LAI	[55]
Modified Triangular VI 2 (MTVI 2)	$\frac{1.5[1.2(R_{798}-R_{550})-2.5(R_{670}-R_{550})]}{\sqrt{(2R_{798}+1)^2-(6R_{798}-5\sqrt{R_{670}})-0.5}}$	LAI	[55]
Simple Ratio (SR)	$\frac{R_{798}}{R_{670}}$	Chlorophylls	[51,64]
Modified Simple Ratio (MSR)	$\left(\frac{R_{798}}{R_{670}}-1\right)/\sqrt{\frac{R_{798}}{R_{670}}+1}$	LAI	[55,65]
Zarco Tejada and Miller (ZTM)	$\frac{R_{750}}{R_{710}}$	Red edge	[66]
Vogelmann Red Edge Index 1 (VOG1)	$\frac{R_{742}}{R_{722}}$	Red edge	[67]

Table 2. *Cont.*

Vegetation Indices	Definition	Commonly Related to	References
Vogelmann Red Edge Index 2 (VOG2)	$\frac{R_{734}-R_{750}}{R_{718}+R_{726}}$	Red edge	[66]
Vogelmann Red Edge Index 3 (VOG3)	$\frac{R_{734}-R_{750}}{R_{718}+R_{722}}$	Red edge	[66]
Red Edge NDVI (RENDVI)	$\frac{R_{754}-R_{702}}{R_{754}+R_{702}}$	Red edge	[68]
Vogelmann's Index (VOI)	$\frac{DR_{718}}{DR_{706}}$	Red edge	[67,69]
Modified Simple Ratio of Derivatives (DMSR)	$\frac{DR_{722}-DR_{502}}{DR_{722}+DR_{502}}$	Chlorophylls	[70]

Note: R denotes the reflectance spectra, DRi denotes the first-order derivative of the reflectance spectra.

2.4. Waveband Selection

Feature selection is an important pre-processing step for hyperspectral data, which can increase the efficiency of the classification model by removing irrelevant and redundant information [71]. Selecting the specific wavebands that are most important for developing more robust classification models is desirable [72,73]. In this study, three waveband selection methods, SDA, CFS and SPA, based on information entropy, correlation, and projection, respectively, were used to find out which wavebands can optimally differentiate mangrove species.

2.4.1. Stepwise Discriminant Analysis

The SDA is a multivariate statistical method, which has been used to discriminate variables. Based on the discriminant analysis, for each step, the variable with the strongest discriminative ability was introduced into the discriminant function, and the variable with the poorest discriminative ability was eliminated. In this manner, the SDA can be used to select a subset of wavebands that had the maximum discriminative ability [74]. In this study, the SDA was used to select the optimal wavebands according to the Wilks' Lambda statistic [51], and implemented by the IBM SPSS Statistics 19 (IBM Inc., Armonk, NY, USA).

2.4.2. Correlation-Based Feature Selection

The CFS is a classic filter method for feature selection [75]. It has proven to be useful for selecting suitable features and facilitating computation. The idea of this algorithm is to calculate the "feature-class" and "feature-feature" correlation matrices from a training set. This algorithm assumes that these features are conditionally independent given the class. In this paper, the best first search algorithm [76] was applied for CFS to select the feature subset with the highest correlation between features and categories, and the lowest correlation between features and features. The CFS and best first search algorithms from the Weka 3.8 attribute selection package were used.

2.4.3. Successive Projections Algorithm

The SPA is a forward variable selection method, which has been used for waveband selection in previous studies [77]. The SPA randomly selects a starting waveband, calculates the maximum projection vector of an unselected waveband, and uses the corresponding waveband as the introduced waveband. After multiple iterations, the characteristic waveband is obtained by a cost function evaluation. This algorithm can effectively eliminate the influence of the collinearity that may exist among wavebands. In this study, the SPA selects the characteristic wavelengths that contain the least redundant information, which are determined by the minimum root mean square error of the prediction [78]. The SPA algorithm was implemented with MATLAB.

2.5. Mangrove Species Classification

We explored four mangrove species classification schemes using the LDA, KNN, RF, and SVM machine-learning classifiers based on the different datasets. These classification models were all executed in MATLAB.

2.5.1. Linear Discriminant Analysis

The LDA is a classic parametric algorithm in the field of data mining and machine learning, which has been widely used in previous classification researches to identify wetland plants, including mangroves [79–81]. The LDA determines the linear discriminant function based on the principle that the distance between classes is the largest and the distance within a class is the smallest, that is, to maximize the ratio of the dispersion between sample classes and minimize the dispersion within a sample class. The LDA makes two assumptions about the normal distribution of the data and the homoscedasticity for which two classes have equal covariance matrices. Finally, the category of unknown samples is determined by the established linear discriminant model.

2.5.2. K-Nearest Neighbor

The KNN is a non-parametric instance-based learning algorithm which has been extensively used for classification and regression [82]. The KNN is based on the assumption that ground objects close in distance are more likely to belong to the same category. The principle of KNN is that the instances within a dataset will generally exist in close proximity to other instances that have similar properties [83]. The distance between the feature vector to be classified and each feature vector in the feature space are calculated. The k nearest neighbor features are then selected. The categories of the testing sample are predicted by the majority vote of its neighbors using the Euclidean distance.

2.5.3. Random Forest

The RF is a non-parametric ensemble-based machine-learning method [84,85], which constructs a multitude of decision trees for learning and predicts the categories of the testing samples based on the average of the predicted values of each decision tree. The RF requires assumptions about independent variables and normality, and it does not need to check the variable interactions and nonlinear effects. The classification model of RF is mainly influenced by two parameters, including the number of decision trees (*ntree*) and the number of variables participating in the classification at the node (*mtry*). It distinguishes classes by individually building decision spaces for each explanatory variable at each node level, and the final classification ultimately depends on the decision spaces at higher nodes.

2.5.4. Support Vector Machine

The SVM is a well-known supervised kernel-based machine-learning method, and has proven to be one of the most widely used and efficient classifiers [86,87]. The SVM aims to find an optimal separating classification hyperplane which assumes that all groups are separable, to maximize the interval between the support planes of each type of data. It has powerful nonlinear and high-dimensional processing capabilities, which can avoid the "dimensionality disaster" caused by high-dimensional sample space and can be applied to small-sample learning. The LIBSVM package developed by Chang and Lin [88] has been widely used to implement the SVM classification model. Considering the nonlinear hyperplane, the radial basis function (RBF) kernel was chosen and two parameters, the cost of constraints (C) and sigma (σ), were determined by a grid search strategy.

2.6. Accuracy Assessment

For each of the classification results, we used a confusion matrix to provide the specific metrics, including the user's accuracy (UA), the producer's accuracy (PA), the overall accuracy (OA) and kappa coefficient (Kappa). The confusion matrix is an effective tool to evaluate the classification performance.

It gives a full description of errors for each category, including the errors of inclusion and errors of exclusion made by the classifiers [89]. The PA denotes the probability of a certain category being correctly recognized. The UA denotes the probability that a sample belongs to a specific category, and the classifier can accurately sort it into this category. The main difference between PA and UA is the cardinality of the accuracy calculation [90]. For PA, the cardinality is the total number of categories by the reference samples. For UA, the cardinality is the total number of categories by the classified samples. PA is concerned with the quality of the method used to produce the classification result, while UA focuses on the credibility of each category in the classification result. The OA is the ratio (%) between the number of correctly classified samples and the number of testing samples [91]. The kappa coefficient is generally used to measure the agreement between the predicted and actual values [92,93].

3. Results

3.1. Spectral Properties of Mangrove Species

The average reflectance curves for the eight mangrove species (Figure 5a) showed typical patterns of vegetation. The trends in these spectral curves were generally similar, which increased continuously from 680 nm and reached a maximum peak around 780 nm, around the red edge region. The differences in the spectral response in the visible light region were indicative of leaf pigments. The mangrove leaves of *C. manghas*, *T. populnea*, and *S. apetala* stands had brighter green colors and consequently higher reflectance within the green reflectance spectral region. The near-infrared signal revealed the multiple scattering within the leaf structure. After the derivative and logarithmic transformations, the spectral variations within the species were reduced, while differences between species were enlarged. Figure 5b–d present the spectral curves of the first-order derivative of the reflectance spectra, the logarithm of the reflectance spectra and its first-order derivative of the leaves from the eight mangrove species, respectively.

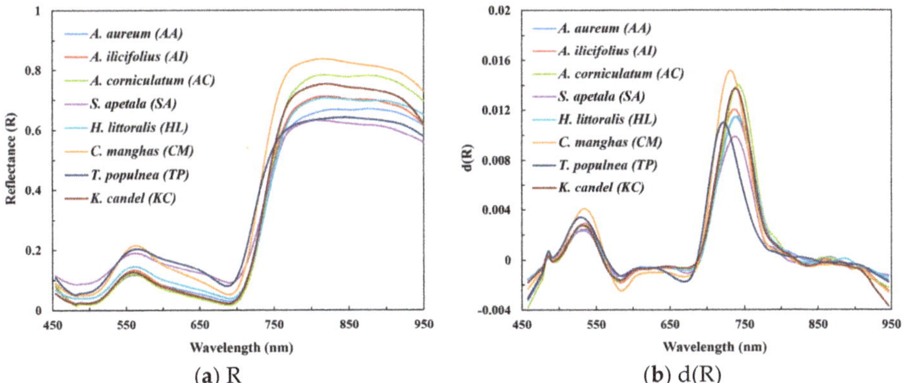

Figure 5. *Cont.*

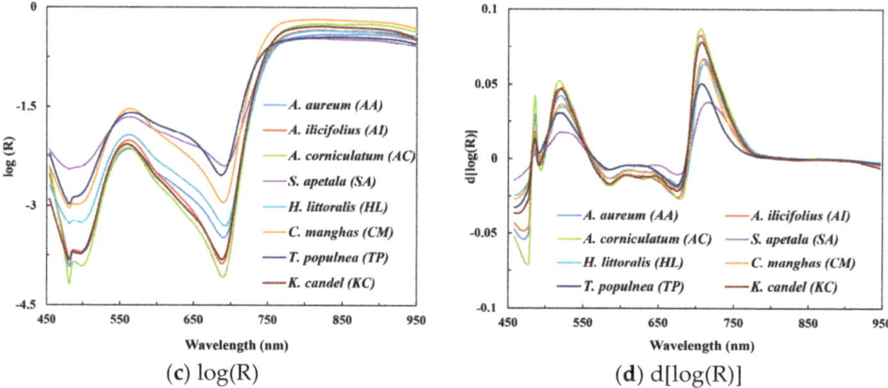

(**c**) log(R) (**d**) d[log(R)]

Figure 5. The average reflectance curves, the spectral curves of derivative and logarithmic transformations of the eight mangrove species. (**a**) R: the average reflectance curves, (**b**) d(R): the average first-order derivative spectral curves, (**c**) log(R): the average log-transformed spectral curves, and (**d**) d[log(R)]: the average first-order derivative of the log-transformed spectral curves.

3.2. Classification Results of the Transformed Datasets

Five hyperspectral datasets were used to identify mangrove species with the four machine-learning classifiers: (a) the reflectance spectra of all 125 wavebands R; (b) the first-order derivative of the reflectance spectra d(R); (c) the logarithm of the reflectance spectra log(R) and (d) its first-order derivative d[log(R)]; and (e) 25 hyperspectral VIs. Table 3 summarizes the classification accuracy assessment results. Based on hyperspectral VIs, the classification accuracies of the four classifiers were all more than 80%. Compared with the other three classifiers, SVM yielded a better overall classification accuracy of 93.54% (Kappa = 0.9253). Both the derivative and logarithmic transformations could improve the mangrove species classification accuracy. The discrimination capabilities of d(R) and log(R) spectral datasets were better than that from the reflectance spectra (Table 4). The d[log(R)]-classification using SVM gave the highest OA of 96.46% (Kappa = 0.9591). The producer's and user's accuracies of the eight mangrove species using SVM were all higher than 90%, especially for the *A. aureum* and *S. apetala* stands. This was mainly because the derivative and logarithmic transformations of reflectance spectra could reduce the multiplicative factors caused by changeable illumination conditions.

Table 3. Classification results in terms of OA (overall accuracy), kappa coefficient, and standard deviation (in bracket) for the eight mangrove species using different spectral datasets and classifiers of the 10-fold cross-validation data.

Spectral Datasets	Performance Metrics	Classifiers			
		LDA	KNN	RF	SVM
Reflectance spectra R	OA(%)	84.17 (5.22)	87.50 (5.29)	87.92 (4.03)	**93.54** (2.68)
	Kappa	0.8179 (0.0595)	0.8562 (0.0607)	0.8609 (0.0460)	**0.9256** (0.0308)
Hyperspectral VIs	OA(%)	85.00 (4.79)	84.58 (5.12)	85.83 (5.62)	**93.54** (2.07)
	Kappa	0.8270 (0.0544)	0.8223 (0.0585)	0.8366 (0.0652)	**0.9253** (0.0244)
First-order derivative spectra d(R)	OA(%)	84.58 (6.45)	90.63 (3.84)	92.71 (3.57)	**95.83** (2.20)
	Kappa	0.8218 (0.0736)	0.8917 (0.0437)	0.9156 (0.0410)	**0.9518** (0.0252)

Table 3. *Cont.*

Spectral Datasets	Performance Metrics	Classifiers			
		LDA	KNN	RF	SVM
Log-transformed spectra log(R)	OA(%)	89.79 (3.73)	88.54 (4.20)	86.04 (4.40)	**93.75** (4.39)
	Kappa	0.8823 (0.0431)	0.8681 (0.0478)	0.8396 (0.0505)	**0.9281** (0.0505)
First-order derivative of log(R) d[log(R)]	OA(%)	85.83 (7.07)	95.00 (3.43)	92.29 (3.55)	**96.46** (2.42)
	Kappa	0.8367 (0.0812)	0.9421 (0.0398)	0.9112 (0.0408)	**0.9591** (0.0280)

Table 4. Summary of classification accuracies and standard deviation (in bracket) of the d[log(R)] spectral dataset using different classifiers of the 10-fold cross-validation data.

Mangrove Species	Classifiers							
	LDA		KNN		RF		SVM	
	PA(%)	UA(%)	PA(%)	UA(%)	PA(%)	UA(%)	PA(/%)	UA(%)
A. aureum (AA)	81.86 (20.30)	92.46 (10.59)	93.65 (8.40)	94.31 (10.89)	93.15 (9.40)	90.06 (11.67)	98.75 (3.95)	100.00 (0.00)
A. ilicifolius (AI)	90.25 (14.26)	88.65 (15.81)	94.67 (11.67)	93.89 (11.25)	93.83 (10.12)	89.08 (8.20)	96.33 (7.77)	95.76 (10.61)
A. corniculatum (AC)	85.33 (17.26)	86.11 (16.47)	93.00 (16.36)	88.99 (8.19)	84.71 (14.72)	85.50 (13.83)	95.89 (9.45)	93.00 (12.01)
S. apetala (SA)	98.33 (5.27)	92.00 (9.76)	98.57 (4.52)	94.72 (9.11)	97.50 (5.27)	97.50 (5.27)	98.57 (4.52)	98.89 (3.51)
H. littoralis (HL)	82.88 (15.53)	93.99 (7.85)	89.40 (16.61)	97.32 (5.66)	87.64 (9.06)	92.05 (10.88)	90.24 (17.83)	97.14 (6.02)
C. manghas (CM)	69.30 (21.82)	67.90 (21.61)	91.65 (9.08)	98.00 (6.32)	89.23 (12.82)	95.00 (8.05)	93.24 (8.83)	96.07 (8.66)
T. populnea (TP)	97.32 (5.66)	90.74 (11.40)	100.00 (0.00)	98.75 (3.95)	98.75 (3.95)	92.08 (13.95)	100.00 (0.00)	97.50 (5.27)
K. candel (KC)	83.17 (19.44)	87.64 (14.97)	96.00 (8.43)	96.67 (10.54)	94.75 (8.70)	98.33 (5.27)	98.00 (6.32)	96.33 (7.77)
OA(%)	85.83 (7.07)		95.00 (3.43)		92.29 (3.55)		**96.46** (2.42)	
Kappa	0.8367 (0.0812)		0.9421 (0.0398)		0.9112 (0.0408)		**0.9591** (0.0280)	

3.3. Optimal Waveband Selection

Considering the high dimensionality of the hyperspectral data and the spectral correlations among different mangrove species, it was necessary to select a few wavebands with lower correlation. For this study, three waveband selection methods, SDA, CFS and SPA, were used to determine the optimal wavebands, and their classification performances were compared. Table 5 shows the results of the effective wavebands selected by the three methods.

Table 5. The selected wavebands, based on the reflectance spectra, using the SDA, CFS and SPA methods.

Methods	Selected Wavebands (nm)
SDA	14 bands: 478–482, 486, 526, 562, 578, 678, 690, 762, 782, 806, 862, 902, 950
CFS	23 bands: 454, 462–466, 474, 502, 554, 578, 598–602, 610, 638, 666, 674, 686, 694–698, 718–722, 734, 870, 902, 946–950
SPA	23 bands: 506–514, 522–534, 542–546, 582, 590, 614, 630, 638, 670, 682–686, 718, 726, 734, 778, 806, 914

To examine the spectral separability of all the wavebands selected by the three methods, the one-way analysis of variance (ANOVA) and multiple significant comparative tests at the 99% confidence level ($p < 0.01$) were performed. As shown in Figure 6, most of the frequencies of occurrence of these selected wavebands were greater than 15 (half of C_8^2) [94,95], which showed their high discriminative capacity for mangrove species. The higher the frequency of occurrence, the more important the waveband, and the stronger the distinguishing ability of the mangrove species. The wavebands with higher frequencies of occurrence at 526 nm, 578 nm, 638 nm, 686 nm, 718 nm, 734 nm, 806 nm, 902 nm and 950 nm, which were selected by two or more waveband selection methods, showed their importance of classifying mangrove species.

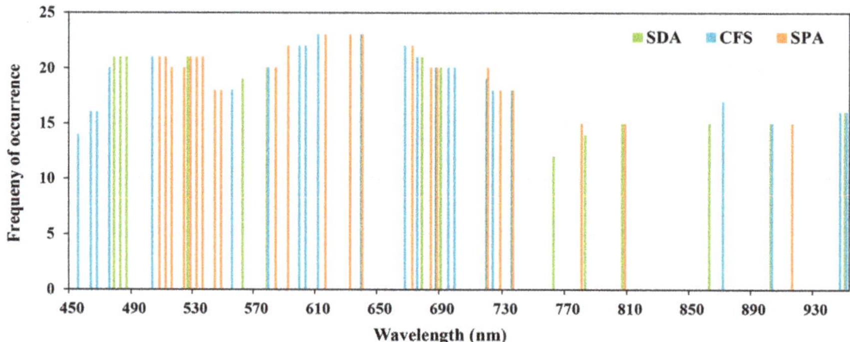

Figure 6. Frequencies of occurrence of the selected wavebands using the SDA, CFS and SPA methods.

3.4. Classification Results with the Selected Wavebands

The selected wavebands were then used as the input data for the LDA, KNN, RF and SVM classifications. Based on the reflectance spectra, the selected wavebands by using the three waveband selection methods (Table 5) and all 125 wavebands were used to identify mangrove species. Table 6 gives the classification results in terms of OA and kappa coefficient for the eight mangrove species identified by the four classifiers. The classification accuracies of these selected wavebands by using the SDA, CFS and SPA methods, were close to, or in some cases, higher than the accuracies when using all 125 wavebands. Among three waveband selection methods, the classification accuracy of the SPA was the highest. Hence, this report only presented the classification results of the wavebands selected by the SPA and the four classifiers (Table 7). Excepted for the LDA, the overall accuracies when using the other three classifiers all reach 80%, where the SVM was more than satisfactory in classifying mangrove species. The classification results indicated that the four classifiers could be easily used to identify the *T. populnea* stands, whose producer's and user's accuracies were more than 90%. For *C. manghas*, the classification accuracy of LDA was the lowest, while SVM could better separate *C. manghas* from other species, with a higher accuracy of 90%.

Table 6. Classification results in terms of OA, kappa coefficient, and standard deviation (in bracket) for the eight mangrove species obtained using different selected wavebands and classifiers of the 10-fold cross-validation data.

Selected Wavebands	Performance Metrics	Classifiers			
		LDA	KNN	RF	SVM
125 wavebands	OA(%)	84.17 (5.22)	87.50 (5.29)	87.92 (4.03)	**93.54** (2.68)
	Kappa	0.8179 (0.0595)	0.8562 (0.0607)	0.8609 (0.0460)	**0.9256** (0.0308)

Table 6. *Cont.*

Selected Wavebands	Performance Metrics	Classifiers			
		LDA	KNN	RF	SVM
14 wavebands (SDA)	OA(%)	74.79 (7.04)	86.46 (5.31)	86.04 (5.38)	**91.46** (4.10)
	Kappa	0.7104 (0.0812)	0.8442 (0.0609)	0.8394 (0.0617)	**0.9017** (0.0471)
23 wavebands (CFS)	OA(%)	81.67 (7.47)	87.92 (4.59)	87.71 (4.33)	**92.29** (2.79)
	Kappa	0.7896 (0.0852)	0.8608 (0.0527)	0.8586 (0.0498)	**0.9112** (0.0321)
23 wavebands (SPA)	OA(%)	83.75 (3.51)	86.88 (4.29)	84.58 (4.93)	**93.13** (1.98)
	Kappa	0.8133 (0.0400)	0.8490 (0.0489)	0.8226 (0.0565)	**0.9208** (0.0227)

Table 7. Summary of classification accuracies and standard deviation (in bracket) using different classifiers of the 10-fold cross-validation data with the 23 wavebands selected by the SPA.

Mangrove Species	Classifiers							
	LDA		KNN		RF		SVM	
	PA(%)	UA(%)	PA(%)	UA(%)	PA(%)	UA(%)	PA(%)	UA(%)
A. aureum (AA)	84.88 (12.70)	77.84 (15.44)	93.99 (7.85)	76.31 (16.02)	86.37 (11.50)	77.55 (14.20)	93.89 (8.11)	84.43 (13.84)
A. ilicifolius (AI)	87.75 (14.55)	92.14 (15.95)	84.57 (15.14)	82.17 (15.95)	75.81 (16.34)	77.89 (11.32)	90.40 (14.09)	89.87 (9.47)
A. corniculatum (AC)	96.67 (10.54)	69.55 (15.29)	72.08 (16.43)	90.50 (12.96)	71.33 (20.25)	80.75 (18.93)	80.71 (15.02)	96.57 (7.35)
S. apetala (SA)	76.62 (21.44)	98.33 (5.27)	98.33 (5.27)	93.57 (8.33)	97.08 (6.23)	96.90 (6.55)	100.00 (0.00)	98.57 (4.52)
H. littoralis (HL)	83.68 (17.68)	80.25 (14.68)	90.07 (14.31)	87.57 (13.23)	84.95 (17.91)	79.92 (13.45)	94.57 (12.95)	88.50 (10.64)
C. manghas (CM)	59.79 (31.60)	73.40 (28.72)	80.90 (22.09)	90.07 (14.31)	83.57 (22.23)	87.64 (12.49)	91.67 (21.15)	96.14 (9.23)
T. populnea (TP)	98.57 (4.52)	98.57 (4.52)	96.67 (7.03)	90.05 (11.09)	95.42 (7.47)	93.17 (9.90)	100.00 (0.00)	98.33 (5.27)
K. candel (KC)	83.99 (15.28)	90.07 (14.31)	75.92 (13.46)	94.25 (9.72)	82.32 (16.24)	92.98 (9.53)	89.17 (14.72)	97.50 (7.91)
OA(%)	83.75 (3.51)		86.88 (4.29)		84.58 (4.93)		**93.13** (1.98)	
Kappa	0.8133 (0.0400)		0.8490 (0.0489)		0.8226 (0.0565)		**0.9208** (0.0227)	

4. Discussion

4.1. Effect of the Optimal Waveband Selection Methods

The selection of effective wavebands can simplify the classification models and reduce the computational cost. By using a fewer number of effective wavebands is possible to achieve or exceed the classification accuracy of the entire waveband dataset [13]. The frequencies of occurrence of the selected wavebands demonstrated their high discriminative capacities for mangrove species (Figure 6), which were further verified by the classification results based on the wavebands selected by the three methods. The SDA method has been widely used for choosing effective wavebands in related research of mangroves [23], while the CFS and SPA methods have been mostly applied to machine-learning-based classifications [10,96]. In this study, the wavebands selected by the SPA showed better performance. There were few overlaps among the wavebands selected by the three methods (Table 5), while neighboring wavebands that had comparable discriminative capacities were selected. The wavebands at 526 nm, 578 nm, 638 nm, 686 nm, 718 nm, 734 nm, 806 nm, 902 nm and 950 nm were frequently selected. The selection results of this study differ from several studies with regards to the selected wavebands for classifying mangrove species [23,97]. This may be expected given the studied species and the specific sensor used in this study.

4.2. Impact of Spectral Datasets With Different Transformations

Spectral transformations and VIs can normally minimize the influence of brightness variations on ground-based spectral measurements. In this study, the classification performance of the selected hyperspectral VIs demonstrated that they were effective for the discrimination of vegetation species. This was consistent with the conclusion of a previous study [32]. Moreover, the derivative and log-transformed datasets manifested better classification results with the overall accuracies of all above 85%, when we employed the four machine-learning classifiers. When the classifications were performed based on the derivative spectra, the classification accuracy of LDA showed a slight improvement from 84.17% to 84.58% (Table 3). The LDA classifier generally requires the assumption of a normal distribution, while the derivative transformation will destroy this distribution [31]. Conversely, the logarithmic transformations can be generally used to normalize the distribution of a dataset. Based on the log-transformed dataset, the classification accuracy increment of LDA was 6.68%.

4.3. Performance of the Machine-Learning Classifiers

The classification performances of four machine-learning classifiers in identifying mangrove species were manifested, where most of the accuracies reached 80%. Compared to the other three non-parametric classifiers, the LDA gave the worst classification performance, especially when using fewer input parameters and the derivative spectra. This may be because the LDA is theoretically limited to parametric datasets and requires the assumption of normal distribution. As shown in Table 7, the lowest PA of 59.79% indicates that the LDA classifier has poor discrimination power for *C. manghas* class, and this class is easy to be mistakenly classified as another class. Conversely, the UA of 73.40% for *C. manghas* class shows that parts of the other classes were misclassified as this class. Previously, the non-parametric classifiers, such as RF and SVM, have proven to be effective for meeting the assumption of normal distribution when used for machine-learning classifications [34,98]. Overall, the SVM outperformed the KNN and RF classifiers, especially when using the selected wavebands and the derivative and log-transformed datasets. However, it should be noted that the classification performances of these machine-learning classifiers may generally depend on the number of features, number of samples, data types and the specific research purposes.

4.4. Applicability of Field Close-range Snapshot Hyperspectral Imaging

Hyperspectral imaging has been widely used to provide excellent detection capabilities for vegetation classification. The use of close-range snapshot hyperspectral imaging provides a new semi-automatic investigation method for hyperspectral measurements with proximal sensing in the field, which enables field surveys more convenient and rapid. The results of this study provided compelling evidence for the application of field close-range snapshot hyperspectral imaging in identifying mangrove species, which also provide a theoretical and practical guidance for monitoring mangrove forests. Compared to previous studies [22–26], which used non-imaging spectrometers, push-broom or staring imaging sensors, this hand-held close-range snapshot hyperspectral imaging, acquiring the spectral and image information at the time of one capture, bridges the gap between point and image data. It also can be considered as the transition from laboratory to field, and further close to realistic application. Furthermore, the snapshot hyperspectral imaging sensor can also be mounted on a UAV platform for use in precision agriculture, such as winter wheat above-ground biomass estimation [99], and vegetation classification [10,100]. The close-range hyperspectral imaging has potential to support the identifying of mangroves at the individual species level, but for field operational applications, there are still several limitations that need to be considered. Many captures need spend much time on preparations (e.g., multiple calibrations with white and dark references) and post-processing (e.g., the selection of samples), and the imaging effects are susceptible to the weather and illumination conditions.

5. Conclusions

In this study, we assessed the feasibility and usefulness of close-range snapshot hyperspectral imaging for mangrove species identification with field hyperspectral measurements. We classified mangrove species using different spectrum-transformed datasets, waveband selection methods, and machine-learning classifiers, and compared the classification results. Our main conclusions include: (1) The SVM proved to be more reliable for identifying mangrove species, when compared with the other three machine-learning classifiers. (2) The classification accuracies of the selected wavebands obtained by the three waveband selection methods, SDA, CFS and SPA, were competitive or comparable to the classification accuracies obtained when using all the wavebands. (3) The derivative and logarithmic transformations and hyperspectral VIs further improve the classification accuracies of mangrove species, especially those susceptible to background contamination and irregular illumination. The results of this study displayed the potential of close-range hyperspectral imaging as a tool in monitoring mangrove forests at the individual species level. The hyperspectral spectra of mangrove canopies acquired by using the snapshot hyperspectral imaging sensor under field conditions can be used to effectively identify mangrove species. The findings of this study can potentially provide further guidance for the application of space-borne and airborne hyperspectral sensors for mangrove forest management and conservation.

Author Contributions: J.C., K.L. and L.L. conceived and designed the experiments; J.C. performed the experiments and analyzed the results; J.C., K.L. and Y.Z. conducted the field investigations; J.C., K.L., L.L., J.L., and Z.H. wrote and revised the manuscript.

Funding: This research was jointly funded by the Science and Technology Planning Project of Guangdong Province (Grant No. 2017A020217003), the Natural Science Foundation of Guangdong (Grant No. 2016A030313261), the National Natural Science Key Foundation of China (Grant No. 41531178), the National Marine Public Welfare Research Project of China (Grant No. 201505012), the National Natural Science Foundation of China (Grant No. 61771496 and 41501368).

Acknowledgments: We would like to thank two doctoral students and three graduate students, including Cuilin Pan, Dashan Wang, Xiang Li, Min Tan and Liheng Peng, from the School of Geography and Planning, Sun Yat-sen University, Guangzhou, China, for their help during the field investigations. We also acknowledge China Oceanic Information Network, National Marine Data and Information Service (NMDIS) for providing the tidal data.

Conflicts of Interest: The authors declare no conflict of interest.

References

1. Kuenzer, C.; Bluemel, A.; Gebhardt, S.; Quoc, T.V.; Dech, S. Remote sensing of mangrove ecosystems: A review. *Remote Sens.* **2011**, *3*, 878–928. [CrossRef]
2. Giri, C. Observation and monitoring of mangrove forests using remote sensing: Opportunities and challenges. *Remote Sens.* **2016**, *8*, 783. [CrossRef]
3. Bahuguna, A.; Nayak, S.; Roy, D. Impact of the tsunami and earthquake of 26th December 2004 on the vital coastal ecosystems of the Andaman and Nicobar islands assessed using RESOURCESAT AWiFS data. *Int. J. Appl. Earth Obs. Geoinf.* **2008**, *10*, 229–237. [CrossRef]
4. Duke, N.C.; Meynecke, J.-O.; Dittmann, S.; Ellison, A.M.; Anger, K.; Berger, U.; Cannicci, S.; Diele, K.; Ewel, K.C.; Field, C.D. A world without mangroves? *Science* **2007**, *317*, 41–42. [CrossRef] [PubMed]
5. Food and Agriculture Organization (FAO). *The World's Mangroves 1980–2005*; FAO: Rome, Italy, 2007.
6. Zhang, H.; Wang, T.; Liu, M.; Jia, M.; Lin, H.; Chu, L.; Devlin, A.T. Potential of combining optical and dual polarimetric SAR data for improving mangrove species discrimination using Rotation Forest. *Remote Sens.* **2018**, *10*, 467. [CrossRef]

7. Pham, T.D.; Bui, D.T.; Yoshino, K.; Le, N.N. Optimized rule-based logistic model tree algorithm for mapping mangrove species using ALOS PALSAR imagery and GIS in the tropical region. *Environ. Earth Sci.* **2018**, *77*, 159. [CrossRef]
8. Zhu, Y.; Liu, K.; Liu, L.; Myint, S.; Wang, S.; Liu, H.; He, Z. Exploring the potential of WorldView-2 red-edge band-based vegetation indices for estimation of mangrove leaf area index with machine learning algorithms. *Remote Sens.* **2017**, *9*, 1060. [CrossRef]
9. Jia, M.; Zhang, Y.; Wang, Z.; Song, K.; Ren, C. Mapping the distribution of mangrove species in the Core Zone of Mai Po Marshes Nature Reserve, Hong Kong, using hyperspectral data and high-resolution data. *Int. J. Appl. Earth Obs. Geoinf.* **2014**, *33*, 226–231. [CrossRef]
10. Cao, J.; Leng, W.; Liu, K.; Liu, L.; He, Z.; Zhu, Y. Object-based mangrove species classification using unmanned aerial vehicle hyperspectral images and digital surface models. *Remote Sens.* **2018**, *10*, 89. [CrossRef]
11. Ruwaimana, M.; Satyanarayana, B.; Otero, V.; Muslim, A.M.; Syafiq, M.; Ibrahim, S.; Raymaekers, D.; Koedam, N.; Dahdouh-Guebas, F. The advantages of using drones over space-borne imagery in the mapping of mangrove forests. *PLoS ONE* **2018**, *13*, e0200288. [CrossRef] [PubMed]
12. Wannasiri, W.; Nagai, M.; Honda, K.; Santitamnont, P.; Miphokasap, P. Extraction of mangrove biophysical parameters using airborne LiDAR. *Remote Sens.* **2013**, *5*, 1787–1808. [CrossRef]
13. Tong, Q.; Zhang, B.; Zheng, L. *Hyperspectral Remote Sensing: Principles, Techniques and Applications*; Higher Education Press: Beijing, China, 2006.
14. Thenkabail, P.S.; Lyon, J.G.; Huete, A. *Hyperspectral Remote Sensing of Vegetation*; CRC Press: Boca Raton, FL, USA, 2016.
15. Kumar, T.; Panigrahy, S.; Kumar, P.; Parihar, J.S. Classification of floristic composition of mangrove forests using hyperspectral data: Case study of Bhitarkanika National Park, India. *J. Coast. Conserv.* **2013**, *17*, 121–132. [CrossRef]
16. Chakravortty, S.; Sinha, D. Analysis of multiple scattering of radiation amongst end members in a mixed pixel of hyperspectral data for identification of mangrove species in a mixed stand. *J. Indian Soc. Remote Sens.* **2015**, *43*, 559–569. [CrossRef]
17. Zhang, C.; Xie, Z. Data fusion and classifier ensemble techniques for vegetation mapping in the coastal everglades. *Geocarto Int.* **2014**, *29*, 228–243. [CrossRef]
18. Green, E.P.; Clark, C.D.; Mumby, P.J.; Edwards, A.J.; Ellis, A. Remote sensing techniques for mangrove mapping. *Int. J. Remote Sens.* **1998**, *19*, 935–956. [CrossRef]
19. Jensen, R.; Mausel, P.; Dias, N.; Gonser, R.; Yang, C.; Everitt, J.; Fletcher, R. Spectral analysis of coastal vegetation and land cover using AISA+ hyperspectral data. *Geocarto Int.* **2007**, *22*, 17–28. [CrossRef]
20. Prasad, K.A.; Gnanappazham, L. Multiple statistical approaches for the discrimination of mangrove species of using transformed field and laboratory hyperspectral data. *Geocarto Int.* **2016**, *31*, 891–912. [CrossRef]
21. Li, S.; Tian, Q. Mangrove canopy species discrimination based on spectral features of Geoeye-1 imagery. *Spectrosc. Spectr. Anal.* **2013**, *33*, 136–141. (In Chinese) [CrossRef]
22. Vaiphasa, C.; Ongsomwang, S.; Vaiphasa, T.; Skidmore, A.K. Tropical mangrove species discrimination using hyperspectral data: A laboratory study. *Estuar. Coast. Shelf Sci.* **2005**, *65*, 371–379. [CrossRef]
23. Zhang, C.; Kovacs, J.M.; Liu, Y.; Flores-Verdugo, F.; Flores-de-Santiago, F. Separating mangrove species and conditions using laboratory hyperspectral data: A case study of a degraded mangrove forest of the Mexican Pacific. *Remote Sens.* **2014**, *6*, 11673–11688. [CrossRef]
24. Yu, X.; Zhang, F.; Liu, Q.; Li, D.; Zhao, D. Analysis of typical mangrove spectral reflectance characteristics. *Spectrosc. Spectr. Anal.* **2013**, *33*, 454–458. (In Chinese) [CrossRef]
25. Yu, X.; Zhao, D.; Zhang, F.; Xiao, Z. Research on mangrove hyperspectrum analysis technology. *J. Binzhou Univ.* **2006**, *22*, 53–56. (In Chinese)
26. Weng, Q.; Lu, C. Research on mangrove canopy apparent spectral reflectance characteristics. *J. Fujian For. Sci. Technol.* **2006**, *33*, 14–19. (In Chinese)
27. Tian, M. Monitoring Winter Wheat Growth Conditions in the Northwest Region of China by Using Hyperspectral Remote Sensing. Ph.D. Dissertation, Northwest A&F University, Shanxi, China, 2017. (In Chinese)
28. Shang, K.; Zhang, X.; Sun, Y.; Zhang, L.; Wang, S.; Zhuang, Z. Sophisticated vegetation classification based on feature band set using hyperspectral image. *Spectrosc. Spectr. Anal.* **2015**, *35*, 1669–1676. [CrossRef]

29. Xiao, B.; Mao, W.; Liang, X.; Zhang, L.; Han, L. Study on varieties identification of Kentucky bluegrass using hyperspectral imaging and discriminant analysis. *Spectrosc. Spectr. Anal.* **2012**, *32*, 1620–1623. (In Chinese) [CrossRef]
30. Gao, J.; Nuyttens, D.; Lootens, P.; He, Y.; Pieters, J.G. Recognising weeds in a maize crop using a random forest machine-learning algorithm and near-infrared snapshot mosaic hyperspectral imagery. *Biosyst. Eng.* **2018**, *170*, 39–50. [CrossRef]
31. Pu, R.; Gong, P. *Hyperspectral Remote Sensing and Its Applications*; Higher Education Press: Beijing, China, 2000.
32. Pu, R.; Gong, P. Hyperspectral remote sensing of vegetation bioparameters. In *Advances in Environmental Remote Sensing: Sensors, Algorithms, and Applications*; CRC Press: Boca Raton, FL, USA, 2011; pp. 101–142.
33. Giri, S.; Mukhopadhyay, A.; Hazra, S.; Mukherjee, S.; Roy, D.; Ghosh, S.; Ghosh, T.; Mitra, D. A study on abundance and distribution of mangrove species in Indian Sundarban using remote sensing technique. *J. Coast. Conserv.* **2014**, *18*, 359–367. [CrossRef]
34. Barrett, B.; Nitze, I.; Green, S.; Cawkwell, F. Assessment of multi-temporal, multi-sensor radar and ancillary spatial data for grasslands monitoring in Ireland using machine learning approaches. *Remote Sens. Environ.* **2014**, *152*, 109–124. [CrossRef]
35. Liao, B.; Wei, G.; Zhang, J.; Tang, G.; Lei, Z.; Yang, X. Studies on dynamic development of mangrove communities on Qi'ao Island, Zhuhai. *J. South China Agric. Univ.* **2008**, *29*, 59–64. (In Chinese)
36. Tang, H.; Liu, K.; Zhu, Y.; Wang, S.; Liu, L.; Song, S. Mangrove community classification based on WorldView-2 image and SVM method. *Acta Sci. Nat. Univ. Sunyatseni* **2015**, *54*, 102–111. (In Chinese)
37. Zhou, F.; Kuang, D.; Jian, Y.; Huang, Q.; Ding, L. Primary study on the composition of mangrove community in Qi'ao Island, Zhuhai. *Ecol. Sci.* **2003**, *22*, 237–241. (In Chinese)
38. Liu, K.; Liu, L.; Liu, H.; Li, X.; Wang, S. Exploring the effects of biophysical parameters on the spatial pattern of rare cold damage to mangrove forests. *Remote Sens. Environ.* **2014**, *150*, 20–33. [CrossRef]
39. Ye, X.; Li, J.; Wang, A. Sedimentary environment and its response to anthropogenic impacts in the coastal wetland of the Qi'ao Island, Zhujiang River Estuary. *Haiyang Xuebao* **2018**, *40*, 79–89. [CrossRef]
40. Aasen, H.; Burkart, A.; Bolten, A.; Bareth, G. Generating 3D hyperspectral information with lightweight UAV snapshot cameras for vegetation monitoring: From camera calibration to quality assurance. *ISPRS-J. Photogramm. Remote Sens.* **2015**, *108*, 245–259. [CrossRef]
41. Savitzky, A.; Golay, M.J.E. Smoothing and differentiation of data by simplified least squares procedures. *Anal. Chem.* **1964**, *36*, 1627–1639. [CrossRef]
42. Vaiphasa, C. Consideration of smoothing techniques for hyperspectral remote sensing. *ISPRS-J. Photogramm. Remote Sens.* **2006**, *60*, 91–99. [CrossRef]
43. Zhao, A.; Tang, X.; Zhang, Z.; Liu, J. Optimizing Savitzky-Golay parameters and its smoothing pretreatment for FTIR gas spectra. *Spectrosc. Spectr. Anal.* **2016**, *36*, 1340–1344. (In Chinese) [CrossRef]
44. Lopatin, J.; Fassnacht, F.E.; Kattenborn, T.; Schmidtlein, S. Mapping plant species in mixed grassland communities using close range imaging spectroscopy. *Remote Sens. Environ.* **2017**, *201*, 12–23. [CrossRef]
45. Zhang, J.; Rivard, B.; Sánchez-Azofeifa, A.; Castro-Esau, K. Intra- and inter-class spectral variability of tropical tree species at La Selva, Costa Rica: Implications for species identification using HYDICE imagery. *Remote Sens. Environ.* **2006**, *105*, 129–141. [CrossRef]
46. Qian, Y.; Yu, J.; Jia, Z.; Yang, F.; Palidan, T. Extraction and analysis of hyperspectral data from typical desert grassland in Xinjiang. *Acta Pratacult. Sin.* **2013**, *22*, 157–166. (In Chinese)
47. Prasad, K.A.; Gnanappazham, L. Species discrimination of mangroves using derivative spectral analysis. *ISPRS Ann. Photogramm. Remote Sens. Spat. Inf. Sci.* **2014**, *2*, 45. [CrossRef]
48. Becker, B.L.; Lusch, D.P.; Qi, J. Identifying optimal spectral bands from in situ measurements of Great Lakes coastal wetlands using second-derivative analysis. *Remote Sens. Environ.* **2005**, *97*, 238–248. [CrossRef]
49. Pearson, R.L.; Miller, L.D. Remote mapping of standing crop biomass for estimation of the productivity of the shortgrass prairie. In Proceedings of the Eighth International Symposium on Remote Sensing of Environment, Ann Arbor, MI, USA, 2–6 October 1972; p. 1355.
50. Sun, Y.; Gong, H. *Quantitative Research on Wetland Plants Based on Hyperspectral Remote Sensing: A Case Study in Honghe Nature Reserve*; China Environmental Science Press: Beijing, China, 2015. (In Chinese)
51. Jain, N.; Ray, S.S.; Singh, J.P.; Panigrahy, S. Use of hyperspectral data to assess the effects of different nitrogen applications on a potato crop. *Precis. Agric.* **2007**, *8*, 225–239. [CrossRef]

52. Zarco-Tejada, P.J.; Berjón, A.; López-Lozano, R.; Miller, J.R.; Martín, P.; Cachorro, V.; González, M.R.; De Frutos, A. Assessing vineyard condition with hyperspectral indices: Leaf and canopy reflectance simulation in a row-structured discontinuous canopy. *Remote Sens. Environ.* **2005**, *99*, 271–287. [CrossRef]
53. Rouse, J.W.; Haas, R.H.; Schell, J.A.; Deering, D.W. *Monitoring Vegetation Systems in the Great Plains with ERTS*; Paper-A20; National Aeronautics and Space Administration (NASA): Washington, DC, USA, 1974; pp. 309–317.
54. Roujean, J.L.; Breon, F.M. Estimating PAR absorbed by vegetation from bidirectional reflectance measurements. *Remote Sens. Environ.* **1995**, *51*, 375–384. [CrossRef]
55. Haboudane, D.; Miller, J.R.; Pattey, E.; Zarco-Tejada, P.J.; Strachan, I.B. Hyperspectral vegetation indices and novel algorithms for predicting green LAI of crop canopies: Modeling and validation in the context of precision agriculture. *Remote Sens. Environ.* **2004**, *90*, 337–352. [CrossRef]
56. Ar, H. A soil-adjusted vegetation index (SAVI). *Remote Sens. Environ.* **1988**, *25*, 295–309. [CrossRef]
57. Baret, F.; Guyot, G. Potentials and limits of vegetation indices for LAI and APAR assessment. *Remote Sens. Environ.* **1991**, *35*, 161–173. [CrossRef]
58. Qi, J.; Chehbouni, A.; Huete, A.R.; Kerr, Y.H.; Sorooshian, S. A modified soil adjusted vegetation index. *Remote Sens. Environ.* **1994**, *48*, 119–126. [CrossRef]
59. Haboudane, D.; Miller, J.R.; Tremblay, N.; Zarco-Tejada, P.J.; Dextraze, L. Integrated narrow-band vegetation indices for prediction of crop chlorophyll content for application to precision agriculture. *Remote Sens. Environ.* **2002**, *81*, 416–426. [CrossRef]
60. Rondeaux, G.; Steven, M.; Baret, F. Optimization of soil-adjusted vegetation indices. *Remote Sens. Environ.* **1996**, *55*, 95–107. [CrossRef]
61. Daughtry, C.S.T.; Walthall, C.L.; Kim, M.S.; Colstoun, E.B.D.; McMurtrey, J.E.I. Estimating corn leaf chlorophyll concentration from leaf and canopy reflectance. *Remote Sens. Environ.* **2000**, *74*, 229–239. [CrossRef]
62. Hernández-Clemente, R.; Navarro-Cerrillo, R.M.; Suárez, L.; Morales, F.; Zarco-Tejada, P.J. Assessing structural effects on PRI for stress detection in conifer forests. *Remote Sens. Environ.* **2011**, *115*, 2360–2375. [CrossRef]
63. Broge, N.H.; Leblanc, E. Comparing prediction power and stability of broadband and hyperspectral vegetation indices for estimation of green leaf area index and canopy chlorophyll density. *Remote Sens. Environ.* **2000**, *76*, 156–172. [CrossRef]
64. Jordan, C.F. Derivation of leaf-area index from quality of light on the forest floor. *Ecology* **1969**, *50*, 663–666. [CrossRef]
65. Chen, J.M. Evaluation of vegetation indices and a modified simple ratio for boreal applications. *Can. J. Remote Sens.* **1996**, *22*, 229–242. [CrossRef]
66. Zarcotejada, P.J.; Miller, J.R.; Noland, T.L.; Mohammed, G.H.; Sampson, P.H. Scaling-up and model inversion methods with narrowband optical indices for chlorophyll content estimation in closed forest canopies with hyperspectral data. *IEEE Trans. Geosci. Remote Sens.* **2001**, *39*, 1491–1507. [CrossRef]
67. Vogelmann, J.E.; Rock, B.N.; Moss, D.M. Red edge spectral measurements from sugar maple leaves. *Int. J. Remote Sens.* **1993**, *14*, 1563–1575. [CrossRef]
68. Gitelson, A.A.; Merzlyak, M.N.; Lichtenthaler, H.K. Detection of red edge position and chlorophyll content by reflectance measurements near 700 nm. *J. Plant Physiol.* **1996**, *148*, 501–508. [CrossRef]
69. Zhang, C.; Chen, K.; Liu, Y.; Kovacs, J.M.; Flores-Verdugo, F.; Santiago, F.J.F.D. Spectral response to varying levels of leaf pigments collected from a degraded mangrove forest. *J. Appl. Remote Sens.* **2012**, *6*, 339–355. [CrossRef]
70. Maire, G.L.; François, C.; Dufrêne, E. Towards universal broad leaf chlorophyll indices using prospect simulated database and hyperspectral reflectance measurements. *Remote Sens. Environ.* **2004**, *89*, 1–28. [CrossRef]
71. Koedsin, W.; Vaiphasa, C. Discrimination of tropical mangroves at the species level with EO-1 Hyperion data. *Remote Sens.* **2013**, *5*, 3562–3582. [CrossRef]
72. Vaiphasa, C.; Skidmore, A.K.; de Boer, W.F.; Vaiphasa, T. A hyperspectral band selector for plant species discrimination. *ISPRS-J. Photogramm. Remote Sens.* **2007**, *62*, 225–235. [CrossRef]
73. Fung, T.; Yan Ma, H.F.; Siu, W.L. Band selection using hyperspectral data of subtropical tree species. *Geocarto Int.* **2003**, *18*, 3–11. [CrossRef]

74. Thenkabail, P.S.; Enclona, E.A.; Ashton, M.S.; Meer, B.V.D. Accuracy assessments of hyperspectral waveband performance for vegetation analysis applications. *Remote Sens. Environ.* **2004**, *91*, 354–376. [CrossRef]
75. Hall, M.A. Feature selection for discrete and numeric class machine learning. In Proceedings of the Seventeenth International Conference on Machine Learning, Stanford, CA, USA, 29 June–2 July 2000; pp. 359–366.
76. Wollmer, M.; Schuller, B.; Eyben, F.; Rigoll, G. Combining long short-term memory and dynamic Bayesian networks for incremental emotion-sensitive artificial listening. *IEEE J. Sel. Top. Signal Process.* **2010**, *4*, 867–881. [CrossRef]
77. Araújo, M.C.U.; Saldanha, T.C.B.; Galvão, R.K.H.; Yoneyama, T.; Chame, H.C.; Visani, V. The successive projections algorithm for variable selection in spectroscopic multicomponent analysis. *Chemom. Intell. Lab. Syst.* **2001**, *57*, 65–73. [CrossRef]
78. Ma, H.; Ji, H.; Lee, W.S. Identification of the citrus greening disease using spectral and textural features based on hyperspectral imaging. *Spectrosc. Spectr. Anal.* **2016**, *36*, 2344–2350. [CrossRef]
79. Clark, M.L.; Roberts, D.A.; Clark, D.B. Hyperspectral discrimination of tropical rain forest tree species at leaf to crown scales. *Remote Sens. Environ.* **2005**, *96*, 375–398. [CrossRef]
80. Feret, J.B.; Asner, G.P. Tree species discrimination in tropical forests using airborne imaging spectroscopy. *IEEE Trans. Geosci. Remote Sens.* **2013**, *51*, 73–84. [CrossRef]
81. Prospere, K.; McLaren, K.; Wilson, B. Plant species discrimination in a tropical wetland using in situ hyperspectral data. *Remote Sens.* **2014**, *6*, 8494–8523. [CrossRef]
82. Cover, T.; Hart, P. Nearest neighbor pattern classification. *IEEE Trans. Inf. Theory* **1967**, *13*, 21–27. [CrossRef]
83. Kotsiantis, S.B. Supervised machine learning: A review of classification techniques. In *Proceedings of the 2007 Conference on Emerging Artificial Intelligence Applications in Computer Engineering: Real Word Ai Systems with Applications in Ehealth, Hci, Information Retrieval and Pervasive Technologies*; IOS Press: Amsterdam, The Netherlands, 2007; pp. 3–24, ISBN 978-1-58603-780-2.
84. Liaw, A.; Wiener, M. Classification and regression by randomforest. *R News* **2002**, *2*, 18–22.
85. Naidoo, L.; Cho, M.A.; Mathieu, R.; Asner, G. Classification of savanna tree species, in the Greater Kruger National Park region, by integrating hyperspectral and LiDAR data in a Random Forest data mining environment. *ISPRS-J. Photogramm. Remote Sens.* **2012**, *69*, 167–179. [CrossRef]
86. Féret, J.-B.; Asner, G.P. Semi-supervised methods to identify individual crowns of lowland tropical canopy species using imaging spectroscopy and LiDAR. *Remote Sens.* **2012**, *4*, 2457–2476. [CrossRef]
87. Tan, Y.; Xia, W.; Xu, B.; Bai, L. Multi-feature classification approach for high spatial resolution hyperspectral images. *J. Indian Soc. Remote Sens.* **2017**, *46*, 9–17. [CrossRef]
88. Chang, C.C.; Lin, C.J. Libsvm: A library for support vector machines. *ACM Trans. Intell. Syst. Technol.* **2011**, *2*, 27. [CrossRef]
89. Stehman, S.V. Selecting and interpreting measures of thematic classification accuracy. *Remote Sens. Environ.* **1997**, *62*, 77–89. [CrossRef]
90. Zhao, Y.; Chen, D.; Yang, L.; Li, X.; Tang, W. *The Principle and Method of Analysis of Remote Sensing Application*; Science Press: Beijing, China, 2003.
91. Abdel-Rahman, E.M.; Mutanga, O.; Adam, E.; Ismail, R. Detecting sirex noctilio grey-attacked and lightning-struck pine trees using airborne hyperspectral data, random forest and support vector machines classifiers. *ISPRS-J. Photogramm. Remote Sens.* **2014**, *88*, 48–59. [CrossRef]
92. Cohen, J. A coefficient of agreement for nominal scales. *Educ. Psychol. Meas.* **1960**, *20*, 37–46. [CrossRef]
93. Cohen, J. Weighted kappa: Nominal scale agreement provision for scaled disagreement or partial credit. *Psychol. Bull.* **1968**, *70*, 213. [CrossRef] [PubMed]
94. Lin, C.; Gong, Z.; Zhao, W.; Fan, L. Identifying typical plant ecological types based on spectral characteristic variables: A case study in Wild Duck Lake wetland, Beijing. *Acta Ecol. Sin.* **2013**, *33*, 1172–1185. [CrossRef]
95. Huang, J.; Sun, Y. Sifting of hyperspectral vegetation indices applying to classification of 5 kinds of plants in Honghe wetlands. *Wetl. Sci.* **2016**, *14*, 888–894. (In Chinese)
96. Zhu, H.; Chu, B.; Zhang, C.; Liu, F.; Jiang, L.; He, Y. Hyperspectral imaging for presymptomatic detection of tobacco disease with successive projections algorithm and machine-learning classifiers. *Sci. Rep.* **2017**, *7*, 4125. [CrossRef] [PubMed]
97. Wang, L.E.; Sousa, W.P. Distinguishing mangrove species with laboratory measurements of hyperspectral leaf reflectance. *Int. J. Remote Sens.* **2009**, *30*, 1267–1281. [CrossRef]

98. Pham, L.T.H.; Brabyn, L. Monitoring mangrove biomass change in Vietnam using spot images and an object-based approach combined with machine learning algorithms. *ISPRS-J. Photogramm. Remote Sens.* **2017**, *128*, 86–97. [CrossRef]
99. Yue, J.; Yang, G.; Li, C.; Li, Z.; Wang, Y.; Feng, H.; Xu, B. Estimation of winter wheat above-ground biomass using unmanned aerial vehicle-based snapshot hyperspectral sensor and crop height improved models. *Remote Sens.* **2017**, *9*, 708. [CrossRef]
100. Ishida, T.; Kurihara, J.; Viray, F.A.; Namuco, S.B.; Paringit, E.C.; Perez, G.J.; Takahashi, Y.; Marciano, J.J., Jr. A novel approach for vegetation classification using UAV-based hyperspectral imaging. *Comput. Electron. Agric.* **2018**, *144*, 80–85. [CrossRef]

Article

An Under-Ice Hyperspectral and RGB Imaging System to Capture Fine-Scale Biophysical Properties of Sea Ice

Emiliano Cimoli [1,*], Klaus M. Meiners [2,3], Arko Lucieer [4] and Vanessa Lucieer [1]

1 Institute for Marine and Antarctic Studies, College of Sciences and Engineering, University of Tasmania, Private Bag 129, Hobart 7001, Tasmania, Australia; vanessa.lucieer@utas.edu.au
2 Australian Antarctic Division, Department of the Environment and Energy, Australian Government, Kingston 7050, Tasmania, Australia; klaus.Meiners@aad.gov.au
3 Australian Antarctic Program Partnership, University of Tasmania, Hobart 7001, Tasmania, Australia
4 Discipline of Geography and Spatial Sciences, School of Technology, Environments and Design, College of Sciences and Engineering, University of Tasmania, Private Bag 76, Hobart 7001, Tasmania, Australia; arko.lucieer@utas.edu.au
* Correspondence: emiliano.cimoli@utas.edu.au; Tel.: +61-0472682108

Received: 26 September 2019; Accepted: 16 November 2019; Published: 2 December 2019

Abstract: Sea-ice biophysical properties are characterized by high spatio-temporal variability ranging from the meso- to the millimeter scale. Ice coring is a common yet coarse point sampling technique that struggles to capture such variability in a non-invasive manner. This hinders quantification and understanding of ice algae biomass patchiness and its complex interaction with some of its sea ice physical drivers. In response to these limitations, a novel under-ice sled system was designed to capture proxies of biomass together with 3D models of bottom topography of land-fast sea-ice. This system couples a pushbroom hyperspectral imaging (HI) sensor with a standard digital RGB camera and was trialed at Cape Evans, Antarctica. HI aims to quantify per-pixel chlorophyll-a content and other ice algae biological properties at the ice-water interface based on light transmitted through the ice. RGB imagery processed with digital photogrammetry aims to capture under-ice structure and topography. Results from a 20 m transect capturing a 0.61 m wide swath at sub-mm spatial resolution are presented. We outline the technical and logistical approach taken and provide recommendations for future deployments and developments of similar systems. A preliminary transect subsample was processed using both established and novel under-ice bio-optical indices (e.g., normalized difference indexes and the area normalized by the maximal band depth) and explorative analyses (e.g., principal component analyses) to establish proxies of algal biomass. This first deployment of HI and digital photogrammetry under-ice provides a proof-of-concept of a novel methodology capable of delivering non-invasive and highly resolved estimates of ice algal biomass in-situ, together with some of its environmental drivers. Nonetheless, various challenges and limitations remain before our method can be adopted across a range of sea-ice conditions. Our work concludes with suggested solutions to these challenges and proposes further method and system developments for future research.

Keywords: sea ice; ice algae; biomass; hyperspectral imaging; fine-scale; photogrammetry; under-ice; underwater; antarctica; structure from motion

1. Introduction

Sea-ice biophysical properties play a central role in controlling primary production and ecosystem function within the polar oceans [1–3]. Primary physical properties of the sea-ice environment include snow depth, ice thickness, sea-ice texture/structure, and under-ice topography. Biological properties often refer to ice algal biomass and include ice algal community composition and physiological condition. Ice algal biomass is strongly dependent on sea-ice physical properties, and both show variability at multiple spatial and temporal scales [4–6].

Ice algal biomass has been observed to display patchiness ranging from the mesoscale to the millimeter-scale and can undergo changes on a daily, weekly, and monthly basis [4,7,8]. The spatio-temporal variability of ice biological properties is determined by some of the sea ice physical properties such as snow depth and ice thickness, governing light availability for the organisms. In addition, ice algal biomass has been linked to sea-ice structure, under-ice roughness, and their complex interplay with the biogeochemical properties of the water column controlled by currents and boundary layer exchange processes [9–13].

A standard proxy for algal biomass in land-fast sea-ice is bottom chlorophyll-a (chl-a) (mg m^{-2}). This has traditionally been derived from melted ice core bottom sections. Typically bottom ice is sampled in 0.03 to 0.1 m long sections, i.e., where most of the biomass is typically found [14].

Capturing and quantifying variability in algal biomass together with some of its associated physical drivers over the full range of spatial scales is extremely challenging. Data for both polar oceans remain sparse in space and time [14–16]. Challenges are in part attributed to the difficulties in conducting fieldwork in polar regions, but also to the spatially limited and invasive nature of traditional point sampling methods such as ice coring. Due to ice algae residing on the underside of sea ice, satellite or airborne remote sensing techniques cannot be used, thereby limiting data collection to field sampling. This has had implications on our capability to properly estimate polar marine primary production, to identify complex under-ice food web dynamics, and assess sea-ice ecosystem responses to environmental change [6,15].

In response to this limitation in sampling methods, under-ice bio-optical methods have emerged as a non-invasive alternative to capture ice algal biomass variability at different spatial scales. These methods are based on the formulation of relationships between spectral radiance or irradiance measurements in the photosynthetically active radiation (PAR, from 400 to 700 nm) range from underneath the ice, and the amount of integrated ice-core chl-a (e.g., [17] or see [4] for a thorough review). Upward looking hyperspectral radiometers mounted on L-shaped deployment arms (or L-arms) have provided means to produce spectra-chl-a relationships by sampling over different spots within an area or non-invasive monitoring of change through time [17–19]. Derived bio-optical relationships can then be applied to datasets obtained from mapping platforms such as remotely operated vehicles (ROVs) [8,20] or instrumented under-ice trawls [7,21]. ROVs permit sampling at the floe-scale area of hundreds of square meters while under-ice trawls are able to cover transects up to two kilometers in length [5]. While these approaches have pushed the spatial boundaries of the surveying, their ability to capture the fine-scale variability of sea bio-physical properties remains limited due to their point sampling nature [22]. Wide solid angles or cosine corrected sensors necessarily integrate over wide surface footprints, particularly when vehicle movements exceed sensor integration times. Large footprints also hinder the effective coupling with the high spatial resolutions achieved by acoustic methods to capture under-ice topography [23], or with photogrammetric methods to capture fine-scale snow depth variability, and sea-ice surface properties [24–26]. Importantly, the obtained resolutions are not always compatible with some of the scales of spatial variability observed for under-ice habitats.

Hyperspectral imaging (HI) has been experimentally tested and proposed as an additional method to look at under-ice biomass variability from cm to sub-mm pixel scales over square-meter areas [27]. Preliminary results suggest that there is potential for HI to be extended to survey tenths of meters transects swaths although until now no in-situ application has been trialed.

From a biogeoscience perspective, HI aims to identify, quantify (measure), and map—chemical, physical, and biological properties—in each of the highly spectrally resolved pixels of the target image. As the technology becomes more portable and accessible, it has found a wide range of applications. A relevant analogous example is HI cameras equipped onto unmanned aerial systems (UAS) which are filling an essential gap between classical ground, full-size aircraft, and satellite sensing systems allowing more mapping at increased resolutions with ease of repeatability [28–31].

Underwater applications of HI are still in a development phase but are presenting opportunities to monitor and map shallow benthic habitats [32,33] and intertidal microphytobenthic environments [34]. HI cameras have also been mounted onto deep-sea ROVs and shown to be a useful taxonomic tool for macrofauna [35] and mapping of manganese nodules [36].

Using HI to investigate processes at the sea ice-water interface presents a new level of technical and logistical challenges. The low temperatures and the difficulty of deploying instruments (and divers) under polar sea-ice are the most obvious. Measuring transmitted light rather than reflected light, however, poses the most constraints. Also, pushbroom HI sensors need to be carefully configured so that the integration time and imaging frequency match the required spatial resolution [28]. Acquired images then typically require a series of radiometric and geometric corrections which are far from trivial for dynamic under-water platforms. Challenges are accentuated in an environment where low, yet variable, downwelling transmitted light availability pushes sensors to their limits. The translucent nature of sea ice would also render the utilization of active light sources, commonly employed in underwater HI applications, as a highly arguable approach. The under-ice realm can be a highly dynamic environment, and where the utilization of common geo-positioning and communication methods employed in typical aerial HI surveys is much more challenging due to the ice cover and viewing geometry [28,37].

This study aims to develop and test the feasibility of the first version of an under-ice sliding hyperspectral imaging system developed to produce in-situ transects several meters long at sub-millimeter spatial resolution. Along with the HI camera, a professional consumer-grade RGB camera was included in the payload for structure from motion (SfM) digital photogrammetry. SfM digital photogrammetry has revolutionized surface topographic mapping by providing a relatively low-cost solution that can provide accurate, high-resolution 3D structures of surfaces of interest through a set of highly overlapping pictures. Particularly relevant is the example of consumer-grade cameras being equipped on UAS to considerably increase the spatial extent of these surveys. For underwater applications, the methodology presents additional challenges which are still a subject of research, but present an equal amount of opportunities [38–41]. Under-ice, few studies have presented the potential of orthomosaic composition from RGB imagery retrieved from underwater vehicles (e.g., [42]), although SfM potential to generate quantitative topography has not been explored before.

Our HI system was tested between November–December 2018 under land fast-sea ice off Cape Evans, Antarctica. The relatively smooth and accessible under-ice surface of land-fast sea ice makes it an appealing first target for testing the technology. The site allows deployment of the system which can slide at a fixed distance underneath the ice. Fast ice also hosts some of the most productive (per volume) microalgal habitats in marine systems [1,43], making it a highly relevant first test target. To the author's knowledge, no published study has applied HI or photogrammetry to the under-ice environment before, nor have HI technologies been tested in polar marine waters.

Overall our study has the following four objectives:

1) To develop and present a novel system capable of capturing fine-scale under-ice biophysical properties based on underwater HI and RGB imagery and photogrammetry.

2) To illustrate the logistical and technical approaches taken for this first in-situ trial.

3) To provide a sample of the primary data outputs of the system and an exploration of the potential data processing workflows aimed to estimate biomass variability and under-ice 3D structure.

4) To present an outlook for the potential of the method, address future system development needs, and highlight the method caveats that require further research.

2. Materials and Methods

2.1. System Design and Sensors

A detailed discussion on the theoretical principles for underwater HI applications can be found in [44] and an extension of such theory from an under-ice perspective can be found in [4]. Here we only discuss the aspects that have driven the design of the under-ice system.

Depending on the camera settings and the desired aims, HI sensors can capture features at different scales ranging from millimeter close-range imagery to continuous swaths of data at the mesoscale. The mapping scale is determined by the sensor distance from the target and the mounting platform. Hyperspectral images are required to be orthorectified to enable extraction of meaningful and accurate metric information of the feature of interest (e.g., distances, shapes, and areas). This is ultimately necessary to compute the biochemical properties of the target [28], and to allow for accurate repeat surveys and co-registration with other datasets.

The modality in which the frame is acquired can be in either pushbroom or as 2D snap-shot imagers. Pushbroom HI line scanners are optimal when it comes to cover large surfaces under dynamic conditions as spectral and spatial information are acquired at the same instance. Pushbroom HI also comes at the best compromise with respect to fundamental sensor properties such as image quality, sensitivity, spectral coverage, and spectral, spatial, and radiometric resolutions [28,45]. However, in order to compose a rectified pushbroom orthoimage, sensors are required to be moving relative to the imaged surface at precisely matched speeds, imaging frequencies (or frame rates), all whilst acquiring a highly stable attitude (pitch, roll, and heading) and distance from the target [28,37,46]. Consequently, pushbroom HI is particularly sensitive when integrated onto dynamic platforms surveying under real environmental conditions and requires the full set of six-position (X, Y, Z) and orientation (pitch, roll, and heading) parameters (pose) assigned for every scan-line. An additional suite of sensors is therefore required to be integrated, and/or additional data products need to be included post-processing for robust HI geometric correction. These include highly precise Global Navigation Satellite Systems (GNSS)/inertial measurements units (IMUs), digital elevation models (DEMs), and orthomosaics of the imaged surface and/or a series of ground control points (GCPs) [28].

Considering that light levels beneath sea-ice are typically very low, ranging from 0.1 to 10% of the incoming solar radiation, HI scans are forced to move at reasonably low speeds so that the signal to noise ratio (SNR) is maximized, requiring integration times and imaging frequency to be optimized (resulting in relatively long integration times and slow imaging frequency required for low-light levels). This makes HI imaging of transmitted under-ice radiance challenging for dynamic underwater conditions and future deployment onto platforms (e.g., ROVs) that are susceptible to continuous buoyancy, speed, drag, and currents adjustments. Also, under-ice navigation and positioning is far from trivial and/or comes at high costs.

The developed approach here aims instead to scan relatively smooth under-ice surfaces by sliding or "skiing" at a predefined fixed distance from the ice at precisely controllable speeds (Figure 1). This enables the scanning movement to remain considerably stable, reducing some of the requirements aforementioned. The transect is prepared to be a pre-defined straight-line between 10 to 40 meters in length, limited in this prototype by the length of tether (Figure 2). Ideally, the set-up is expected to permit stable scanning speeds matched to the low-light levels experienced and the need of pushbroom HI orthorectification to be suppressed (or minimized). To achieve a steady, slow, and controllable movement, two WG1500 manual worm gear winches (Dutton Lainson, NE, USA) were established at each end-point of the surveyed transects (Figure 2). Stainless steel wires were attached from each winch to the respective end of the aluminum frame legs of the payload rig, which allowed the system to precisely slide back and forth through controlled winch rotations (Figure 2).

Such a sliding concept is only possible on under-ice surfaces which are relatively flat—a common feature of land-fast sea ice in both the Arctic [20,47] and in Antarctica [48]. Fast-ice is not only a relevant target for first tests of the technology, but also provides a relatively simpler optical set-up where algae are mostly residing at the bottom of the ice, at least during spring [14]. Under rougher under-ice surfaces (e.g., pack ice, platelet ice, ice fissures, and cracks caused by pressure ridges or medium to large brinicles) the scanning advancement of the system result could be impeded with such a skies-based concept.

Figure 3 displays the core components of the internal payload that were fitted in the system enclosure. An overview of all sensors, equipment specifications, and their purpose for this first test can be found in Table 1. Detailed information of technical design and software employed to operate the payload can be found Appendix A. The appendix also includes a schematization of power supply and data transmission paths from the surface elements to the enclosure interior and the external payload (Figure A1).

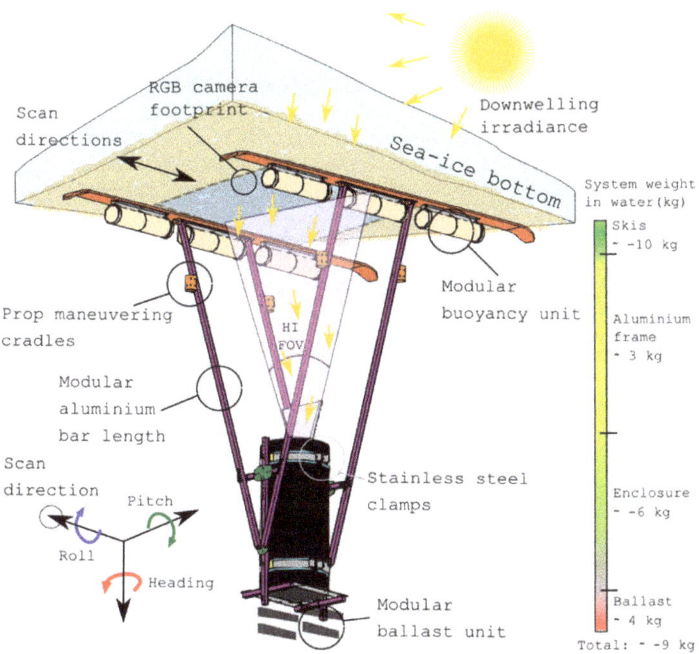

Figure 1. Concept design of the under-ice hyperspectral and RGB imaging system to capture fine-scale biophysical properties of sea ice. The system is designed to retrieve bio-optical relationship from downwelling sea-ice transmitted radiance. The sliding system aims to smoothly scan transects tenths of meters. It has a variable ski span of 0.82 to 1.2 m, a ski length of 1.48 m and a height of approximately 2 m. Its modular buoyancy system allows adjustment of the upward push against the ice and stabilizes the structure under different payload set-ups. The figure also shows the payload attitude reference system relative to the sensors orientation (heading, roll, and pitch). HI refers to Hyperspectral Imaging and FOV to Field of View.

To select an appropriate distance between the imaging sensors and the ice, we considered the trade-off between HI and RGB imaging specifications together with a series of environmental and logistical constraints (see [4] for a trade-offs overview). For example, spatial resolution and image footprint are inversely correlated since increased distances from the ice yields a larger footprint at the cost of pixel size. Increasing the distance from the ice also enlarges the depth of field (DOF), which is an important factor to consider for close-range optical HI and RGB imaging applications. The DOF should be large enough to cover at least the sea-ice skeletal layer where most of the algal biomass is concentrated. Nonetheless, while gaining distance from the ice seems appealing to increase survey area, it increases logistical and technical problems which are relevant to the deployment of a large sliding platform beneath thick ice cover. Such problematics add up to the known effects of the water column on measured light intensity and spectral composition in the visible range [49]. Overall, the increased costs of deploying optical sensors underwater need to be considered together with the additional challenges of geometric and chromatic correction of underwater images associated to the diverse refractive indices across the seawater-glass-air interface [50–52]. Such aberrations are not trivial to correct and depend on multiple factors such as the sensors optical parameters and settings, deployment mode (e.g., distance from the ice and field of view (FOV) inclinations), water optical properties and the underwater housing lens design (e.g., flat versus dome), and material (e.g., thickness of the acrylic window).

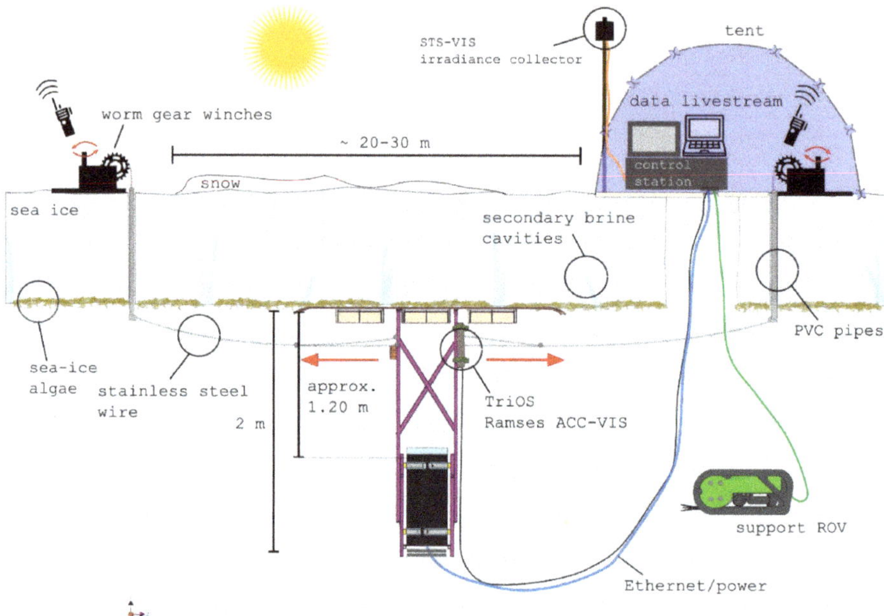

Figure 2. Field deployment and operation concept for the under-ice hyperspectral imaging and RGB scanning system. Two worm gear winches provide highly controllable slow movement back and forth along predefined transect. Movement commands are provided via radio communication and manual winching. The support remotely operated vehicle (ROV) is used to establish a tow-line between the deployment hole and the opposite transect endpoint. The deployment and operation require at least three people. Figure is not to scale.

For this prototype test, we found that an enclosure with a flat-port fitted with sensors separated approximately one meter from the ice would be a good compromise considering our equipment, deployment capabilities, and the spatial variability of the target (sea-ice algae) that we were surveying (Figure 3). The custom-built and low-cost aluminum frame that set the distance from the sensor to the ice was approximately 1.20 ± 0.10 m in length (variable by changing the angle of the legs and steel clamps position). It also allowed the legs to be modified to any desired length if required (Figures 1 and 2). The span between the 1.48-m-long skies ranged from 0.82 to 1.2 meters. It was confirmed that no components of the frame or skis interfered with the sensors FOVs and that FOVs of both sensors largely overlapped for coherent HI and 3D data interpretation.

Since the system travels at a fixed distance from the target, the horizontal and vertical footprint of the sensors can be estimated for the entire transect using standard imaging formulas (e.g., see Appendix B in [24]). Nonetheless, a flat-port causes magnification of images due to the multiple refractions at the air-acrylic-water interfaces, thus reducing the apparent FOV [51]. The amount of magnification is generally ≤1.33 and can be theoretically obtained using Snell's law. However, such calculations are not straightforward and require a series of sensor optical parameters and sensor specifications, not always easily retrievable. Some include entrance pupil distance relative to the port and imaging object, underwater focus distance and port thickness, among others. To precisely calculate the sensor footprint on the ice, a simpler way is to image objects of known length from which we can retrieve pixel size and derive horizontal and vertical footprint thereafter.

Finally, it is important to consider that miniaturization of remote sensing payloads is always preferable but is inevitably associated with increased cost and/or complexity [28,29]. We must then consider logistical and technical constraints as significant factors that could impede the deployment of a cost-effective solution. It was also preferable to use commercially available and off-the-shelf components when possible, to foster ease of replicability. For example, it was considered mandatory for the system to be surface powered and to be able to stream data to operator and change sensors acquisition parameters based on observed circumstances in real-time. The latter is not straightforward, considering a large amount of data is generated over the multiple high-frequency imaging processes. Costly underwater fiber optic connectors and tethers were avoided by allocating an internal digital processing unit (DPU) within the enclosure, which directly interfaced with the multiple sensors and allowed for on-board data storage (Figure 3). Power and communication with the surface was enabled through an ethernet/power cable permitting for virtual network computing (VNC). Altogether, these design features come at the cost of payload volume, and the entire payload was fitted into a cylindrical enclosure with an internal diameter of 0.23 m and a length of 0.6 m (Figure 3).

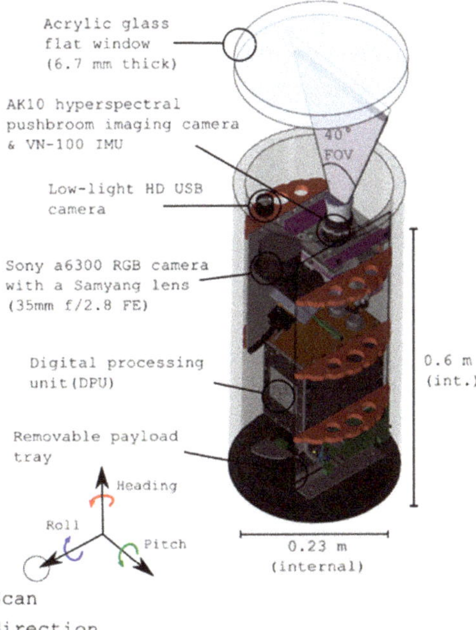

Figure 3. An overview of the payload main internal components, their allocation within the enclosure and volume required to host the payload. AK10 stands for AISA Kestrel 10. The figure also includes the payload attitude reference system relative to the sensors orientation (heading, roll and pitch).

Table 1. Summary of all optical sensors utilized in the internal and external components of the developed system together with their specifications (top part). The table also includes specifications of other components required to run the system (bottom part). Field of view (FOV)h and FOVv stand for the vertical and horizontal field of view. Underwater FOV is only an ≤ estimate approximation based on simplified theoretical formulas. FWHM refers to full width to half maximum.

Sensor	Fore-Optics/Lens	Field of View (FOV_h/FOV_v)	Number of Spatial Pixels	Spectral Range	Spectral Resolution and FWHM
Specim AISA Kestrel 10 pushbroom sensor	aperture: F/2.4 focal length: 35.375	40°/0.0388° in air ~29.88°/0.029° underwater	2048 × 1 or 1024 × 1 (binned)	400–1000 nm	1.75 / 3.5 / 7 nm/pixel (depending on binning)
Sony a6300 with Samyang AF 35 mm FE	max aperture F/2.8 focal length: 35 mm	37.2°/25.12° in air ~27.5°/18.7° underwater	6000 × 4000	Visible	RGB
Low Light USB HD cam	focal length: 2.97 mm	80°/64° in air ~57.3°/46.6° underwater	1920 × 1080	Visible	RGB
TriOS Ramses ACC	cosine corrected diffuser	Cosine response	Point sampling	320–950 nm	3.3 nm/pixel
STS-VIS	CC-3-DA cosine corrected diffuser	Cosine response	Point sampling	350–800 nm	3.0 nm/pixel (50 μm slit version)
Other components					
Digital Processing Unit (DPU)	Used to interface and operate all internal sensors/cameras with the surface PC using VNC and has custom electronics from Specim (see Appendix A). Specifications are: Windows 7 Pro, Intel Core i5, 64 bit, 8GB RAM, PIXCI EB1 frame grabber, CameraLink converter, 500 GB HyperX SATA SSD.				
VN-100 IMU	Measures system attitude used for future geo-rectification of HI imagery (see Appendix A). Operated through the DPU via a Python script. Specifications: 0.5° Static Pitch/Roll, 1.0° Dynamic Pitch/Roll, 5°/hr Gyro In-Run Bias (typ.), 800 Hz IMU Data, ±16 g Accelerometer Range, ±2000°/sec Gyroscope Range, no GPS unit is included in this model.				
Garmin GPS 18x LVC	Used for above surface GPS lock and time-stamp synchronization (see Appendix A). Specifications: 12-channel GPS receiver tracks, up to 12 satellites, one-pulse-per-second logic-level output with a rising edge aligned to within 1 microsecond of UTC. 1 Hz, output data in NMEA 0183 format.				
Lumen Subsea Lights (LEDs)	Four units attached as external payload. Intensity manipulated from above surface using a custom build-control. Specifications: Max brightness of 1500 lumens dimmable, beam angle of 135 deg in water and color temperature of 6200 Kelvin.				

The overall height of the system (including the frame legs and skis—Figure 1) was approximately 2 m which required a well-regulated buoyancy to keep the system vertical and pushing against the ice with moderate upward pressure to allow for smooth scanning. This was achieved through modular buoyancy and ballast units that regulated the system's vertical buoyancy and stability based on local conditions as displayed in Figure 1.

One benefit of the system's frame size was that it allowed the incorporation of external sensors in the future. For example, for our first tests, we included an upward-looking TriOS Ramses ACC-VIS spectrally resolved irradiance sensor near the ice-water interface to measure light directly exiting the sea-ice matrix (seen in Figure 2 and specified in Table 1).

2.2. Field Site and Transect Preparation

First trials of the system occurred during November–December 2018 under highly productive Antarctic land-fast sea ice off Cape Evans (77.6371733°S, 166.4018691°E) [13,14,48]. As seen in Figure 4c, we did not experience any platelet ice during the period of our surveys, contrary to what was experienced over the same site during other studies [23,48]. The area was characterized by a relatively homogenous sea-ice thickness of approximately 1.8 ± 0.01 m, except for ridged or cracked areas, and this was confirmed by our sampling. The area was also largely snow-free due to wind-induced snowdrift and displacement. An ice hole site was selected from which three transects with variable surface conditions could be surveyed. Transect directions pointed towards northwest (NW), west (W), and southwest (SW). In this study, we provide only a data sample from the western transect as this paper aims to describe the technical performance of the payload and its potential for research applications (see objectives). The analysis of the remaining transects and biophysical investigation of the under-ice habitat at Cape Evans will be presented in a later study.

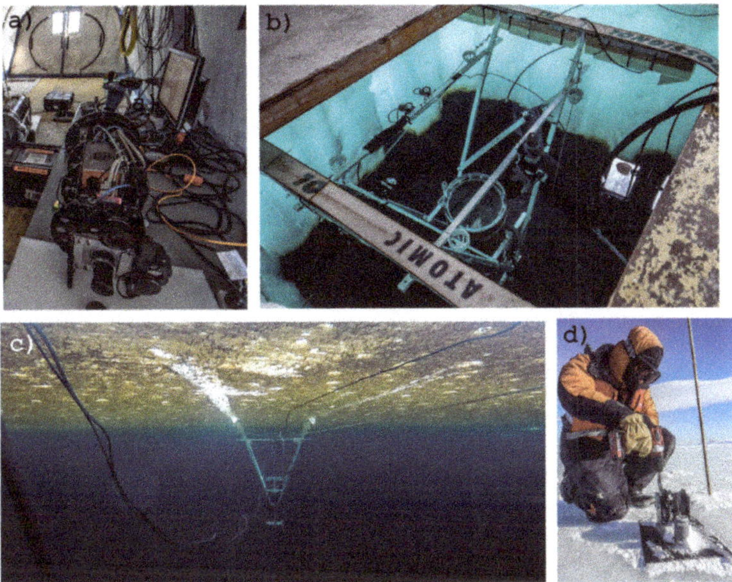

Figure 4. Field pictures of the first deployment at Cape Evans, Antarctica. **a**) The system control station together with the removable payload tray. **b**) The system deployed in the water prior to under-ice immersion. Visible is the external payload composed of the TriOS Ramses ACC and a set of four Lumen Subsea LEDs, and the prop maneuvering cradles. **c**) The system scanning over the selected transect underneath the highly productive fast-ice of Cape Evans. **d**) One of the worm gear winches at the opposite side of the transect in speed-up mode using a drill adapter.

The 2 m by 1.8 m ice-hole was made through a combination of 6″ Jiffy auger holes and hot-water drilling. A polar haven tent was erected on top of the hole to maintain a safe and constant-temperature working environment for the equipment. To create a tow-line for the winch system, a 6″ Jiffy auger hole was drilled at the end-side of each targeted transect. From this hole, a rope with a deadweight was immersed and rendered visible from the under-ice. A Seabotix LBV-300 ROV (Teledyne Marine, Seabotix, California, USA) equipped with a grabber arm was deployed from the central hole to grab and retrieve the tow-line from the smaller hole at the end of the transect (Figure 2). Following the installation of the winches, the rope was replaced with the winch wire and this was attached to the under-ice sled.

2.3. Deployment and Data Acquisition

The AK10 only allows for manual focus, and the system does not currently have the capability for remote focusing. The focus distance was required to be set to the predefined scanning distance of the system of approximately 1.2 m. Nonetheless, we need to consider that the focal distance and DOF have the potential to change underwater under a flat port set-up to ultimately affect image sharpness. We, therefore, used an underwater focusing target immerged in the ice hole together with a dummy acrylic glass port to focus the camera under dry conditions while mimicking the underwater optical set-up. The Sony a6300 interface allowed for remote autofocus.

We selected sunny and completely cloud-free days for our deployments to maximize under-ice transmitted light (and thus HI SNR). Before deployment, the enclosure was vacuumed using a standard vacuum pump and PREVCO vacuum kit manifold assembly to an internal pressure of −15 in.-Hg in gauge for leak testing and to reduce internal condensation risks. Although the air in Antarctica is typically very dry, this process is important to avoid any condensation within the enclosure due to the considerable heat produced by sensors and equipment compared to the exterior temperature.

Due to its voluminous shape and weight, the system required two to three people to be manually deployed into the ice hole. The system was then manually pushed below the 1.8 m thick sea ice by two people using rods inserted into the incorporated cradles (see Figures 1 and 4b). The system can then be rotated into the desired transect direction (e.g., western).

Once under-ice, the system was winched three to four meters away from the hole and the tent to avoid interference in the light conditions beneath the ice. We were able to speed up the worm gear winches (designed to be slow for data acquisition) using a winch adapted electric drill as seen in Figure 4d to move the system into the right position for data collection. An initial assessment of the HI signal intensity from directly under-ice was then performed. The optimal traveling speed and HI and RGB imaging settings were then maximized for both SNR and image quality.

The AK10 data storing and imaging settings, including integration time, imaging frequency, spatial, and spectral binning were controlled in real-time using the Lumo Recorder software (Specim Spectral Imaging, Oulu Finland). For HI, the spatial and spectral dimensions were binned to 1024 spatial pixels across track, and a spectral resolution of 3.5 nm (178 bands), respectively. Whilst the spectral dimension could have been further binned to 7.5 nm for increasing the signal; this was avoided as too coarse spectral resolutions are known to hamper the application of some of the HI processing methods for ice algae [27]. The HI frequency was set to 10 Hz and an exposure time of 99 ms (maximum setting available). The ideal sled system speed for these settings was found to be around 0.008 ms-1 corresponding roughly to one rotation of our worm gear winch per second. The read-out frequency of the IMU was also set to 10 Hz aiming for HI and IMU data time-stamp synchronization at the decisecond (ds) level. The survey distance of 1.18 m between the HI sensor and the ice resulted in a HI footprint width on the ice of approximately 0.61 m and a pixel size of 0.00625 m. The Lumo Recorder software was programmed to acquire 100 samples of a dark frame image with the shutter closed at the end of each acquired hyperspectral image. Dark frame images were taken for the subsequent radiometric correction of the imagery through the removal of dark current noise.

The Sony a6300 is operated through the Sony Imaging Edge software "Remote" feature. The software allows live streaming the camera view and permits exposure control, ISO, time-lapse shooting interval, and AF settings to be modified. We found that at the selected winch speed, an imaging interval of 0.1 Hz was sufficient to guarantee abundant forward overlap (>90%). This relatively large sampling interval, together with the slow movement allowed the camera to be set to AF, which resulted in sharp and focused images. The ISO was set to 250; aperture maximized to f/2.8 and shutter speed set to 1/250 sec for most of the circumstances. The altitude of the camera was around 1.2 m, which yielded an estimated footprint width of 0.586 m in water and a resolution of 0.0001 m. All images were captured in the Sony RAW format (.ARW) to allow for any eventual image pre-processing approaches (e.g., see appendix in [24]).

The radiometrically calibrated Ramses ACC-VIS was synchronized to acquire an under-ice irradiance sample at the same time as each Sony a6300 RGB image (0.1 Hz) was taken. In this way, it is possible to link every image to a Ramses ACC-VIS radiometric irradiance sample and locate images spatially across the transect through the retrieved camera positions following SfM digital photogrammetry.

The STS-VIS radiometer was set-up to acquire a measurement of incoming downwelling solar irradiance every minute considering the highly stable conditions during the surveys and the relatively low variability in sun angle.

Following system retrieval, HI, RGB imagery, and IMU navigation data files were downloaded directly from the SATA SSD within the DPU. VNC allows for direct data transfer from the payload to the surface, but the operation is time-consuming for large files such as the HI imagery data files.

2.4. Data Processing

Both hyperspectral image analysis and SfM photogrammetry are active research topics for many land-based applications. The adaptation of established terrestrial procedures to novel under-ice applications requires targeted studies aiming to identify, test, and evaluate their performance in an under-ice context. Here we present only preliminary data outputs of the developed system and assess their quality and potentials from a biophysical perspective. We do this by looking exclusively at the western transect and selecting a successful subsample for hyperspectral image analysis and processing (Figure 5), namely block B. For the RGB imagery and photogrammetry, we retrieve for the first time a high-resolution orthomosaic and DEM of the under-ice using commercially available software. For HI, we adapt some of the known methods in under-ice bio-optical literature to the hyperspectral images and illustrate potential new ones.

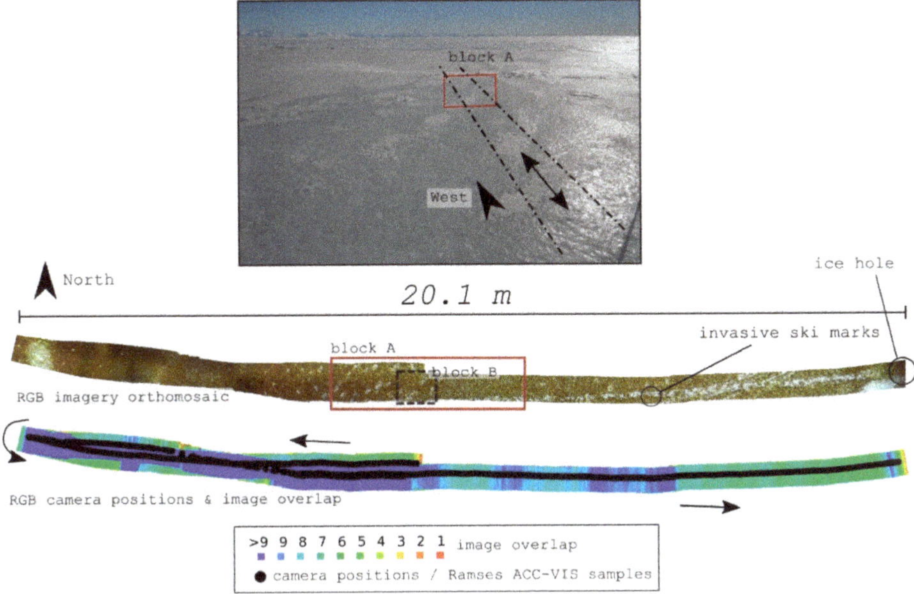

Figure 5. Overview of the surveyed western transect produced with structure from motion (SfM) digital photogrammetry using the RGB imagery. Camera positions and Ramses ACC irradiance samples were synchronized to the same sampling frequency, so they match in space. Blocks A and B within the transect were selected for further image analysis. On top is a photograph of the transect direction viewed from above the surface. Displaying the typical survey conditions (little to zero snow) of the study area.

2.4.1. RGB Imagery and SfM Digital Photogrammetry

It is well known that image quality and poor camera network geometries can considerably affect SfM model's reconstruction and the extraction of accurate metric information. Image quality in non-metric cameras is influenced by the camera sensor, lens quality, mechanical stability, and the overall image acquisition process under dynamic conditions. Poor camera network geometry refers to the lack of forward or side overlap in the imagery and/or lack of oblique imagery. Underwater, SfM photogrammetry is further challenged when using flat-ports due to the multiple refraction processes that magnify FOV, affect the focal length and produce a series of geometrical (e.g., radial distortion) and chromatic aberrations in the images directly affecting camera calibration algorithms in SfM, which ultimately affect the reconstructed model.

While image quality, per se, was not considered problematic in our transect dataset, the flat port did cause non-negligible effects on the imagery (e.g., noticeable pincushion distortion). To solve such aberrations and obtain an accurate camera calibration one can formulate the complex mathematical models of the imaging process in water [53,54] or perform a rigorous camera calibration using underwater targets with precisely known geometry [55]. Another option is to rely on camera self-calibration, which refers to the calibration process using only image point correspondences for large and well-composed datasets [56,57]. However, self-calibration is challenging in our dataset as camera network geometry is particularly weak when dealing with elongated strips with only nadir images and no side overlap and/or oblique imagery [57]. Systematic errors produced in such datasets can cause bending and non-linear deformations in the photogrammetric models as confirmed by our tests [57,58]. Here we apply a simple preliminary solution to the camera calibration problem using a constrained self-calibration approach by taking advantage of the flat under-ice surface, the known transect lengths and a series of identifiable reference points that were also measured from above the surface.

Prior to photogrammetric processing, 733 Sony RAW images acquired for the western transect were first imported into Adobe Lightroom where an initial lens correction and manual batch compensation for pincushion distortion was performed. Lightroom considers camera lens profiles into its corrections, and this empirical "trial-and-error" approach is simply to partially reduce bending of the model to a near straight level. Duplicate images were discarded as labeled repetitions during sled idle times, and the remaining images were exported from Lightroom as .JPG files for further SfM processing.

The 3D reconstruction of the under-ice surface was created using Agisoft Metashape (previously Photoscan), is a software package which has been extensively used for 3D modeling and photogrammetry over a wide range of geoscience applications [59,60]. The workflows for under-ice DEM and orthomosaic generation are described here. Photo alignment accuracy was selected as medium (for computational reasons) and provided a first estimate of camera calibration parameters and the reconstructed scene. The produced sparse point cloud model at this stage was noticeably bent and deformed. We proceeded to filter outlier's and low accuracy points using the gradual selection tools. Due to the smooth nature of the surface (Figure 4c), we assumed that all the surface areas with little algal cover were level with a reference height of 0.0 m, and created a dense and well-distributed network of reference level markers with a Z position (altitude) 0.0 m. We also added the known transect length as a scale bar length reference together with a series of points that were identifiable and could be referenced to above surface positions whose relative position could be measured with a measuring tape. For our entire western transect, we allocated 32 of these reference points, termed ground control points (GCPs) [61,62].

All these level reference GCPs are assigned with a high marker accuracy of 0.002 m in Metashape reference settings options. The model is then processed using the optimization of camera alignment feature where non-linear deformations can be removed by optimizing the estimated point cloud and camera calibration parameters based on these known reference marker coordinates [59]. During this optimization, Metashape adjusts estimated point coordinates and camera parameters minimizing the sum of reprojection error and reference coordinate misalignment error.

The Metashape workflow is then followed by dense cloud reconstruction (medium quality and aggressive depth filtering), 3D mesh from the dense cloud (Arbitrary surface type, medium quality, enabled interpolation, and aggressive depth filtering), texture mapping (orthophoto mapping mode and mosaic blending mode), and finally DEM and Orthophoto production. The scaled orthomosaic and DEM were exported in .TIF format to QGIS and the DEM was processed with a hillshade function for visualization purposes.

2.4.2. Hyperspectral Imaging and Radiometer Data

The retrieved HI images of block A and B consisted of a three-dimensional (x, y, λ) data cube where x and y represent the spatial dimensions, and λ the spectral dimension. The first two steps of the HI processing workflow include radiance conversion of digital numbers (DN) and pushbroom image rectification. The system was designed so that little to no geometric rectification and IMU data integration is required. This was the case for block A and B of the analyzed transect (Figure 5).

Per-pixel radiance conversion was done using Specim Caligeo PRO software (Spectral Imaging, Specim Ltd., Finland) which addresses noise and geometric aberrations inherent to the sensor and performs the conversion of DN into downwelling spectral radiance L_d (λ, mW m^2 sr^{-1} nm^{-1}) using the in-situ acquired dark current frames and the associated calibration files. For the present study, spectral bands <400 nm and >700 nm were considerably noisy and outside the range of interest, therefore spectral subsetting was applied reducing the data to a total of 89 bands.

The block B HI subsamples are then smoothed using a Savitzy-Golay low-pass filter with a polynomial order of three and frame length of nine aiming to reduce noise in the transmitted signals without hindering the retrieval of fine spectral features [63,64].

Following this procedure, we adapted methodologies previously applied to track biomass variability from under-ice spectra such as normalized difference indices (NDIs) and principal component analysis (PCA) (also known as EOF) [5,17,19,27]. Every pixel within the HI subsample was integral-normalized to reduce the amplitude component of spectral variability and to focus on differences in spectral shape, a pre-processing standardization method previously applied in sea-ice bio-optical literature [8,19,48].

PCA for hyperspectral remote sensing is typically employed for dimensionality reduction, to reveal complex relationships among spectral features or for the identification of prevalent spectral characteristics. PCA has been widely used in optical oceanography for extracting information about seawater constituents from spectral data (e.g., [65,66]). In our case, PCA was applied to the spectral dimension of block B data cube to explore and highlight the most variable features and relationships across all pixels in the block B image [27,67].

Spectral indices, such as NDIs, have been linearly correlated to the logarithm of sampled chl-a in multiple sea-ice studies [5,19,48]. Since we have not developed a specific spectra-biomass relationship for our site that applies to the developed HI payload yet, a couple of identified optimal NDIs from the land-fast sea-ice of Davis Station and McMurdo Sound, Antarctica by [48] were selected and utilized as a proxy of biomass. Before index implementation, block B was spatially binned to two by two pixels, reducing the spatial resolution from 0.624 mm to 1.2 mm, but boosting per pixel signal. The following NDI equation was then applied to every pixel in the image:

$$\mathrm{NDI}(\lambda_1, \lambda_2) = \frac{L_d(\lambda_1) - L_d(\lambda_2)}{L_d(\lambda_1) + L_d(\lambda_2)} \tag{1}$$

where λ_1 and λ_2 are wavelength bands selected across the sensor spectral range and L_d (λ, mW m^2 sr^{-1} nm^{-1}) is the solar downwelling radiance transmitted through the ice. From [48], we selected 441:426 nm and 648:567 nm as two different NDIs in different areas of the spectrum and applied the NDI equation to every pixel in the block B image. In this study, we used radiance to compute the indexes rather than under-ice radiance normalized to surface irradiance (or transflectance [68]). Changes in above surface illumination conditions (e.g., solar geometry and atmospheric effects) within the block A and B image subsample were considered negligible.

In addition to adapting PCA and NDIs to under-ice HI, we also tested for the use of an index called Area under curve Normalized to Maximal Band depth between 650–700 nm ($\mathrm{ANMB}_{650\text{–}700}$) of the continuum removed spectrum [69]. $\mathrm{ANMB}_{650\text{-}700}$ has been successfully applied for chl-a and chl-b mapping using HI of Norwegian spruce trees [69] and Antarctic moss beds [70], and here we use it as a proxy of chl-*a* or ice algal biomass.

For this index, we applied the same Savitzky-Golay low-pass filter and the two by two spatial binning factor, but no integral normalization is performed. Instead, the entire image is normalized by the highest spectrum intensity within the block, which corresponds to an algal free cavity in the ice visible in the image (shown later in the results section). This provides a proxy of light transmittance over roughly the last 5 to 15 cm of ice bottom and enhances visibility of the absorption peak of chl-a at 670 nm of each pixel spectrum. The continuum removal transformation on the spectrum is a fundamental pre-processing step to enhance and standardize the specific absorption features of biochemical constituents [71]. It allows for the normalization of the transmittance spectra so that individual absorption features can be compared from a common baseline. Following a localized continuum removal, we can calculate the Area Under Curve in the range between 650 and 700 nm ($AUC_{650–700}$) where chl-a attains one of its absorption peaks:

$$AUC_{650\text{–}700} = \frac{1}{2}\sum_{j=1}^{n-1}(\lambda_{j+1} - \lambda_j)(\rho_{j+1} + \rho_j) \tag{2}$$

where ρ_j and ρ_{j+1} are values of the continuum-removed transmitted spectra at the j and j+1 bands, λ_j and λ_{j+1} are wavelengths of the j and j+1 bands, and n is the number of the used spectral bands. We can then calculate the $ANMB_{650\text{-}700}$ index as:

$$ANMB_{650\text{–}700} = \frac{AUC_{650\text{–}700}}{MBD_{650\text{–}700}} \tag{3}$$

where $MBD_{650\text{–}700}$ is a Maximal Band Depth of the continuum-removed reflectance, generally at one of the spectrally stable wavelengths of strongest chl-a absorption around 670–680 nm. Normalization of $AUC_{650\text{–}700}$ by $MBD_{650\text{–}700}$ is a crucial step for strengthening the relationship between $ANMB_{650\text{–}700}$ and the chl-a content for higher chl-a concentrations. The logic behind this spectral index is exploiting well-known changes of the transmittance signature shapes produced within these wavelengths mainly by the changes in algal chl-a content.

In order to validate the robustness of the HI data compared to traditional means of acquiring under-ice spectra, hyperspectral irradiance variability measured with the Ramses ACC-VIS across the entire transect (samples shown as black dots in Figure 5) was computed and compared with spectra of every pixel in block B. The Ramses ACC-VIS data further allow us to gain an estimate of downwelling irradiance intensity exiting the ice-water interface and was used to gain an insight of the light levels experienced under-ice. These can then be used to baseline the signal quality of the data achieved using our HI system under those specific conditions. The TriOS Ramses ACC-VIS was radiometrically calibrated using the factory provided calibration files (traceable within international standards) during the data acquisition process.

3. Results

3.1. Deployment and Operation Performance

The system was successfully deployed and retrieved for the three targeted transects (NW, W, and SW). For the western transect analyzed here, a total of 736 RGB images and Ramses ACC-VIS irradiance samples were acquired, in both forward and backward directions (Figure 5). The overall scanning operation lasted approximately 2.5 hours, not including system set-up. Considering the air (−5 to 5 °C) and water (−1.8 °C) temperatures experienced, the electronics in the housing functioned well under the challenging environmental conditions and were kept above freezing point by heat produced from the multiple electronics. HI-sensor temperature sensors indicated that temperature was maintained at around 17 °C over the entire western transect.

As shown in Figure 6, the system was able to produce natively well-composed pushbroom hyperspectral images without the need for any rectification methods and/or additional attitude and navigation data (e.g., see block A in Figure 6c).

However, occasional lagging instances in the sled-motion during scanning of some sections of the transect hampered smooth pushbroom HI data acquisition. Sometimes these lags were long enough (0.5–3 seconds instances) that data collection had to be interrupted and the sled system to be forwarded until the movement was smooth again. In other cases, they were acceptable and could eventually be corrected through the integration of the IMU data algorithms and image correction filters (e.g., Figure 6 lagging instance). Transect blocks requiring rigorous geometric rectification and post-processing are out of the scope of this study and will be investigated in the future through the development of targeted geometric HI correction algorithms.

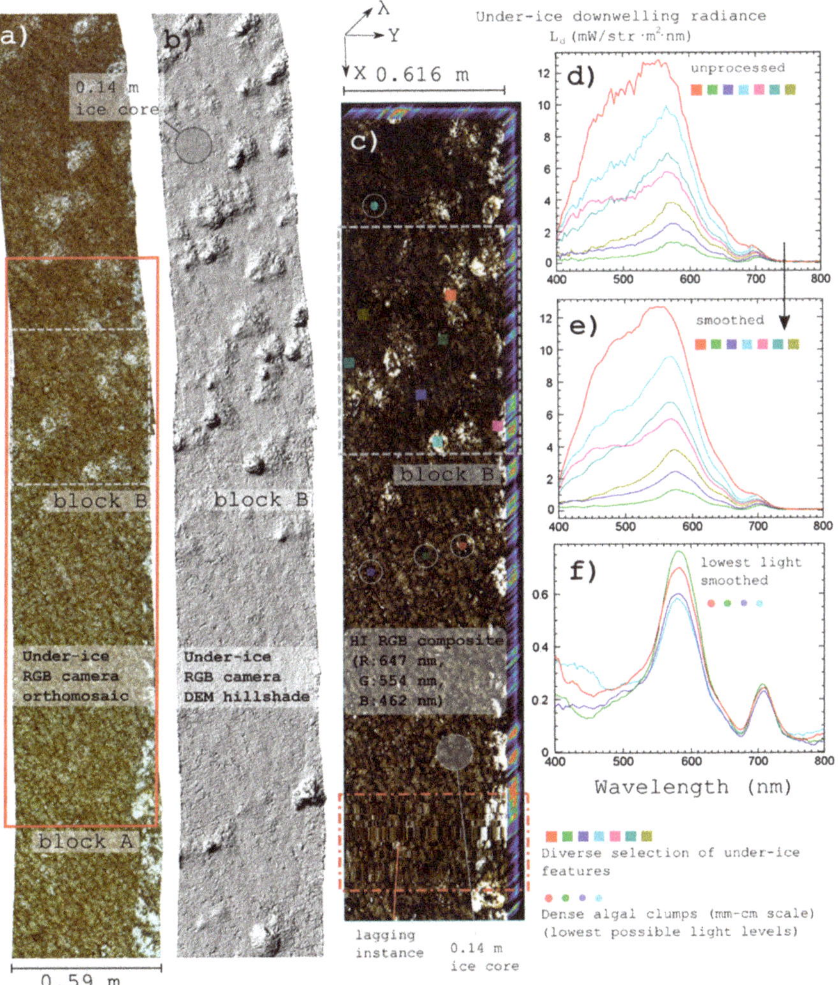

Figure 6. Display of the main data products of the developed under-ice payload. Block A and block B refer to two different subsections within the western transect that were selected for further analyses. **a**) Under-ice orthomosaic produced from the RGB imagery. **b**) Hillshade of the SfM derived digital elevation model (DEM) illustrating relief structure produced by the large cavities. **c**) Visual representation of the hyperspectral data cube for block A including block B as an RGB composite. Panel **d**) and **e**) display the high variability of radiance spectra for a selected variety of spots within block B (both unprocessed and smoothed with a Savitzky-Golay filter respectively). Panel **f**) display four of the darkest pixels within the image associated to extremely dense algal clumps. For all plots, spectrum shows a × 4 pixels spectral average which corresponds to approx. 1.2 mm pixel size. Native pixel size is 0.624 mm.

Transects also did not always followed a straight line, but instead, the trajectory displayed a slight bend as can be seen from Figure 5. This means that the system showed changes in heading according to its attitude reference system (heading, roll, and pitch) shown in Figures 1 and 3. Transect bending is only noticeable when considering long distances rather than over the shorter accomplished HI scans. However, this track deviation did have an impact on the imaged transect as forward, and backward travels did not perfectly overlap in some instances producing unnatural invasive marks such as the visible ski tracks in Figure 5.

3.2. RGB Imagery and Photogrammetry

For the western transect, 615 camera positions were aligned successfully, and optimization produced an overall flat 3D model of the under-ice surface (Figures 5 and 6a). Dense reconstruction of the model resulted in a rich and well-composed dense point cloud (100,199,561 points). The first estimation of the total area covered was 13 m^2 for the western transect. The final resolution of the displayed orthomosaic was 0.0994 mm/pixel and for the DEM 0.821 mm/pixel with a point density of 1.48 points mm^{-2}. The total RMSE of the Euclidean distance between the generated reference level markers and the corresponding estimated points in the reconstructed 3D model was 0.0762 m (0.0623 m X error, 0.0322 m Y error, and 0.0297 m Z error). While this error does not reflect a rigorous accuracy assessment of the absolute geometric accuracy of the model, our interest in these first trials was in the ability to retrieve complex topographic features. The relative (within model) accuracy and point density are sufficiently high for this purpose.

The RGB orthomosaic illustrates the high level of algal biomass under the land-fast sea ice of Cape Evans. This encompasses both gentle changes in illumination and also different shades of brown and green coloration over the full 20.1 m transect (Figure 5). Zooming into block A, Figure 6a displays complex networks of ice algal aggregations and patches together with the presence of large bright cavities embodying large secondary brine channels [72]. The DEM hillshade in Figure 6b shows that while at first sight, the under-ice at Cape Evans seems like a featureless surface, it has high levels of relief complexity attributed mainly to an extensive network of secondary pore spaces [72]. Looking at Figure 5, they appear to occur in specific areas of the western transect. These pore cavities range widely in size and depth and are believed to be a result of a series of sea ice thermodynamic processes of brine flushing and merging of channels during the advancement of the summer season (e.g., [72,73]). An ice core footprint of 0.14 m in diameter is provided as a reference scale for these large brine pores in Figure 6a,c. However, the total depth of the cavities is difficult to capture with digital photogrammetry, and we could only image and reconstruct up to a certain depth depending on their width. Smaller subtle relief and undulations of the under-ice surface are also observable from the DEM hillshade (Figure 6b). Since these are not recognizable as white spots from the imagery itself, they are perhaps not strictly related to brine release processes but rather ice undulations of yet unknown origin. The DEM hillshade also captures micro-rugosity in the 3D model attributed to protrusion of dense algal clumps mostly formed by the diatom species *Berkeleya adeliensis* (F. Kennedy pers. communication). The hillshade map also displays a specific orientation pattern assumed to be driven by the underlying water currents. *Berkeleya adeliensis* was found to be the predominant species together with the interstitial diatom *Nitzschia stellata* from microscopic observations.

Current-driven orientation of algae strands and the biophysical complexity of the under-ice habitat were also observed in the high-resolution Sony a6300 RGB images shown in Figure 7. These images not only display the native quality of the RGB imagery but also show additional important biophysical properties of the under-ice habitat such as the sea-ice skeletal layer and its crystal orientation (Figure 7a) [72,74]. Figure 7a was taken nearby the ice hole, and the difference between what appears to be the hanging *Berkeleya adeliensis* and interstitial diatom species is clearly visible. Later into the transect in Figure 7b, a certain degree of algal orientation can also be observed together with some of the large secondary brine channels. Zooming in on Figure 7b, we also observed high concentrations of oxygen bubbles produced by the photosynthesizing algae. Also, several types of under-ice fauna were visible along the high-resolution imagery dataset such as ctenophores (Figure 7a) and amphipods (Figure 7b).

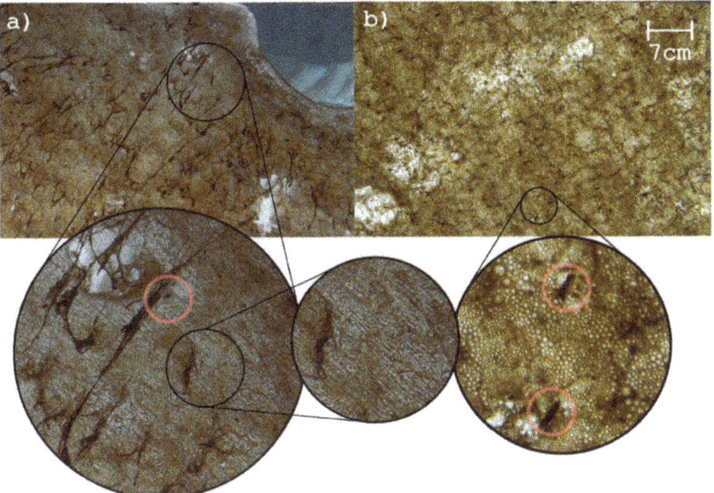

Figure 7. Two upward looking RGB image samples taken from the Sony a6300 camera dataset shown at full resolution. Both images display some examples of spotted under-ice feeders (circled). Left image shows a ctenophore (comb jelly) and right image shows a couple of circled amphipods. **a**) Image taken nearby the visible deployment ice hole. The image zooms into a large brine channel and further on the highly detailed under-ice skeletal layer. **b**) Image taken midway on the transect displaying the high concentration of oxygen bubbles produced by the photosynthesizing ice algae.

3.3. Hyperspectral Imaging and Radiometric Data

A visual representation of the block A hyperspectral data cube within the western transect is shown in Figure 6c. The quality of the image composition shows minimal geometric noise and a robust geometrical resemblance with the RGB orthomosaic for the entire block A subsample. The cube also shows an example of one of the lagging instances in the sliding sled system as previously noted.

The right-hand plots in Figure 6 display the quality of the measured spectral signatures in terms of overall intensity for the under-ice downwelling radiance L_d (λ, mW m^2 sr^{-1} nm^{-1}) unprocessed (Figure 6d) and smoothed (Figure 6e). It is clear that high variability of light intensity and spectral shape can be found across a series of features within the <1 m^2 area of block B. Such variability can change up to one order of magnitude and is mostly ruled by the presence of the secondary brine channels together with the drastic differences in algal concentrations and aggregations, but also due to the different algal species/morphotypes (e.g., hanging versus interstitial) among other factors. Despite the highly contrasting under-ice light regime induced by the large brine features, the camera dynamic range allowed to optimize settings to the lower light areas (e.g., algal patches) without saturating the pixels over the secondary brine pores.

Absorption by algal associated chl-a is easily observable over almost all pixels in the image as a reduction in intensity over the 440 ± 20 nm and 680 ± 10 nm bands. Higher ice algal biomass reduces transmitted radiance in the blue part of the spectrum and produces a compressed curve in the green part of the spectrum [75]. Absorption features by ice algae tend to drastically decrease nearby and within the secondary brine channels (e.g., red spectrum in Figure 6d–e) except in circumstances where we find dense algal webs hanging in the middle of these cavities (e.g., celeste spectrum in Figure 6d) or highly concentrated algal clumps scattered around these cavities. From the entire block A image, we also selected some of the lowest light pixels we could find, and their spectrum can be seen in Figure 6f.

The SNR noticeably decreases for such targets, and the blue region (400 to 500 nm) seems to be noise dominated. Nonetheless, the spectrum still displays strong chl-a signatures in the 680 ± 10 nm band curve and an overall meaningful signal.

The mean irradiance spectrum ± standard deviation (sd) measured with the Ramses ACC-VIS for the length of the whole 20.1 meters transect is shown in Figure 8a. The total irradiance energy integrated over the PAR range (400–700 nm), $E_{d,PAR}$ (λ, W m^2) averaged 0.35 (λ, W m^2), with a 0.20 sd, and a total range of 0.07–1.5 (λ, W m^2). Figure 8a also helps to characterize the spectrum variability across the entire transect. Interestingly, a similar degree of variability (although in terms of radiance) is experienced within the <1 m^2 block B subsample as seen in Figure 8b showing the mean spectrum ± sd of all pixels of block B. Figure 8c displays the integral-normalized mean spectrum of all the pixels of the block B hyperspectral data cube overlaid by the integral-normalized mean spectrum of the entire western transect using the Ramses-ACC-VIS. Figure 8d displays all pixels of the block B image normalized by the highest light intensity pixel in the images which is attributed to the light exiting one of the secondary large brine channels or cavities (seen Figure 6). This plot indicates properties of the transmitted light over the bottom layer of the ice where >98% of the biomass thrives. The normalized spectra were used to compute the $ANMB_{650—700}$ index. The normalization greatly accentuates the absorption features of chl-a in the blue area centered at 450 nm and the red peak centered around 670 nm.

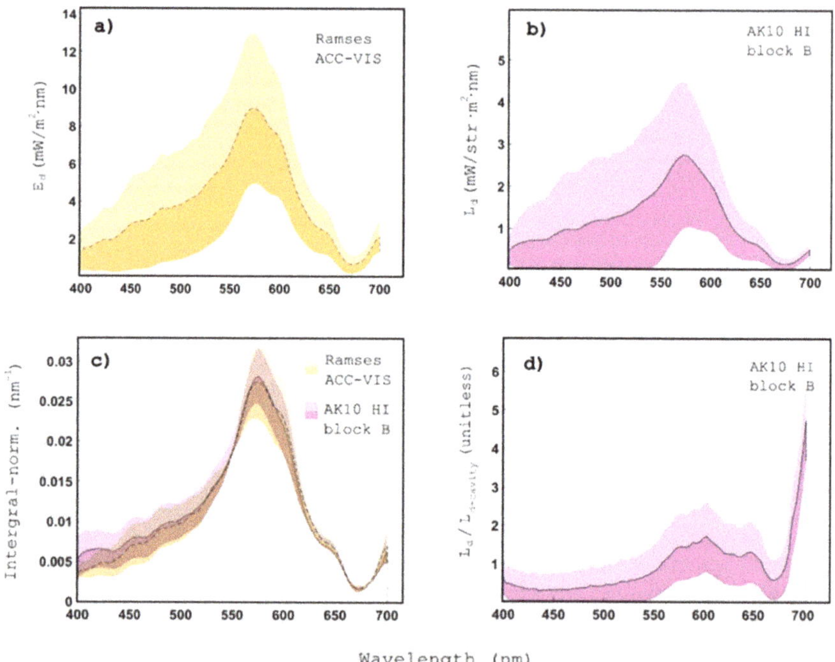

Figure 8. **a**) Mean ± one standard deviation of downwelling under-ice irradiance (E_d) spectra from the TriOS RAMESES ACC-VIS located near the ice water interface for the full 20.1 m transect. **b**) Mean ± one standard deviation of under-ice downwelling radiance spectra (L_d) from all the pixels of block B hyperspectral image subsample from the AK10. **c**) Mean ± one standard deviation of under-ice irradiance and radiance spectra normalized by area under curve for the Ramses ACC-VIS over all the transect and for all pixels of block B AK10 hyperspectral image. **d**) Mean ± one standard deviation of under-ice downwelling radiance (L_d) normalized by the maximum radiance pixel of all block B and corresponding to one of the cavities or secondary brine channels seen in the image ($L_{d\text{-}cavity}$).

PCA results are shown in Figure 9. The loadings of the first nine principal components explaining >99.54 % of spectral variability within the image are shown for completeness. Figure 8 also displays the loading scores applied to each pixel of block B for the first three principal components (PC1, PC2, and PC3) together with an RGB composite of block B. PCA results show well resolved and coherent principal components similar to what was reported previously in the literature employing PCA (or EOF) using under-ice radiance and irradiance sensors in-situ [5,19], or for HI in artificial sea-ice simulation tanks [27]. The PC1 loadings account mainly for variability in light intensity attributed to a mixture of factors and embody the general trend of the under-ice light spectrum. PC2 seems to be more influenced by the two contrasting dip areas around 440 ± 20 nm and 680 ± 10 nm suggesting a possible correlation with algal chl-a pigments. Nonetheless, PCA at this stage serves as an exploratory tool and it remains difficult to assess the nature of PC3 and the remaining PCs without further analyses of pigment composition e.g., through high-performance liquid chromatography (HPLC) (e.g., [43,48]). The PCA score images also evidence some subtle line artifact features across the scanning direction of the hyperspectral image (Figure 9). These are attributed to small vibrations or micro-lagging instances whose visibility is enhanced following integral-normalization and PCA processing.

The results of the NDI (648:567 nm) and $ANMB_{650–700}$ indices applied as relative proxies of biomass variability to block B are presented in Figure 10a,b, respectively. Interestingly, Figure 10 suggests that both indices provide a similar result in terms of biomass distribution patterns and capture spatial scales previously unprecedented. However, NDIs seems to produce nosier images compared to $ANMB_{650–700}$. It might be argued that the $ANMB_{650–700}$ is based on the curve shape information of the light transmitted through the algal layer and such normalization was not applied to compute NDIs. However, both methods were tested and showed that using quantitative changes of transmitted radiance intensity produced less noisy images in case of NDIs.

NDIs applied to block B over the blue/violet area (441:426 nm) were also tested and provided similar results although with slightly noisier imagery (not shown). A final interesting observation is the anisotropic noise pattern observed in both NDIs (648:567 nm) (Figure 10a) and PCA across the scanning direction of the image (Figure 9). This is observable as noisier zones at the top and bottom of the image.

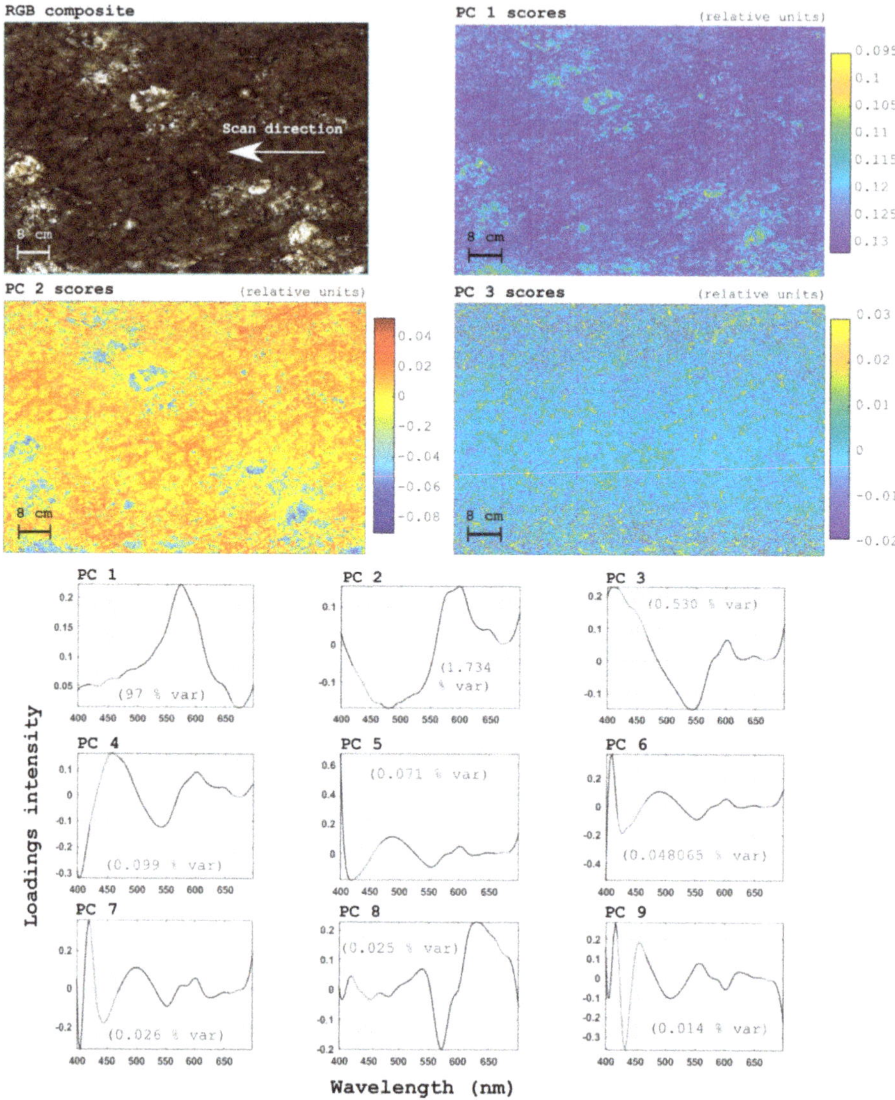

Figure 9. Results of principal component analysis (PCA, also known as EOF), applied to the spectral dimension of block B (hyperspectral image subsample of the western transect). Top images display the first three PC scores applied to every pixel of the image using corresponding loadings for each component. Bottom plots display the loadings for each wavelength for each principal component. Plot display as well the proportion of variance explained by each corresponding component. Light grey areas highlight the maximum chl-a absorption regions at 440 and 670 nm. Spatial resolution for PCA was maintained to a native 0.625 mm.

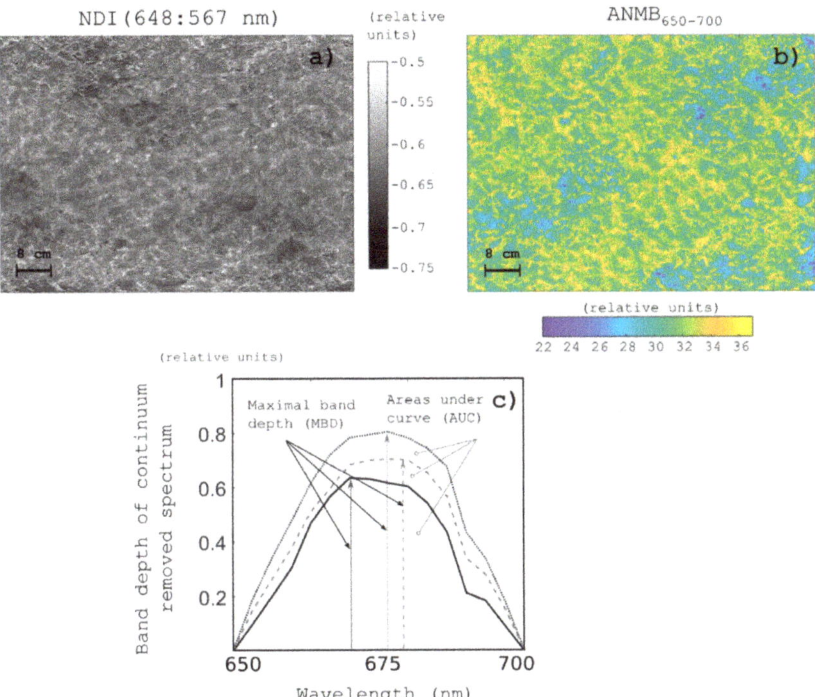

Figure 10. Application of spectral indexes as proxies of chl-a distribution over block b HI subsample. **a**) Results from the application of a commonly used index in sea-ice bio-optical literature, the normalized difference index (NDI), applied for wavelengths 648:567 nm on block B hyperspectral image subsample. **b**) Application of a novel index to sea-ice bio-optical literature, the area under curve normalized to maximal band depth (ANMB) between wavelengths 650 to 700, applied to the same block B. **c**) Plot of continuum removed spectrum of three random pixels within block B to help visualizing the ANMB 650–700 concept and its association with chl-a absorption. For the color bars, higher values (towards red) correspond to higher expected biomass. Spatial resolution for the indices was binned to 1.2 mm.

4. Discussion

4.1. Under-Ice Hyperspectral Imaging Data Quality and Processing

The present study outlines a novel platform incorporating two emerging underwater optical methods for capturing fine-scale biophysical properties of the under-ice habitat non-invasively. Passive HI and digital photogrammetry were tested for the first time to observe the ice-water interface and were deployed using a relatively simple under-ice sled. The sliding concept took advantage of the fixed and smooth surface of land-fast sea ice to minimize costly set-ups and yielded geometrically coherent hyperspectral imagery without the need of georectification. To the authors knowledge, only three underwater HI payload designs have been documented before. The Ecotone UHI (Ecotone, Trondheim, Norway) is a commercial solution designed for deep or shallow ROV-based seafloor observations and utilizes active light sources [36,76]. The Ecotone UHI has also been equipped onto unmanned seafloor vehicles (USV) for shallow seafloor mapping [33]. The other two are documented in [32,34] and comprise of stationary time-lapse observations and a diver-operated set-up.

In terms of data processing, the aim was to provide a preliminary outlook of the system's data outputs and its potentials. The preliminary results presented here indicates that it is possible to apply simple, yet effective, algorithms to retrieve chl-a per surface area on a sub-mm per pixel basis over tenths-of-meters-long transects. Figure 8 shows that the under-ice spectral signatures of traditional and novel sensors are comparable. They are also comparable with studies over similar Antarctic land-fast sea-ice areas (e.g., [48]). Established under-ice bio-optical methods for retrieving sea-ice biomass proxies in-situ (e.g., NDIs or PCA models) were also successfully adapted to the acquired hyperspectral imagery (Figures 9 and 10). NDIs values outputted are observed to match the range of values over the same or similar sea-ice areas [48] and PCA loadings shown strong similarities in shape if compared with results from other studies both in real sea ice and in artificial ice tanks [7,27]; the difference being that in this study they were retrieved on a sub-mm per-pixel basis.

PCA results retrieved chl-a signatures over its PC2 component and reaffirm the utility of PCA for explorative analyses. For example, the pronounced "shoulder" deviation towards 470 nm in PC2 loadings is likely associated with a higher concentration of accessory algal pigments such as fucoxanthin [77,78]. PCA analyses also suggest the possibility to retrieve PC/EOF based regression models to develop chl-a-spectra relationships, algorithms that have been proven successful for a wide range of sea-ice conditions [5,19,27].

The use of the NDIs positioned at wavelengths 648:567 and 441:426 nm was also tested with meaningful per-pixel biomass proxy representations although images were characterized by consistent pixel noise, particularly for the blue region of the spectrum (Figure 10). This is probably attributed to the lower SNR inherent to mm-scale hyperspectral resolution image pixels compared to wide-footprint radiometric sensors. SNR changes due to variations in intensity and shape of the retrieved spectra, which varies as the target constituent concentrations change and as the noise changes depending on sensor settings and specifications. The high ice algal biomass typically found at Cape Evans (see [1] for biomass ranges), favors algal associated spectral shapes, but heavily reduced light availability and consequently per pixel SNR on the overall spectrum, particularly in the blue region where chl-a attains one of its major absorption ranges (e.g., for the NDI 441:426, see Figure 8b,d). In fact, from Figure 10a, we can observe how noise is drastically reduced over the high light intensity brine channel areas. Studies [48] and [22] also highlighted how in general NDIs were producing poor relationships at the Cape Evans site. However, this might be because of different reasons such as the presence of platelet ice (which we did not experience during our study), the consequent poor spatial variability in biomass at the measured scale, or perhaps the difficulty in ice-coring and sampling chl-a from sloughing platelet ice [22].

The $ANMB_{650-700}$ index explored here is directly linked to the absorption properties of chl-a in the red region of the spectrum. It takes the advantage of hyperspectral data to finely integrate over the narrow absorption peak of chl-a in the 650 to 700 nm range. While it is not guaranteed that a meaningful quantitative relationship with sampled chl-a will be retrieved, the index performed better than NDIs for our case by providing less noisy and coherent images (Figure 10b,c). Increases in chl-a concentration (with absorption maximum around 665 to 680 nm) causes chl-a absorption feature to deepen at the 680 ± 10 nm dip. While the spectrum of the transmitted radiance in this range can show signs of saturation, the adjacent wavebands at longer wavelengths remain sensible to changes as the peak broadens and thus extending the area under curve (Figure 10c) [69]. The index was also designed to reduce the impact of other confounding factors of the imaged target within its complex 3D environment [69], and this might also supported the index performance in our case. A continuum-removed integrative index could also have worked better than a band ratio (e.g., NDIs) under this high biomass case (and therefore less light and SNR) as it integrates a larger area (AUC) hence providing a stronger signal per-pixel (Figure 10c). In fact, the performance of ANMB and similar indices is expected to deteriorate under low chlorophylls (chl-a and chl-b) amounts [69,79].

Future work in this area will explore the performance and comparison of these indices for the Cape Evans site, and to work on the retrieval of quantitative correlations tailored to our encountered sea-ice conditions that are suitable to be applied to HI data. It was also noticed how different pre-processing, normalization, and standardization techniques (not all shown here) affected the visualization of indexes applied to the images and the performance of exploration methods such as PCA. Such observations prompt for the investigation of optimal workflows to process and analyze under-ice HI data.

4.2. System Performance and Future Developments

While most of the transect could be scanned as planned, some issues were experienced during the scanning process such as invasive ski marks and occasional lagging which hampered pushbroom image composition (Figures 6 and 8). Nonetheless, the low gear winch system was capable of delivering extremely slow speeds in a stable manner as observed in the imagery. The observed angular deviations are comparable to data for professional gimbal stabilization systems for UAV applications [80]. They had negligible effects on the HI image composition and RGB imagery in our case due to the close-range set-up and the extremely slow speeds. The only trade-off of the system is that the winches had to be manually rotated which is a time-consuming and personnel demanding process. For future deployments, we plan to automate and motorize the winch system. The changes in transect heading are likely attributed to a combination of small-scale ice irregularities, inhomogeneous surface drag and/or the effect of intermittent currents observed from our underwater footage. There is also the possibility of a loosened ski frame support which went unnoticed. The cause of the lagging could be attributed to these roll changes but could not be precisely identified either. Investigation of the RGB imagery did not point to a particular ice condition that could have induced the lagging. A too strong buoyancy force against the ice (−9 kg in water, Figure 1) might have increased surface drag to a counterproductive level.

These aspects can firstly be improved by developing an improved sliding system and refining its technical design. However, greater advantage is envisaged in exploring manual or automatic pushbroom HI rectification techniques through the incorporation of overlapping RGB orthomosaics, also known as co-registration [81–83]. This approach co-registers the hyperspectral imagery based on a reference RGB orthomosaic through image matching procedures (e.g., feature matching and transformation based on matching points [28]). The only requirements for co-registration are spatially similar and overlapping HI and RGB imagery and good accuracy for the RGB orthomosaic reference. Advances in camera calibration and triangulation procedures permit the generation of RGB orthomosaics with high geometric fidelity using a limited amount of GCPs and/or consumer-grade navigation data [60,61,84]. Although a more accurate assessment is still required, the western transect RGB orthomosaic resulted in a highly resolved and metrically scaled photogrammetric model which could be used for co-registration for example (Figures 5 and 6a,b). This was possible as our camera calibration and model reconstruction heavily benefitted from a constant sea-ice thickness and imaging altitude which allowed to impose an artificial network of GCPs of precisely known positions in the 3D space. The same approach would not be possible under highly heterogeneous topographies or would not be as effective for highly dynamic imaging conditions. Under these sub-optimal circumstances, the options could be to retrieve an accurate camera model using underwater calibration targets [39,55], to estimate it through its mathematical formulation and/or to implement the use of dome ports [51]. Another option remains the addition of physical GCPs. Compared to the seafloor, the sea-ice can be used as an opportunistic reference surface where ground control points visible below and above the ice can be allocated (e.g., Nicolaus and Katlein, 2013). GCP positioning can then be accomplished using conventional GNSS devices and manual measuring or by referencing them in a local reference system. This is advantageous as positioning underwater typically requires the acquisition of acoustic data, which may depend on information from the under-ice vehicle/platform to a research vessel through a network of deployed transponders [8,85–87]. This process requires considerably more effort and resources and would arguably suit the precision required by line scanning orthorectification methods.

By taking advantage of the referenceable sea-ice surface and co-registration methods we could then theoretically develop algorithms analogous to aerial HI algorithms based on the scaled RGB orthomosaics, the partially rectified HI scans and the acquired consumer-grade IMU data [37,81,82]. These future developments will aim to support the geometric correction of distortions caused by the dynamics of the HI frame, such as the lagging instances (Figure 6c). In addition, robust geometric correction will pave the way for a more independent system that can operate under rougher under-ice topographies and at increased distances from the ice. The system needs to strive towards increased distance from the ice, and ease of operability under diverse under-ice conditions. As the technology develops, there is also potential to drastically reduce the weight and volume of the payload. Eventually, this may allow the development of HI payloads for ROVs or unmanned underwater vehicles (AUVs) to drastically increase the spatial extent of the surveys, although there are physical and technical challenges associated which are briefly discussed in the last sub-section.

4.3. Potential Applications of Under-Ice Hyperspectral and RGB Imaging Payloads

Compared to standard imagery or multispectral imagery, HI provides narrow spectral resolutions, high bit depths, and actual radiometric and referenceable units. Higher spectral fidelity sensors with reasonable spectral resolution would not only be beneficial to produce quantitative estimates of fine scale sea-ice biophysical properties, but also to develop tailored relationships for each study area and move towards more universal approaches and algorithms [4]. The complex under-ice perspective will undoubtedly pose new challenges and constraints. However, several additional indices or machine learning approaches coupled with radiative transfer modelling efforts could be tested and adapted to produce more robust and universal relationships to retrieve diverse biophysical properties. Some examples can be found in forestry and agriculture [29,79,88], ocean color [89,90], chemometrics [67], and other environments [34,91].

Low-cost imagery sensors—such as RGB, near-infrared (NIR) or multispectral—have also served well in multiple close-range remote sensing applications to retrieve qualitative and quantitative information from biological targets [92–94]. For the under-ice environment, RGB imagery has been used to qualitatively assess the spatial distribution of algae [9,95,96]. Therefore, RGB or multispectral imagery should be considered from a cost-benefit analysis perspective based on desired research aims and available resources.

In theory, hyperspectral resolution data has the potential to resolve beyond pure biomass estimates towards more sophisticated biological traits such as ice algal photophysiology [91,97,98], species composition [97,99,100], pigment detection [101–103], and feature classification and mapping [35,76,104]. An interesting field is also being explored in the retrieval of primary production estimates from spectral data in combination with in-vitro photosynthetic parameters for ice algae [7,105] or with PAM fluorometry for microphytobenthic communities [106].

Compared to point sampling radiometers, the main advantage of imaging payloads is the possibility to capture the information at ultra-high spatial resolutions (in this case sub-mm scales) in a non-invasive manner (e.g., [34]). Under sea ice, this will allow future studies to investigate multi-scale ice-algal dynamics and how they covary with environmental drivers over space and time [4,11,47,107]. With little additional effort, the RGB imagery and close-range digital photogrammetry provided an accessible tool to producing ultra-high resolution orthomosaic and 3D models of the under-ice surface.

Surface topography is a well-known factor driving spatial distributions in many marine ecosystems (e.g., [108]). Under-ice, the potential of high-resolution HI and 3D data fusion could contribute to new opportunities to monitor some of the sea-ice biophysical interactions which were previously difficult to capture. The effects of under-ice topography on sea-ice algal biomass distributions has long been queried and investigated [10,109,110]. Recent studies have further observed and inquired about the role of under-ice topography and underlying currents on algal biomass distribution at multiple spatial scales [11,96,111]. Hydrodynamic shadows can foster the accumulation of diatoms, algal aggregates, and may also provide shelter for under-ice fauna [9,112,113]. The RGB imagery not only can provide under-ice roughness but it could also serve to gain further insight into grazing dynamics by sympagic fauna (Figure 7).

The effects of sea-ice structure and physical properties also go beyond effects on biomass distribution and are known to influence algal photophysiology, species composition, and production [1,3,9,114]. Although intrinsically different from some of the Arctic examples cited above, the dataset presented here clearly illustrates a complex biophysical scene for Antarctic land-fast sea ice even within a square meter area (Figures 8 and 9). For example, we found large secondary brine channels to characterize specific areas of the scanned transect (see Figure 5). These cavities augmented transmitted light conditions that showed localized maxima of up to one order of magnitude (Figure 6). The question arises whether these under-ice features have an impact on algal distribution, species composition, and/or photophysiology, or if they play any role in hydrodynamic regimes and under-ice grazing dynamics. The presented methodology may contribute to a better understanding of some of these complex biophysical interactions.

4.4. Caveats and Future Challenges

Our sliding system has been designed for deployments over relatively smooth under-ice bottoms. Nonetheless, the principles of operation of HI and digital photogrammetry remain applicable to any ice type, provided that under-ice light levels are sufficient. In cases where the sliding concept is not applicable (e.g., rough pack ice), platforms will need to be equipped with sensors to accurately trace HI sensor attitude and dynamics.

In this study, the HI payload was operated under thick (1.8 m) and almost snow free fast ice (Figure 5). To account for low under-ice irradiance levels (0.35 ± 0.20 W m^{-2}) the system was operated at extremely slow scanning speeds (0.008 ms^{-1}). These irradiance values are comparable to under-ice light levels and variability for Arctic fast-ice during spring [68], and help to provide a baseline for the range of under-ice irradiances intensities for which our payload could acquire meaningful HI signals. However, many other sea-ice conditions remain to be explored (e.g., with deep snow packs) and which may pose significant technical challenges. Low light levels will push sensors to their sensitivity limits, necessarily affect SNR and hinder the integration of pushbroom HI payloads onto more efficient and dynamic underwater platforms, such as ROVs and AUVs. A series of studies have already employed pushbroom HI sensors for seafloor mapping using ROVs [35,36,76], diver operated systems [32], or unmanned surface systems [33]. A first study has also discussed HI feasibility onto AUVs [115]. However, these applications positively benefitted from artificial light sources that illuminate the imaged scan line, or were performed in shallow, clear tropical waters. For mapping under-ice environments, there is a trade-off between sensor integrations times, imaging frequencies, and platform dynamicity under low light conditions that will need further investigation [4].

The inclusion of underwater IMUs, relative positioning systems, and implementation of targeted under-ice pushbroom HI orthorectification methods will open up new avenues for this type of research. While active lights sources could be eventually considered for under-ice mapping, the resulting mixture between reflected and transmitted light through a complex and translucent medium would render data processing and interpretation extremely challenging. In fact, our system features a set of artificial light sources as shown in Figure 4b and schematized in Figure A1. Using a custom-built control (Figure A1), the LEDs were tested and observed to provide a slight increase in the measured signal. However, it was preferred for the scans here presented to avoid their use to avoid complicated data interpretations. The effect of strong LEDs on relatively low-light adapted algal communities could also question the invasiveness of the methodology.

Additional challenges arise due to the complex nature of sea-ice optical properties and the resulting anisotropic under-ice light field [116,117]. The anisotropic light field is shaped by the lamellar sea-ice features funneling light in the downwards direction creating a forward peaked light field. Lamellar structures associated with columnar ice were clearly observed in our site (e.g., Figure 7). Analogous above surface HI applications (e.g., equipped onto UASs) have acknowledged the impact of an anisotropic leaving reflectance on the retrieval of biochemical parameters using spectral data [118–120]. In this study, we experienced noise artefacts over block B sample processed images as an increase in noise patterns at the upper and bottom edges of the image. This is most likely inherent to camera optical design and sensitivity heterogeneity across the spatial dimension, but it could also be in part attributed to light-field anisotropy. A forward peaked light field could mean a stronger signal at the center of the line scan and a decreasing signal towards the edges of our ~30 FOV. However, other possible causes should be taken into consideration (e.g., data processing artifacts) or the dense oxygen bubble layer causing multiple refraction effects (Figure 7). Eventually, the impact of an anisotropic under-ice light field, or other particular environmental conditions (e.g., oxygen bubbles), on HI data will need to be further assessed, and corrections developed towards improved estimates and interpretations.

We did not apply any corrections for the water column effects to either the RGB imagery or to the HI data processing workflow. This is acceptable as the water column in between the ice and the enclosure was <1.1 m and our site was characterized by exceptionally clear waters (see Figure 4c). Antarctic surface waters are generally considered to have low particle loads with low backscattering (e.g., [121]). Nonetheless, as we increase sensor distance from the target, or in case of consistent under-ice phytoplankton abundance (e.g., [122]), the impact of the water column should be addressed with standard color correction approaches for RGB imagery [44,123,124] and for the water column correction of hyperspectral radiometric data if possible [44,125].

Our sea-ice site also benefitted from optically "favorable" conditions where biomass was high and resided mostly in the bottom 3 cm of the ice. Due to the scattering properties of sea ice, the bottom 3 cm algal layer can be considered as an evenly illuminated "thin" sheet that was scanned with our payload. While ice algal biomass is known to be generally concentrated at the ice bottom, where organisms enjoy more favorable living conditions [1,16], there are many circumstances of vertically variable distributions [16]. Future applications of HI for sea ice with a certain degree of vertical biomass variability (e.g., in Antarctic pack ice) will need to consider these effects. Due to the scattering nature of sea ice, biomass in the sea-ice interior will probably have a negative impact on discernible spatial resolutions and image interpretation. Larger protruding algal filaments that are only loosely attached to the subsurface of the ice could also be a problem under dynamic currents for both HI and RGB imagery. In our case, filaments did not represent a significant problem as they were relatively short and under-ice currents during scanning seemed monodirectional, thus providing a relatively still scene (Figure 7). Finally, the feasibility and performance of HI to capture biomass variability under the extremely different biomass ranges found in the sea ice need to be assessed. A compilation of biomass ranges found in sea ice can be found in Arrigo (2017). A previous experimental study has shown HI to be able to discern biomass ranges as low as 0.036–2.72 mg m^{-2} [27], but much more work is required to investigate the impact of different concentrations on per-pixel SNR and regression algorithms performance.

5. Conclusions and Outlook

Sea-ice biophysical properties can exhibit high spatio-temporal variability at very fine scales (e.g., <1 m^2) which are difficult to capture and quantify using traditional methodologies. In response to some of these limitations, this study has presented a proof-of-concept for using hyperspectral imaging and digital photogrammetry for under-ice habitat mapping using a modular, low-speed sliding platform. The particular "inverted" under-ice perspective poses new challenges and limitations to HI which were thoroughly discussed in this study. We demonstrate that the new system was able to map a ~20 m-long transect with geometrically consistent pushbroom hyperspectral imagery, together with overlapping digital elevation models of the under-ice surface, at sub-mm spatial resolution.

Despite the low light irradiance levels experienced ($E_{d,PAR}$ = 0.35 ± 0.20 (λ, W m^{-2})), our HI payload attained suitable per-pixel under-ice signals for employing established bio-optical approaches without the need of active light sources. Minor issues with the sliding system where experienced (e.g., occasional lagging and ski marks), but these seem to be addressable with feasible system modifications and/or data processing techniques.

Future work will aim to address system performance and technological capabilities to increase the spatial extent of the surveys, data acquisition under rougher ice types and investigate pushbroom image rectification approaches based on RGB imagery. RGB imagery and digital photogrammetry were shown to provide ultra-high-resolution DEMs and orthomosaics of the under-ice habitat. RGB imagery presents diverse opportunities to qualitatively and quantitatively map and investigate under-ice objects of interest together with highly detailed under-ice roughness.

For under-ice HI, several aspects remain to be investigated prior to method standardization. We think light limitation, the complex under-ice light field, and vertically variable biomass distributions should be further investigated and their impacts on HI methods assessed. For digital photogrammetry, efficient camera calibration approaches should be tested in an under-ice context to assess and improve DEMs accuracy.

Finally, we need to move from relative indices proxies to actual quantitative per-pixel biomass estimates. In this study we only underlined the potential of applying spectral indexes and dimensionality reduction methods to retrieve biomass proxies. We then validated them by comparison with values documented in other studies. The next step is to develop a targeted biomass-spectra calibration suitable and tailored for our sensor payload and study area. This will be eventually applied to the full set of images transect imagery so that we can gain a better understanding of the under-ice habitat characteristics at Cape Evans, Antarctica. We expect that HI systems will contribute to fill a niche gap in the mechanistic understanding of some of the complex under-ice biophysical interactions.

Author Contributions: Conceptualization, E.C., K.M.M., A.L. and V.L.; methodology, E.C. and V.L.; software, E.C.; validation, E.C.; formal analysis, E.C.; investigation, E.C.; resources, V.L. and K.M.M.; data curation, E.C.; writing—original draft preparation, E.C.; writing—review and editing, E.C., K.M.M., A.L. and V.L.; visualization, E.C.; supervision, K.M.M., A.L. and V.L.; project administration, V.L.; funding acquisition, V.L.

Funding: This research was supported under Australian Research Council's Special Research Initiative for Antarctic Gateway Partnership (Project ID SR140300001), and the New Zealand Antarctic Research Institute (NZARI) under project number K043-1819-A. Emiliano Cimoli is supported by the Antarctic Gateway Partnership and the University of Tasmania's Ph.D. program.

Acknowledgments: We are grateful for the support of Antarctica New Zealand staff and the K043 team for field-based operation and logistics. We gratefully acknowledge the support of Richard Ballard for circuitry production, and technical support during stages of this project. We are also thankful to several members of the TerraLuma team for valuable advice and support.

Conflicts of Interest: The authors declare no conflict of interest.

Appendix A Technical Design and Specifications

The core equipment of the developed system consisted of an AISA Kestrel 10 pushbroom HI camera (AK10) (Specim Spectral Imaging Ltd., Oulu, Finland), a DPU (Specim Spectral Imaging Ltd., Oulu, Finland) and a Sony a6300 mirrorless digital camera together with a Samyang 35 mm prime lens. Accessories include a Low-Light HD USB camera (Blue Robotics Inc., California, USA) a VN-100 Inertial Measurement Unit (IMU) (VectorNav Technologies, LLC, Dallas, USA) and a Garmin 18x LVC GPS (Garmin, USA). In our independent external payload, we included a TriOS Ramses ACC hyperspectral cosine corrected spectroradiometer (TriOS Mess- und Datentechnik GmbH, Rastede, Germany) and a set of four daisy-chained Subsea Lumen Lights (Blue Robotics Inc., California, USA).

Figure A1 illustrates the power supply and data transmission paths from the surface elements to the enclosure interior and the external payloads. The internal payload is fitted inside an off-the-shelf black anodized cylindrical aluminum enclosure manufactured by PREVCO (PREVCO Subsea, Fountain Hills, USA) that seals via two nitrile O-rings for each end plate and is rated to a depth of 100 m (Figure 3). The end-cap is fitted with a single underwater connector and a pressure release/vacuum hole. The connector used for external communication and power supply with the surface was a 13 contacts circular SubConn power/Ethernet.

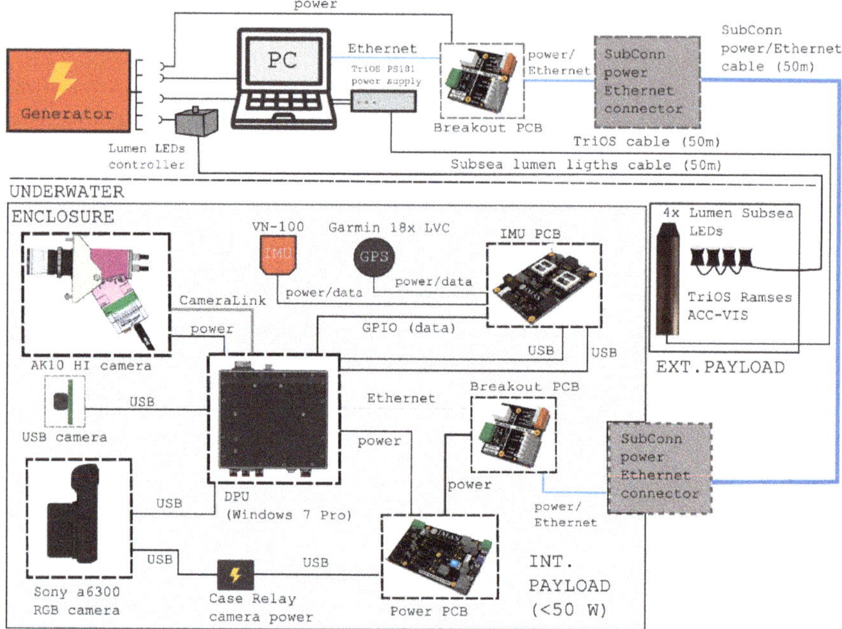

Figure A1. Schematics of the electronic power and communication streams for the internal and the additional external under-ice payloads.

All components of the payload were mounted around a custom-made vertical aluminum tray (20.32 by 60.96 cm) that hangs from the enclosure end cap for full swift removal and insertion of the payload (Figures 3 and 4a). The umbilical used is a 13 contacts SubConn power/ethernet (Type: D-P-P4TP24#/4C18#, 50 m long) and is received on the surface by another circular 13 contacts connector. Data and power streams were then divided within the enclosure and above the surface from the connectors by unregulated breakout PCBs (Figure 4). The ethernet stream within the enclosure connects directly to the DPU, allowing for VNC from the above surface PC. In our set-up, we used the freely available TightVNC software (https://www.tightvnc.com/) for this purpose. Most sensors were interfaced and powered through the DPU using their respective data/power cables, as shown in Figure 4 and operated through their own software. Only the Sony a6300 was powered using a different route. The power stream within the enclosure went through a power regulator PCB that fed the DPU directly and the Sony a6300 camera via Tether Tools Case Relay (Tether Tools, Phoenix, USA), and a Relay camera coupler for the Sony a6300.

An alternative low-cost solution was designed to synchronize and time stamp pushbroom frames with attitude (roll, pitch, and heading) data from the IMU. The IMU PCB in Figure 4 collected NMEA strings from the GPS and sent them through USB serial connection to the DPU for local time synchronization. The VN-100 IMU uses a 1 PPS sync input from the GPS as a trigger and reference to start its internal timer and to synchronize it to the GPS clock. This means that for each packet of data that the IMU outputs, it stamps the packet with an accurate GPS timestamp. Within the DPU the data is recorded using a Python script (run through Eclipse IDE and PyDev). The script takes the IMU packet producing a file with the IMU data (heading, pitch, and roll) and the NMEA string (position and GPS time) from the GPS and then correlates the IMU's internal "stopwatch time" to GPS time. The script also adds local PC time for reference with other sensors. GPS lock was performed before deployment (above the surface). Once underwater, the Garmin 18x LVC transmitted NMEA and 1PPS signals even when it could not see satellites. The NMEA times continued to update, but according to the internal real-time clock on the GPS. This, however, means that a short drift may happen in the GPS clock reference time over long iterations.

The AK10, DPU, IMU, and GPS together have a power consumption of <42 W. The Sony a6300 power consumption was estimated to be <7.5 W. The Low-Light USB camera was only used as additional visual support and was run through iSPY open-source software (https://www.ispyconnect.com/) for live-stream footage and video recording and had a power consumption of <1 W. The total power requirement of the internal payload is estimated to be < 50 W which can be easily powered by conventional generators. The total VNC data rate oscillates well below 1 Gbit/s (up to 75 m) supported by the SubConn cable, which leaves enough space for additional sensor streaming and data transfer.

The external payload components were operated separately using their respective cables by standard means. The TriOS Ramses ACC VIS was set-up with a connecting cable (50 m) and a TriOS PS101 power supply operated through the TriOS MSDA_XE software. The four Lumen Lights LEDs location were powered and dimmed through their separate 50 m lumen cables.

References

1. Arrigo, K.R. Sea ice as a habitat for primary producers. In *Sea Ice*; Thomas, D.N., Ed.; John Wiley & Sons, Ltd.: Hoboken, NJ, USA, 2017; pp. 352–369.
2. Kohlbach, D.; Graeve, M.; Lange, B.A.; David, C.; Schaafsma, F.L.; van Franeker, J.A.; Vortkamp, M.; Brandt, A.; Flores, H. Dependency of Antarctic zooplankton species on ice algae-produced carbon suggests a sea ice-driven pelagic ecosystem during winter. *Glob. Chang. Biol.* **2018**, *24*, 4667–4681. [CrossRef] [PubMed]
3. Van Leeuwe, M.A.; Tedesco, L.; Arrigo, K.R.; Assmy, P.; Campbell, K.; Meiners, K.M.; Rintala, J.M.; Selz, V.; Thomas, D.N.; Stefels, J. Microalgal community structure and primary production in Arctic and Antarctic sea ice: A synthesis. *Elem. Sci. Anthr.* **2018**, *6*, 4. [CrossRef]
4. Cimoli, E.; Meiners, K.M.; Lund-Hansen, L.C.; Lucieer, V. Spatial variability in sea-ice algal biomass: An under-ice remote sensing perspective. *Adv. Polar Sci.* **2017**, *28*, 268–296.
5. Lange, B.A.; Katlein, C.; Nicolaus, M.; Peeken, I.; Flores, H. Sea ice algae chlorophyll a concentrations derived from under-ice spectral radiation profiling platforms. *J. Geophys. Res. Ocean.* **2016**, *121*, 8511–8534. [CrossRef]
6. Miller, L.A.; Fripiat, F.; Else, B.G.; Bowman, J.S.; Brown, K.A.; Collins, R.E.; Ewert, M.; Fransson, A.; Gosselin, M.; Lannuzel, D.; et al. Methods for biogeochemical studies of sea ice: The state of the art, caveats, and recommendations. *Elem. Sci. Anthr.* **2015**, *3*, 000038. [CrossRef]
7. Lange, B.A.; Katlein, C.; Castellani, G.; Fernández-Méndez, M.; Nicolaus, M.; Peeken, I.; Flores, H. Characterizing Spatial Variability of Ice Algal Chlorophyll a and Net Primary Production between Sea Ice Habitats Using Horizontal Profiling Platforms. *Front. Mar. Sci.* **2017**, *4*, 349. [CrossRef]
8. Meiners, K.M.; Arndt, S.; Bestley, S.; Krumpen, T.; Ricker, R.; Milnes, M.; Newbery, K.; Freier, U.; Jarman, S.; King, R.; et al. Antarctic pack ice algal distribution: Floe-scale spatial variability and predictability from physical parameters. *Geophys. Res. Lett.* **2017**, *44*, 7382–7390. [CrossRef]
9. Fernández-Méndez, M.; Olsen, L.M.; Kauko, H.M.; Meyer, A.; Rösel, A.; Merkouriadi, I.; Mundy, C.J.; Ehn, J.K.; Johansson, A.M.; Wagner, P.M.; et al. Algal Hot Spots in a Changing Arctic Ocean: Sea-Ice Ridges and the Snow-Ice Interface. *Front. Mar. Sci.* **2018**, *5*, 75. [CrossRef]
10. Krembs, C.; Tuschling, K.; Juterzenka, K.V. The topography of the ice-water interface – its influence on the colonization of sea ice by algae. *Polar Biol.* **2002**, *25*, 106–117. [CrossRef]
11. Lund-Hansen, L.C.; Hawes, I.; Nielsen, M.H.; Sorrell, B.K. Is colonization of sea ice by diatoms facilitated by increased surface roughness in growing ice crystals? *Polar Biol.* **2016**, *40*, 593–602. [CrossRef]
12. Monti, D.; Legendre, L.; Therriault, J.C.; Demers, S. Horizontal distribution of sea-ice microalgae: Environmental control and spatial processes (southeastern Hudson Bay, Canada). *Mar. Ecol. Prog. Ser.* **1996**, *133*, 229–240. [CrossRef]
13. Ryan, K.G.; Hegseth, E.N.; Martin, A.; Davy, S.K.; O'Toole, R.; Ralph, P.J.; McMinn, A.; Thorn, C.J. Comparison of the microalgal community within fast ice at two sites along the Ross Sea coast, Antarctica. *Antarct. Sci.* **2006**, *18*, 583. [CrossRef]

14. Meiners, K.M.; Vancoppenolle, M.; Carnat, G.; Castellani, G.; Delille, B.; Delille, D.; Dieckmann, G.S.; Flores, H.; Fripiat, F.; Grotti, M.; et al. Chlorophyll- a in Antarctic Landfast Sea Ice: A First Synthesis of Historical Ice Core Data. *J. Geophys. Res. Ocean.* **2018**, *123*, 8444–8459. [CrossRef]
15. Leu, E.; Mundy, C.J.; Assmy, P.; Campbell, K.; Gabrielsen, T.M.; Gosselin, M.; Juul-Pedersen, T.; Gradinger, R. Arctic spring awakening–Steering principles behind the phenology of vernal ice algal blooms. *Prog. Oceanogr.* **2015**, *139*, 151–170. [CrossRef]
16. Meiners, K.M.; Vancoppenolle, M.; Thanassekos, S.; Dieckmann, G.S.; Thomas, D.N.; Tison, J.L.; Arrigo, K.R.; Garrison, D.L.; McMinn, A.; Lannuzel, D.; et al. Chlorophyll a in Antarctic sea ice from historical ice core data. *Geophys. Res. Lett.* **2012**, 39. [CrossRef]
17. Mundy, C.J.; Ehn, J.K.; Barber, D.G.; Michel, C. Influence of snow cover and algae on the spectral dependence of transmitted irradiance through Arctic landfast first-year sea ice. *J. Geophys. Res.* **2017**, 112. [CrossRef]
18. Campbell, K.; Mundy, C.J.; Barber, D.G.; Gosselin, M.; Giguère, N. Remote Estimates of Ice Algae Biomass and Their Response to Environmental Conditions during Spring Melt. *ARCTIC* **2014**, *67*, 375. [CrossRef]
19. Melbourne-Thomas, J.; Meiners, K.M.; Mundy, C.J.; Schallenberg, C.; Tattersall, K.L.; Dieckmann, G.S. Algorithms to estimate Antarctic sea ice algal biomass from under-ice irradiance spectra at regional scales. *Mar. Ecol. Prog. Ser.* **2015**, *536*, 107–121. [CrossRef]
20. Lund-Hansen, L.C.; Juul, T.; Eskildsen, T.D.; Hawes, I.; Sorrell, B.; Melvad, C.; Hancke, K. A low-cost remotely operated vehicle (ROV) with an optical positioning system for under-ice measurements and sampling. *Cold Reg. Sci. Technol.* **2018**, *151*, 148–155. [CrossRef]
21. Van Franeker, J.A.; Flores, H.; Van Dorssen, M. The surface and under ice trawl (SUIT). Frozen Desert Alive- Role Sea Ice Pelagic Macrofauna Its Predat. Ph.D. Thesis, University of Groningen, Groningen, The Netherlands, 2009; pp. 181–188.
22. Forrest, A.L.; Lund-Hansen, L.C.; Sorrell, B.K.; Bowden-Floyd, I.; Lucieer, V.; Cossu, R.; Lange, B.A.; Hawes, I. Exploring Spatial Heterogeneity of Antarctic Sea Ice Algae Using an Autonomous Underwater Vehicle Mounted Irradiance Sensor. *Front. Earth Sci.* **2019**, *7*, 1–13. [CrossRef]
23. Lucieer, V.; Nau, A.; Forrest, A.; Hawes, I. Fine-Scale Sea Ice Structure Characterized Using Underwater Acoustic Methods. *Remote Sens.* **2016**, *8*, 821. [CrossRef]
24. Cimoli, E.; Marcer, M.; Vandecrux, B.; Bøggild, C.E.; Williams, G.; Simonsen, S.B. Application of Low-Cost UASs and Digital Photogrammetry for High-Resolution Snow Depth Mapping in the Arctic. *Remote Sens.* **2017**, *9*, 1144. [CrossRef]
25. Irvine-Fynn, T.D.L.; Sanz-Ablanedo, E.; Rutter, N.; Smith, M.W.; Chandler, J.H. Measuring glacier surface roughness using plot-scale, close-range digital photogrammetry. *J. Glaciol.* **2014**, *60*, 957–969. [CrossRef]
26. Li, T.; Zhang, B.; Cheng, X.; Westoby, M.J.; Li, Z.; Ma, C.; Hui, F.; Shokr, M.; Liu, Y.; Chen, Z.; et al. Resolving Fine-Scale Surface Features on Polar Sea Ice: A First Assessment of UAS Photogrammetry Without Ground Control. *Remote Sens.* **2019**, *11*, 784. [CrossRef]
27. Cimoli, E.; Lucieer, A.; Meiners, K.M.; Lund-Hansen, L.C.; Kennedy, F.; Martin, A.; McMINN, A.; Lucieer, V. Towards improved estimates of sea-ice algal biomass: Experimental assessment of hyperspectral imaging cameras for under-ice studies. *Ann. Glaciol.* **2017**, *58*, 68–77. [CrossRef]
28. Aasen, H.; Honkavaara, E.; Lucieer, A.; Zarco-Tejada, P.J. Quantitative Remote Sensing at Ultra-High Resolution with UAV Spectroscopy: A Review of Sensor Technology, Measurement Procedures, and Data Correction Workflows. *Remote Sens.* **2018**, *10*, 1091. [CrossRef]
29. Adão, T.; Hruška, J.; Pádua, L.; Bessa, J.; Peres, E.; Morais, R.; Sousa, J. Hyperspectral Imaging: A Review on UAV-Based Sensors, Data Processing and Applications for Agriculture and Forestry. *Remote Sens.* **2017**, *9*, 1110. [CrossRef]
30. Jaud, M.; Le Dantec, N.; Ammann, J.; Grandjean, P.; Constantin, D.; Akhtman, Y.; Barbieux, K.; Allemand, P.; Delacourt, C.; Merminod, B. Direct Georeferencing of a Pushbroom, Lightweight Hyperspectral System for Mini-UAV Applications. *Remote Sens.* **2018**, *10*, 204. [CrossRef]
31. Lucieer, A.; Malenovský, Z.; Veness, T.; Wallace, L. HyperUAS-Imaging Spectroscopy from a Multirotor Unmanned Aircraft System. *J. Field Robot.* **2014**, *31*, 571–590. [CrossRef]
32. Chennu, A.; Färber, P.; De'ath, G.; de Beer, D.; Fabricius, K.E. A diver-operated hyperspectral imaging and topographic surveying system for automated mapping of benthic habitats. *Sci. Rep.* **2017**, *7*, 7122. [CrossRef]
33. Mogstad, A.A.; Johnsen, G.; Ludvigsen, M. Shallow-Water Habitat Mapping using Underwater Hyperspectral Imaging from an Unmanned Surface Vehicle: A Pilot Study. *Remote Sens.* **2019**, *11*, 685. [CrossRef]

34. Chennu, A.; Färber, P.; Volkenborn, N.; Al-Najjar, M.A.; Janssen, F.; de Beer, D.; Polerecky, L. Hyperspectral imaging of the microscale distribution and dynamics of microphytobenthos in intertidal sediments: Hyperspectral imaging of MPB biofilms. *Limnol. Oceanogr. Methods* **2013**, *11*, 511–528. [CrossRef]
35. Dumke, I.; Purser, A.; Marcon, Y.; Nornes, S.M.; Johnsen, G.; Ludvigsen, M.; Søreide, F. Underwater hyperspectral imaging as an in situ taxonomic tool for deep-sea megafauna. *Sci. Rep.* **2018**, *8*, 12860. [CrossRef] [PubMed]
36. Dumke, I.; Nornes, S.M.; Purser, A.; Marcon, Y.; Ludvigsen, M.; Ellefmo, S.L.; Johnsen, G.; Søreide, F. First hyperspectral imaging survey of the deep seafloor: High-resolution mapping of manganese nodules. *Remote Sens. Environ.* **2018**, *209*, 19–30. [CrossRef]
37. Yeh, C.-K.; Tsai, V.J.D. Direct georeferencing of airborne pushbroom images. *J. Chin. Inst. Eng.* **2015**, *38*, 653–664. [CrossRef]
38. Friedman, A.; Pizarro, O.; Williams, S.B.; Johnson-Roberson, M. Multi-Scale Measures of Rugosity, Slope and Aspect from Benthic Stereo Image Reconstructions. *PLoS ONE* **2012**, *7*, e50440. [CrossRef]
39. Maas, H.-G. On the Accuracy Potential in Underwater/Multimedia Photogrammetry. *Sensors* **2015**, *15*, 18140–18152. [CrossRef]
40. McCarthy, J.; Benjamin, J. Multi-image Photogrammetry for Underwater Archaeological Site Recording: An Accessible, Diver-Based Approach. *J. Marit. Archaeol.* **2014**, *9*, 95–114. [CrossRef]
41. Raoult, V.; David, P.A.; Dupont, S.F.; Mathewson, C.P.; O'Neill, S.J.; Powell, N.N.; Williamson, J.E. GoPros^{TM} as an underwater photogrammetry tool for citizen science. *PeerJ* **2016**, 4. [CrossRef]
42. Johnsen, G.; Norli, M.; Moline, M.; Robbins, I.; von Quillfeldt, C.; Sørensen, K.; Cottier, F.; Berge, J. The advective origin of an under-ice spring bloom in the Arctic Ocean using multiple observational platforms. *Polar Biol.* **2018**, *41*, 1197–1216. [CrossRef]
43. Arrigo, K.R.; Brown, Z.W.; Mills, M.M. Sea ice algal biomass and physiology in the Amundsen Sea, Antarctica. *Elem. Sci. Anthr.* **2014**, *2*, 000028. [CrossRef]
44. Johnsen, G.; Volent, Z.; Dierssen, H.; Pettersen, R.; Van Ardelan, M.; Søreide, F.; Fearns, P.; Ludvigsen, M.; Moline, M. Underwater hyperspectral imagery to create biogeochemical maps of seafloor properties. In *Subsea Optics and Imaging*; Elsevier: Amsterdam, The Netherlands, 2013; pp. 508e–540e.
45. Huang, H.; Liu, L.; Ngadi, M. Recent Developments in Hyperspectral Imaging for Assessment of Food Quality and Safety. *Sensors* **2014**, *14*, 7248–7276. [CrossRef] [PubMed]
46. Ramirez-Paredes, J.-P.; Lary, D.J.; Gans, N.R. Low-altitude Terrestrial Spectroscopy from a Pushbroom Sensor. *J. Field Robot.* **2016**, *33*, 837–852. [CrossRef]
47. Lund-Hansen, L.C.; Hawes, I.; Sorrell, B.K.; Nielsen, M.H. Removal of snow cover inhibits spring growth of Arctic ice algae through physiological and behavioral effects. *Polar Biol.* **2014**, *37*, 471–481. [CrossRef]
48. Wongpan, P.; Meiners, K.M.; Langhorne, P.J.; Heil, P.; Smith, I.J.; Leonard, G.H.; Massom, R.A.; Clementson, L.A.; Haskell, T.G. Estimation of Antarctic Land-Fast Sea Ice Algal Biomass and Snow Thickness from Under-Ice Radiance Spectra in Two Contrasting Areas. *J. Geophys. Res. Ocean.* **2018**, *123*, 1907–1923. [CrossRef]
49. Morel, A.; Maritorena, S. Bio-optical properties of oceanic waters- A reappraisal. *J. Geophys. Res.* **2001**, *106*, 7163–7180. [CrossRef]
50. Bryson, M.; Johnson-Roberson, M.; Pizarro, O.; Williams, S.B. Colour-Consistent Structure-from-Motion Models using Underwater Imagery. In *Robotics: Science and Systems*; MIT Press: Cambridge, MA, USA, 2012; pp. 1–8.
51. Menna, F.; Nocerino, E.; Fassi, F.; Remondino, F. Geometric and Optic Characterization of a Hemispherical Dome Port for Underwater Photogrammetry. *Sensors* **2016**, *16*, 48. [CrossRef]
52. Telem, G.; Filin, S. Photogrammetric modeling of underwater environments. *ISPRS J. Photogramm. Remote Sens.* **2010**, *65*, 433–444. [CrossRef]
53. Łuczyński, T.; Pfingsthorn, M.; Birk, A. The Pinax-model for accurate and efficient refraction correction of underwater cameras in flat-pane housings. *Ocean Eng.* **2017**, *133*, 9–22. [CrossRef]
54. Treibitz, T.; Schechner, Y.; Kunz, C.; Singh, H. Flat Refractive Geometry. *IEEE Trans. Pattern Anal. Mach. Intell.* **2012**, *34*, 51–65. [CrossRef]
55. Shortis, M. Calibration Techniques for Accurate Measurements by Underwater Camera Systems. *Sensors* **2015**, *15*, 30810–30826. [CrossRef] [PubMed]

56. Oniga, V.-E.; Pfeifer, N.; Loghin, A.-M. 3D Calibration Test-Field for Digital Cameras Mounted on Unmanned Aerial Systems (UAS). *Remote Sens.* **2018**, *10*, 2017. [CrossRef]
57. Piazza, P.; Cummings, V.J.; Lohrer, D.M.; Marini, S.; Marriott, P.; Menna, F.; Nocerino, E.; Peirano, A.; Schiaparelli, S. Divers-operated underwater photogrammetry: Applications in the study of antarctic benthos. *Int. Arch. Photogramm. Remote Sens. Spat. Inf. Sci.* **2018**, *42*, 885–892. [CrossRef]
58. Menna, F.; Nocerino, E.; Remondino, F. Flat versus hemispherical dome ports in underwaterphotogrammetry. *Int. Arch. Photogramm. Remote Sens. Spat. Inf. Sci.* **2017**, *42*, 481–487. [CrossRef]
59. *Agisoft Metashape User Manual - Professional Edition, Version 1.5*. Available online: www.agisoft.com/pdf/metashape-pro_1_5_en.pdf (accessed on 2 December 2019).
60. Fonstad, M.A.; Dietrich, J.T.; Courville, B.C.; Jensen, J.L.; Carbonneau, P.E. Topographic structure from motion: A new development in photogrammetric measurement. *Earth Surf. Process. Landf.* **2013**, *38*, 421–430. [CrossRef]
61. Tonkin, T.N.; Midgley, N.G. Ground-Control Networks for Image Based Surface Reconstruction: An Investigation of Optimum Survey Designs Using UAV Derived Imagery and Structure-from-Motion Photogrammetry. *Remote Sens.* **2016**, *8*, 786. [CrossRef]
62. Westoby, M.J.; Brasington, J.; Glasser, N.F.; Hambrey, M.J.; Reynolds, J.M. "Structure-from-Motion" photogrammetry: A low-cost, effective tool for geoscience applications. *Geomorphology* **2012**, *179*, 300–314. [CrossRef]
63. Savitzky, A.; Golay, M.J. Smoothing and Differentiation of Data by Simplified Least Squares Procedures. *Anal. Chem.* **1964**, *36*, 1627–1639. [CrossRef]
64. Schafer, R.W. What Is a Savitzky-Golay Filter? [Lecture Notes]. *IEEE Signal Process. Mag.* **2011**, *28*, 111–117. [CrossRef]
65. Craig, S.E.; Jones, C.T.; Li, W.K.; Lazin, G.; Horne, E.; Caverhill, C.; Cullen, J.J. Deriving optical metrics of coastal phytoplankton biomass from ocean colour. *Remote Sens. Environ.* **2012**, *119*, 72–83. [CrossRef]
66. Lubac, B.; Loisel, H. Variability and classification of remote sensing reflectance spectra in the eastern English Channel and southern North Sea. *Remote Sens. Environ.* **2007**, *110*, 45–58. [CrossRef]
67. Amigo, J.M.; Babamoradi, H.; Elcoroaristizabal, S. Hyperspectral image analysis. A tutorial. *Anal. Chim. Acta* **2015**, *896*, 34–51. [CrossRef] [PubMed]
68. Nicolaus, M.; Petrich, C.; Hudson, S.R.; Granskog, M.A. Variability of light transmission through Arctic land-fast sea ice during spring. *Cryosphere* **2013**, *7*, 977–986. [CrossRef]
69. Malenovsky, Z.; Ufer, C.; Lhotáková, Z.; Clevers, J.G.; Schaepman, M.E.; Albrechtová, J.; Cudlín, P. A new hyperspectral index for chlorophyll estimation: Area under curve normalised to maximal band depth between 650-725 nm. *EARSeL eProc.* **2006**, *5*, 12.
70. Malenovský, Z.; Lucieer, A.; King, D.H.; Turnbull, J.D.; Robinson, S.A. Unmanned aircraft system advances health mapping of fragile polar vegetation. *Methods Ecol. Evol.* **2017**, *8*, 1842–1857. [CrossRef]
71. Kokaly, R.F.; Clark, R.N. Spectroscopic determination of leaf biochemistry using band-depth analysis of absorption features and stepwise multiple linear regression. *Remote Sens. Environ.* **1999**, *67*, 21. [CrossRef]
72. Petrich, C.; Eicken, H. Overview of sea ice growth and properties. In *Sea Ice*; Thomas, D.N., Ed.; John Wiley & Sons, Ltd.: Chichester, UK, 2016; pp. 1–41.
73. Polashenski, C.; Perovich, D.; Courville, Z. The mechanisms of sea ice melt pond formation and evolution: Mechanisms of melt pond evolution. *J. Geophys. Res. Oceans* **2012**, 117. [CrossRef]
74. Weeks, W.F.; Gow, A.J. Preferred Crystal Orientations in the Fast Ice Along the Margins of the Arctic Ocean. *J. Geophys. Res.* **1978**, *83*, 5105–5121. [CrossRef]
75. Legendre, L.; Gosselin, M. In situ spectroradiometric estimation of microalgal biomass in first-year sea ice. *Polar Biol.* **1991**, *11*, 113–115. [CrossRef]
76. Foglini, F.; Grande, V.; Marchese, F.; Bracchi, V.A.; Prampolini, M.; Angeletti, L.; Castellan, G.; Chimienti, G.; Hansen, I.M.; Gudmundsen, M.; et al. Application of Hyperspectral Imaging to Underwater Habitat Mapping, Southern Adriatic Sea. *Sensors* **2019**, *19*, 2261. [CrossRef]
77. Kirk, J.T.O. *Light and Photosynthesis in Aquatic Ecosystems*, 3rd ed.; Cambridge University Press: Cambridge, UK; New York, NY, USA, 2011.
78. Lund-Hansen, L.C.; Markager, S.; Hancke, K.; Stratmann, T.; Rysgaard, S.; Ramløv, H.; Sorrell, B.K. Effects of sea-ice light attenuation and CDOM absorption in the water below the Eurasian sector of central Arctic Ocean (>88°N). *Polar Res.* **2015**, *34*, 23978. [CrossRef]

79. Malenovský, Z.; Homolová, L.; Zurita-Milla, R.; Lukeš, P.; Kaplan, V.; Hanuš, J.; Gastellu-Etchegorry, J.P.; Schaepman, M.E. Retrieval of spruce leaf chlorophyll content from airborne image data using continuum removal and radiative transfer. *Remote Sens. Environ.* **2013**, *131*, 85–102. [CrossRef]
80. Arroyo-Mora, J.P.; Kalacska, M.; Inamdar, D.; Soffer, R.; Lucanus, O.; Gorman, J.; Naprstek, T.; Schaaf, E.S.; Ifimov, G.; Elmer, K.; et al. Implementation of a UAV–Hyperspectral Pushbroom Imager for Ecological Monitoring. *Drones* **2019**, *3*, 12. [CrossRef]
81. Fang, H.; Hu, B.; Yu, Z.; Xu, H.; He, C.; Li, A.; Liu, Y. Semi-automatic geometric correction of airborne hyperspectral push-broom images using ground control points and linear features. *Int. J. Remote Sens.* **2018**, *39*, 4115–4129. [CrossRef]
82. Habib, A.; Han, Y.; Xiong, W.; He, F.; Zhang, Z.; Crawford, M. Automated Ortho-Rectification of UAV-Based Hyperspectral Data over an Agricultural Field Using Frame RGB Imagery. *Remote Sens.* **2016**, *8*, 796. [CrossRef]
83. Turner, D.; Lucieer, A.; Malenovský, Z.; King, D.; Robinson, S. Spatial Co-Registration of Ultra-High Resolution Visible, Multispectral and Thermal Images Acquired with a Micro-UAV over Antarctic Moss Beds. *Remote Sens.* **2014**, *6*, 4003–4024. [CrossRef]
84. Marcer, M.; Stentoft, P.A.; Bjerre, E.; Cimoli, E.; Bjørk, A.; Stenseng, L.; Machguth, H. Three Decades of Volume Change of a Small Greenlandic Glacier Using Ground Penetrating Radar, Structure from Motion, and Aerial Photogrammetry. *Arct. Antarct. Alp. Res.* **2017**, *49*, 411–425. [CrossRef]
85. Nicolaus, M.; Katlein, C. Mapping radiation transfer through sea ice using a remotely operated vehicle (ROV). *Cryosphere* **2013**, *7*, 763–777. [CrossRef]
86. Cazenave, F.; Zook, R.; Carroll, D.; Flagg, M.; Kim, S. Development of the Rov Scini and deployment in Mcmurdo sound, Antarctica. *J. Ocean Technol.* **2011**, *6*, 20.
87. Williams, G.; Maksym, T.; Wilkinson, J.; Kunz, C.; Murphy, C.; Kimball, P.; Singh, H. Thick and deformed Antarctic sea ice mapped with autonomous underwater vehicles. *Nat. Geosci.* **2014**, *8*, 61–67. [CrossRef]
88. Wang, L.; Zhou, X.; Zhu, X.; Dong, Z.; Guo, W. Estimation of biomass in wheat using random forest regression algorithm and remote sensing data. *Crop J.* **2016**, *4*, 212–219. [CrossRef]
89. Blondeau-Patissier, D.; Gower, J.F.R.; Dekker, A.G.; Phinn, S.R.; Brando, V.E. A review of ocean color remote sensing methods and statistical techniques for the detection, mapping and analysis of phytoplankton blooms in coastal and open oceans. *Prog. Oceanogr.* **2014**, *123*, 123–144. [CrossRef]
90. Matus-Hernández, M.Á.; Hernández-Saavedra, N.Y.; Martínez-Rincón, R.O. Predictive performance of regression models to estimate Chlorophyll-a concentration based on Landsat imagery. *PLoS ONE* **2018**, *13*, e0205682. [CrossRef]
91. Malenovský, Z.; Turnbull, J.D.; Lucieer, A.; Robinson, S.A. Antarctic moss stress assessment based on chlorophyll content and leaf density retrieved from imaging spectroscopy data. *New Phytol.* **2015**, *208*, 608–624. [CrossRef] [PubMed]
92. Näsi, R.; Viljanen, N.; Kaivosoja, J.; Alhonoja, K.; Hakala, T.; Markelin, L.; Honkavaara, E. Estimating Biomass and Nitrogen Amount of Barley and Grass Using UAV and Aircraft Based Spectral and Photogrammetric 3D Features. *Remote Sens.* **2018**, *10*, 1082. [CrossRef]
93. Shen, X.; Cao, L.; Yang, B.; Xu, Z.; Wang, G. Estimation of Forest Structural Attributes Using Spectral Indices and Point Clouds from UAS-Based Multispectral and RGB Imageries. *Remote Sens.* **2019**, *11*, 800. [CrossRef]
94. Taghizadeh, M.; Gowen, A.A.; O'Donnell, C.P. Comparison of hyperspectral imaging with conventional RGB imaging for quality evaluation of Agaricus bisporus mushrooms. *Biosyst. Eng.* **2011**, *108*, 191–194. [CrossRef]
95. Ambrose, W.G.; von Quillfeldt, C.; Clough, L.M.; Tilney, P.V.R.; Tucker, T. The sub-ice algal community in the Chukchi sea: Large- and small-scale patterns of abundance based on images from a remotely operated vehicle. *Polar Biol.* **2005**, *28*, 784–795. [CrossRef]
96. Katlein, C.; Fernández-Méndez, M.; Wenzhöfer, F.; Nicolaus, M. Distribution of algal aggregates under summer sea ice in the Central Arctic. *Polar Biol.* **2015**, *38*, 719–731. [CrossRef]
97. Jesus, B.; Mouget, J.-L.; Perkins, R.G. Detection of Diatom Xanthophyll Cycle Using Spectral Reflectance. *J. Phycol.* **2008**, *44*, 1349–1359. [CrossRef]
98. Perkins, R.G.; Williamson, C.J.; Brodie, J.; Barillé, L.; Launeau, P.; Lavaud, J.; Yallop, M.L.; Jesus, B. Microspatial variability in community structure and photophysiology of calcified macroalgal microbiomes revealed by coupling of hyperspectral and high-resolution fluorescence imaging. *Sci. Rep.* **2016**, *6*, 22343. [CrossRef] [PubMed]

99. Mehrubeoglu, M.; Teng, M.; Zimba, P. Resolving Mixed Algal Species in Hyperspectral Images. *Sensors* **2013**, *14*, 1–21. [CrossRef] [PubMed]
100. Xi, H.; Hieronymi, M.; Röttgers, R.; Krasemann, H.; Qiu, Z. Hyperspectral Differentiation of Phytoplankton Taxonomic Groups: A Comparison between Using Remote Sensing Reflectance and Absorption Spectra. *Remote Sens.* **2015**, *7*, 14781–14805. [CrossRef]
101. Blackburn, G.A. Wavelet decomposition of hyperspectral data: A novel approach to quantifying pigment concentrations in vegetation. *Int. J. Remote Sens.* **2007**, *28*, 2831–2855. [CrossRef]
102. Pettersen, R.; Johnsen, G.; Bruheim, P.; Andreassen, T. Development of hyperspectral imaging as a bio-optical taxonomic tool for pigmented marine organisms. *Org. Divers. Evol.* **2014**, *14*, 237–246. [CrossRef]
103. Taylor, B.B.; Taylor, M.H.; Dinter, T.; Bracher, A. Estimation of relative phycoerythrin concentrations from hyperspectral underwater radiance measurements—-A statistical approach. *J. Geophys. Res. Ocean.* **2013**, *118*, 2948–2960. [CrossRef]
104. Caras, T.; Karnieli, A. Ground-Level Classification of a Coral Reef Using a Hyperspectral Camera. *Remote Sens.* **2015**, *7*, 7521–7544. [CrossRef]
105. Müller, S.; Vähätalo, A.V.; Uusikivi, J.; Majaneva, M.A.M.; Majaneva, S.; Autio, R.; Rintala, J.M. Primary production calculations for sea ice from bio-optical observations in the Baltic Sea. *Elem Sci Anthr.* **2016**, *4*, 000121. [CrossRef]
106. Méléder, V.; Jesus, B.; Barnett, A.; Barillé, L.; Lavaud, J. Microphytobenthos primary production estimated by hyperspectral reflectance. *PLoS ONE* **2018**, *13*, e0197093. [CrossRef]
107. Campbell, K.; Mundy, C.J.; Barber, D.G.; Gosselin, M. Characterizing the sea ice algae chlorophyll a–snow depth relationship over Arctic spring melt using transmitted irradiance. *J. Mar. Syst.* **2015**, *147*, 76–84. [CrossRef]
108. Dustan, P.; Doherty, O.; Pardede, S. Digital Reef Rugosity Estimates Coral Reef Habitat Complexity. *PLoS ONE* **2013**, *8*, e57386. [CrossRef] [PubMed]
109. Gutt, J. The occurrence of sub-ice algal aggregations off northeast Greenland. *Polar Biol.* **1995**, *15*, 247–252. [CrossRef]
110. Krembs, C.; Mock, T.; Gradinger, R. A mesocosm study of physical-biological interactions in artificial sea ice: Effects of brine channel surface evolution and brine movement on algal biomass. *Polar Biol.* **2001**, *24*, 356–364. [CrossRef]
111. Lange, B.A.; Michel, C.; Beckers, J.F.; Casey, J.A.; Flores, H.; Hatam, I.; Meisterhans, G.; Niemi, A.; Haas, C. Comparing Springtime Ice-Algal Chlorophyll a and Physical Properties of Multi-Year and First-Year Sea Ice from the Lincoln Sea. *PLoS ONE* **2015**, *10*, e0122418. [CrossRef]
112. Hop, H.; Pavlova, O. Distribution and biomass transport of ice amphipods in drifting sea ice around Svalbard. *Deep Sea Res. Part II Top. Stud. Oceanogr.* **2008**, *55*, 2292–2307. [CrossRef]
113. Werner, I. Grazing of Arctic under-ice amphipods on sea-ice algae. *Mar. Ecol. Prog. Ser.* **1997**, *160*, 93–99. [CrossRef]
114. Arrigo, K.; Dieckmann, G.; Gosselin, M.; Robinson, D.; Fritsen, C.; Sullivan, C. High resolution study of the platelet ice ecosystem in McMurdo Sound, Antarctica:biomass, nutrient, and production profiles within a dense microalgal bloom. *Mar. Ecol. Prog. Ser.* **1995**, *127*, 255–268. [CrossRef]
115. Sture, Ø.; Ludvigsen, M.; Aas, L.M.S. Autonomous underwater vehicles as a platform for underwater hyperspectral imaging. In Proceedings of the OCEANS 2017-Aberdeen, Aberdeen, UK, 19–22 June 2017; pp. 1–8.
116. Katlein, C.; Perovich, D.K.; Nicolaus, M. Geometric Effects of an Inhomogeneous Sea Ice Cover on the under Ice Light Field. *Front. Earth Sci.* **2016**, *4*, 1–10. [CrossRef]
117. Katlein, C.; Nicolaus, M.; Petrich, C. The anisotropic scattering coefficient of sea ice. *J. Geophys. Res. Ocean.* **2014**, *119*, 842–855. [CrossRef]
118. Aasen, H.; Bolten, A. Multi-temporal high-resolution imaging spectroscopy with hyperspectral 2D imagers—From theory to application. *Remote Sens. Environ.* **2018**, *205*, 374–389. [CrossRef]
119. Buchhorn, M.; Raynolds, M.K.; Walker, D.A. Influence of BRDF on NDVI and biomass estimations of Alaska Arctic tundra. *Environ. Res. Lett.* **2016**, *11*, 125002. [CrossRef]
120. Zhao, F.; Li, Y.; Dai, X.; Verhoef, W.; Guo, Y.; Shang, H.; Gu, X.; Huang, Y.; Yu, T.; Huang, J. Simulated impact of sensor field of view and distance on field measurements of bidirectional reflectance factors for row crops. *Remote Sens. Environ.* **2015**, *156*, 129–142. [CrossRef]

121. Piazza, P.; Cummings, V.; Guzzi, A.; Hawes, I.; Lohrer, A.; Marini, S.; Marriott, P.; Menna, F.; Nocerino, E.; Peirano, A.; et al. Underwater photogrammetry in Antarctica: Long-term observations in benthic ecosystems and legacy data rescue. *Polar Biol.* **2019**, *42*, 1061–1079. [CrossRef]
122. Arrigo, K.R.; Perovich, D.K.; Pickart, R.S.; Brown, Z.W.; Van Dijken, G.L.; Lowry, K.E.; Mills, M.M.; Palmer, M.A.; Balch, W.M.; Bahr, F.; et al. Massive Phytoplankton Blooms Under Arctic Sea Ice. *Science* **2012**, *336*, 1408. [CrossRef]
123. Åhlén, J.; Sundgren, D.; Bengtsson, E. Application of underwater hyperspectral data for color correction purposes. *Pattern Recognit. Image Anal.* **2007**, *17*, 170–173. [CrossRef]
124. Bryson, M.; Johnson-Roberson, M.; Pizarro, O.; Williams, S.B. True Color Correction of Autonomous Underwater Vehicle Imagery. *J. Field Robot.* **2015**, *33*, 853–874. [CrossRef]
125. Yang, C.; Yang, D.; Cao, W.; Zhao, J.; Wang, G.; Sun, Z.; Xu, Z.; Ravi Kumar, M.S. Analysis of seagrass reflectivity by using a water column correction algorithm. *Int. J. Remote Sens.* **2010**, *31*, 4595–4608. [CrossRef]

Article

High-Spatial Resolution Monitoring of *Phycocyanin* and *Chlorophyll-a* Using Airborne Hyperspectral Imagery

Jong Cheol Pyo [1], Mayzonee Ligaray [1], Yong Sung Kwon [1], Myoung-Hwan Ahn [2], Kyunghyun Kim [3], Hyuk Lee [4], Taegu Kang [4], Seong Been Cho [5], Yongeun Park [6,*] and Kyung Hwa Cho [1,*]

1 School of Urban and Environmental Engineering, Ulsan National Institute of Science and Technology, Ulsan 689-798, Korea; jcp01@unist.ac.kr (J.C.P.); mayzonee@gmail.com (M.L.); wizkys@unist.ac.kr (Y.S.K.)
2 Department of Atmospheric Science and Engineering, Ewha Womans University, Ewha-Yeodae-Gil 52, Seodaemoon-Gu, Seoul 03760, Korea; terryahn65@gmail.com
3 Yeongsan River Environmental Research Center, National Institute of Environmental Research, Gwangju 61011, Korea; matthias@korea.kr
4 Water Quality Assessment Research Division, National Institute of Environmental Research, Environmental Research Complex, Incheon 22689, Korea; ehyuk72@korea.kr (H.L.); taegu98@korea.kr (T.K.)
5 GEOSTROY Inc., Hwa-gok-lo 68-82, Gangseo-Gu, Seoul 07566, Korea; losttemple85@geostory.co.kr
6 School of Civil and Environmental Engineering, Konkuk University, Neumgdong-Ro 120, Gangjin-Gu, Seoul 05029, Korea
* Correspondence: yepark@konkuk.ac.kr (Y.P.); khcho@unist.ac.kr (K.H.C.); Tel.: +82-2-2049-6106(Y.P.); +82-52-217-2829 (K.H.C.)

Received: 14 June 2018; Accepted: 21 July 2018; Published: 26 July 2018

Abstract: Hyperspectral imagery (HSI) provides substantial information on optical features of water bodies that is usually applicable to water quality monitoring. However, it generates considerable uncertainties in assessments of spatial and temporal variation in water quality. Thus, this study explored the influence of different optical methods on the spatial distribution and concentration of *phycocyanin* (PC), *chlorophyll-a* (Chl-a), and total suspended solids (TSSs) and evaluated the dependence of algal distribution on flow velocity. Four ground-based and airborne monitoring campaigns were conducted to measure water surface reflectance. The actual concentrations of PC, Chl-a, and TSSs were also determined, while four bio-optical algorithms were calibrated to estimate the PC and Chl-a concentrations. Artificial neural network atmospheric correction achieved Nash-Sutcliffe Efficiency (NSE) values of 0.80 and 0.76 for the training and validation steps, respectively. Moderate resolution atmospheric transmission 6 (MODTRAN 6) showed an NSE value >0.8; whereas, atmospheric and topographic correction 4 (ATCOR 4) yielded a negative NSE value. The MODTRAN 6 correction led to the highest R^2 values and lowest root mean square error values for all algorithms in terms of PC and Chl-a. The PC:Chl-a distribution generated using HSI proved to be negatively dependent on flow velocity (p-value = 0.003) and successfully indicated cyanobacteria risk regions in the study area.

Keywords: Hyperspectral image; atmospheric correction; bio-optical algorithm; *phycocyanin*; *chlorophyll-a*

1. Introduction

Severe algal blooms, mainly caused by anthropogenic effects, are an ongoing cause of water quality problems in inland waters globally [1–6]. Massive nutrient loads from both point and non-point sources accelerate the growth and biomass production of algae [7]. In Korea, increases in water retention times because of the construction of multi-functional weirs contributes to the frequent formation of cyanobacterial blooms [8–10]. Baekje Weir along the Geum River, for example, has recently received increased attention because of water quality issues caused by frequent outbreaks of severe cyanobacterial blooms [11,12]. These have caused water quality degradation in the weir, which can lead to adverse effects on human health [13,14].

Remote-sensing techniques are useful in the detection of algal blooms because they can detect algae over large areas at a high time resolution [15–27]. Specifically, many bio-optical algorithms that use remotely sensed data have been developed to estimate the concentrations of algal pigments such as *chlorophyll-a* (Chl-a) and *phycocyanin* (PC) [28–32]. PC and Chl-a concentrations have been estimated using various apparent optical property (AOP) algorithms [33–36] and inherent optical property (IOP) algorithms [30,31,37–39]. AOP algorithms utilize water surface reflectance to estimate the algal concentration using multiple reflectance bands. The authors of [17] and [28] introduced a two-band ratio algorithm and three-band ratio algorithm for Chl-a and PC estimation, respectively. IOP algorithms use absorption and the back scattering coefficient for estimation of algal pigments. The authors of [29,30] estimated Chl-a and PC concentration using the ratio of the absorption coefficient and specific absorption coefficient.

Hyperspectral imagery (HSI) provides a spatially detailed information map of high spectral resolution. This high resolution allows hyperspectral images to be used for the identification and analysis of sophisticated spatial and spectral information [40–44]. Accurate retrieval of algal biomass from a hyperspectral image requires an atmospheric correction to remove atmospheric interference. Commercial atmospheric software packages such as atmospheric and topographic correction (ATCOR) [45] are often used to correct the images. Thiemann and Kaufmann [46] implemented atmospheric correction of hyperspectral images using ATCOR to generate maps of Secchi disk transparency and Chl-a concentration. Alternatively, users can perform the atmospheric correction themselves using the Moderate resolution atmospheric transmission (MODTRAN) software, which provides atmospheric correction parameters [47]. Giardino et al. [48] used MODTRAN to perform atmospheric correction of HSI to retrieve the Chl-a of an inland water. Furthermore, machine learning techniques have been introduced to correct atmospheric effects using observed atmospheric parameters [49–51].

Previous studies have used various atmospheric correction methods either to achieve good correction performance [22,52–56] or simply to estimate the target [44,57–60]. However, few studies have quantitatively analyzed the dependence of the atmospheric correction performance on the correction method [61]. In particular, algal detection studies, which consider the effect of atmospheric correction on PC and Chl-a concentration estimates, have been rarely completed. Moreover, when preprocessed hyperspectral images provide an algal distribution map, the concentration level and spatial distribution of algae are influenced by environmental factors such as water temperature, nutrients, and water retention time. The authors of [9,62,63] showed the distribution of algae is mainly affected by influencing factors such as hydrodynamic patterns. Thus, identifying the cause of an algal bloom is important after producing an algal distribution map.

Therefore, the objectives of this study were to (1) implement atmospheric correction of hyperspectral images using MODTRAN 6, ATCOR 4, and Artificial Neural Network (ANN); (2) develop bio-optical algorithms to estimate PC and Chl-a concentration using the corrected hyperspectral images; (3) identify the influence of the atmospheric correction on PC and Chl-a quantification; and (4) evaluate the algal distribution with hydrodynamic patterns.

2. Materials and Methods

2.1. Study Area

The Baekje Reservoir (36°32′N 126°94′E) is an artificial weir in the Geum River in South Korea (Figure 1). The Geum River has a total length of 395 km. It has three man-made weirs (the Baekje, Gongju, and Saejong reservoirs) and one dam (the Daechung dam). The distance between the Baekje and Gongju weirs along the river is 23 km, over which the width is 50 m and the average water depth is 4 m. The Baekje Weir has a length of 331 m and height of 5.3 m, and a total storage capacity of 24×10^6 m^3. The stored water is used for both domestic and agricultural purposes.

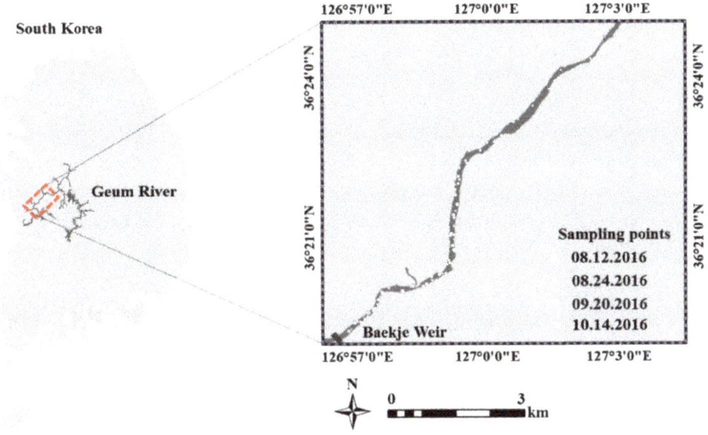

Figure 1. Study site in the Baekje Reservoir region.

2.2. Remote Sensing of PC and Chl-a Pigment

Figure 2 shows the research scheme employed in this study. The diagram is divided into four parts: (a) data collection from field and laboratory measurements were conducted; (b) atmospheric correction of hyperspectral images was performed using MODTRAN 6 and ATCOR 4 software; (c) atmospheric correction parameters generated by MODTRAN 6 were used in an ANN-simulated atmospheric correction; and (d) the corrected reflectance values from MODTRAN 6, ATCOR 4, and ANN were applied to build bio-optical algorithms for the estimation of PC and Chl-a concentrations. Finally, the bio-optical algorithms were used to generate the spatial and temporal distribution of the PC and Chl-a concentrations and identify the influence of the atmospheric correction on the PC and Chl-a quantification.

2.2.1. Water Sampling and Experimental Work

Field campaigns were conducted on 12 August, 24 August, 20 September, and 14 October 2016 (Table 1). During the sampling period, optical data and water samples were collected at 74 monitoring stations.

The measured radiance and irradiance data were used to calculate the remote-sensing reflectance while the PC and Chl-a were quantified in the water samples. Cyanobacteria contain the PC pigment, which harvests light through photosynthesis [64]. The representative absorption band for PC is around 620 nm [65]. For PC extraction, water samples were concentrated using a phytoplankton net with a 20-µm mesh size. The pre-concentration water volume varied from 10 L to 45 L. The concentrated water was stored in a 100-mL wide-necked bottle and kept cool in a box with ice. The water samples were analyzed within 24 h in the laboratory. PC was extracted by applying the freezing and thawing method [66]. The detailed experimental procedures of PC extraction are described in [38]. Two liters of water samples were collected for Chl-a analysis and analyzed within 24 h. The extraction of Chl-a followed a standard method [67]. Specific information regarding Chl-a analysis is presented in [38]. A standard method was used for the analysis of total suspended solid (TSS) concentration. A glass microfiber filter (GF/C, WHATMAN Inc., Piscataway, NJ, USA) was preferentially washed with deionized water and dried in a desiccator. Before filtration, the weight of the filter was measured using an analytical balance (EX224, OHAUS Inc., Parsippany, NJ, USA). After the filtration of the water samples, the used filter papers were placed in a drying oven (DO-150, HYSC Inc., Seoul, Korea) for 2 h. Finally, the dried filters were weighed using the analytical balance. The TSS concentration was calculated using the following equation:

$$\text{Total Suspended Solid } (\text{mg L}^{-1}) = \frac{(F_a - F_b) \times 1000}{V} \quad (1)$$

where, F_a is the weight of the filter after filtration (mg), F_b is the weight of the filter before filtration (mg), and V is the volume (mL) of the sample.

The absorption coefficient of phytoplankton was measured using light transmission measurement. This measurement was able to obtain the phytoplankton absorption coefficient without the signal of a non-algal particle by performing bleaching processing. A more detailed method of absorption coefficient measurement was followed by [38].

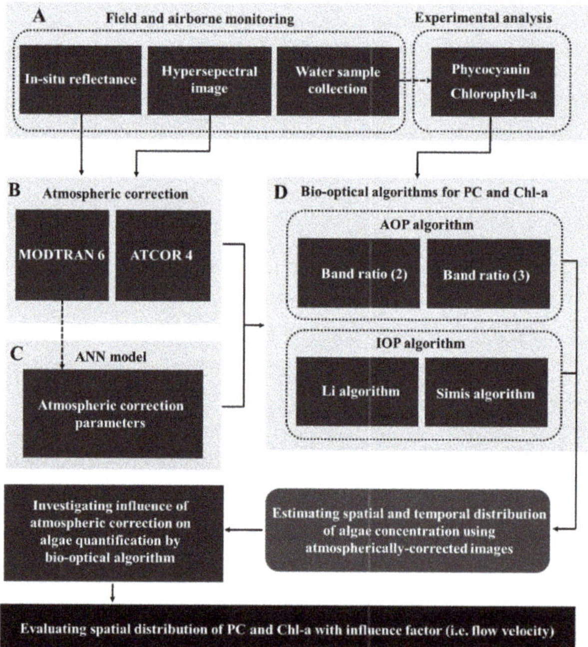

Figure 2. Schematic diagram for identifying the influence of the atmospheric correction and hydrodynamic pattern on algal quantification. (**A**) shows the field and airborne monitoring and experimental analysis, (**B**,**C**) show the atmospheric correction with commercial models and the Artificial Neural Network (ANN) model, and (**D**) shows the bio-optical algorithm calibration and application. PC = *phycocyanin*; AOP = apparent optical property; IOP = inherent optical property; MODTRAN = moderate resolution atmospheric transmission; ATCOR = atmospheric and topographic correction

2.2.2. Field Optical and Hyperspectral Reflectance Data

A FieldSpec HandHeld2 hand-held spectroradiometer (ASD, Inc., Longmont, CO, USA) was used to collect radiance and irradiance data. The spectroradiometer has a band range of 325 nm to 1075 nm and a 3-nm spectral resolution. The device has a 25-degree field-of-view. This spectroradiometer collects sky irradiance (E_r) using a cosine detector fore-optic, and water surface radiance (L_w) and sky radiance (L_s) using a bare fiber fore-optic. Optical sampling requires a specific position of the spectroradiometer with 130–135 degrees of azimuth angle and 40–45 degrees of zenith angle. This position minimizes ambient interferences such as the sun glint and shading effects [29]. The remote-sensing reflectance uses the ratio of irradiance to radiance as follows:

$$R_{rs}(\lambda, 0+) = \frac{L_w(\lambda, 0+) - 0.025L_s(\lambda)}{Er(\lambda)} \quad (2)$$

where $L_w(\lambda, 0+)$ is the water leaving radiance, $L_s(\lambda)$ is the downwelling sky radiance, $E_r(\lambda)$ is the downwelling sky irradiance, $R_{rs}(\lambda, 0+)$ is the remote-sensing reflectance, and 0+ denotes the water surface. Detailed information on the remote-sensing spectra is presented in [38].

Table 1 presents a summary of the field campaigns. Four airborne monitoring campaigns along the Geum River were implemented simultaneously with the ground-based monitoring. The airborne campaigns were performed by ASIA Aero Survey co., Ltd. (Seoul, Korea) using an AISA Dual airborne hyperspectral sensor. The sensor direction is perpendicular to the ground. The altitude of the aircraft was 3 km and the flying time was between two and three hours, beginning at 8:30 a.m. The hyperspectral image has 127 wavelength bands from 404 nm to 996 nm. The image has a spectral resolution of 4 nm to 5 nm and a spatial resolution of 2 m × 2 m. This study applied atmospheric correction using ATCOR 4, MODTRAN 6, and an ANN.

Table 1. Monitoring and experimental data acquisition.

Date	Point	Airborne Campaign	Min/Max PC Concentration *	Min/Max Chl-a Concentration *	Min/Max PC:Chl-a	Min/Max TSS Concentration **
12 August 2016	18	Implemented	6.25/150.90	14.19/111.40	0.32/1.91	6.27/40.14
24 August 2016	19	Implemented	12.48/100.00	25.95/61.44	0.28/2.72	10.13/23.33
20 September 2016	17	Implemented	0.83/1.64	11.85/60.88	0.025/0.089	11.47/19.33
14 October 2016	20	Implemented	0.19/0.88	13.74/46.17	0.0062/0.047	13.60/19.60

* Unit of PC and *chlorophyll-a* (Chl-a) concentration is mg m^{-3} and ** unit of total suspended solids (TSS) is mg L^{-1}.

2.2.3. Atmospheric Correction

ATCOR is a commercial software package that was developed during the 1990s [45,68]. The main features of this software are correction of the topographic and adjacency effect and spectral smoothing [45]. ATCOR 4 is a user-friendly software for atmospheric correction of HSI. Its ease of use stems from the straightforward and fast simulation [69].

MODTRAN was developed by Spectral Science, Inc. (Burlington, MA, USA) and the Air Force Research Laboratory (AFRL) [47]. The MODTRAN code solves the radiative transfer function to generate physical parameters related to atmospheric correction such as transmittance and spherical albedo. MODTRAN version 6 has a graphical user interface (GUI), making this software user friendly.

This study implemented atmospheric correction using the ANN to simulate ρ_{surf}. Detailed descriptions of the atmospheric correction using ATCOR 4, MODTRAN 6, and ANN are presented in Appendix A in the Supplementary Material.

2.2.4. Bio-Optical Algorithms for Determination of PC and Chl-a Concentration

This study estimated PC and Chl-a concertation using four bio-optical algorithms, two AOP algorithms, and two IOP algorithms, as follows.

AOP Algorithm

The AOP algorithm is based on the following remote-sensing reflectance:

$$R_{rs}(\lambda) = 0.54\left(\frac{f}{q}\frac{b(\lambda)}{a(\lambda)+b(\lambda)}\right) \tag{3}$$

where $R_{rs}(\lambda)$ is the remote-sensing reflectance on the water surface, f is the geometric light factor, q is the light distribution factor, $b(\lambda)$ is the backscattering coefficient, and $a(\lambda)$ is the absorption coefficient.

This study used either two or three bands to estimate PC and Chl-a concentration. The first algorithm used was the two-band ratio algorithm [17,37,70–72], which is referred to as band ratio (2) in this text. Band ratio (2) estimates the PC concentration as follows:

$$PC\ (\text{mg m}^{-3}) \propto \frac{R_{rs}(708)}{R_{rs}(622)} \tag{4}$$

$$Chl-a\ (\text{mg m}^{-3}) \propto \frac{R_{rs}(708)}{R_{rs}(660)} \tag{5}$$

where $R_{rs}(708)$ is the reflectance at 708 nm, $R_{rs}(660)$ is the reflectance at 660 nm, and $R_{rs}(622)$ is the reflectance at 622 nm.

The second algorithm used was the three-band ratio algorithm [28], which is referred to as the band ratio (3) in this manuscript. PC concentration was estimated by band ratio (3) using the following equations:

$$PC\ (\text{mg m}^{-3}) \propto \left(\frac{1}{R_{rs}(622)} - \frac{1}{R_{rs}(708)}\right)\cdot R_{rs}(755) \tag{6}$$

$$Chl-a\ (\text{mg m}^{-3}) \propto \left(\frac{1}{R_{rs}(660)} - \frac{1}{R_{rs}(708)}\right)\cdot R_{rs}(755) \tag{7}$$

where $R_{rs}(755)$ is the reflectance at 755 nm.

IOP Algorithm

The IOP algorithm mainly uses absorption and $b(\lambda)$ by rearranging Equation (4) in terms of $a(\lambda)$. Many studies have used the ratio form to retrieve $a(\lambda)$ because this allows the removal of geometric and ambient light effects, assuming these effects are independent of wavelength [29]. The $a(\lambda)$ equation is classified according to whether $b(\lambda)$ is wavelength-dependent or not. Simis et al. [31] and Duan et al. [37] suggested the following formulation for the $a(\lambda)$ equation with a wavelength-independent $b(\lambda)$:

$$a(\lambda_a) = \frac{R_{rs}(\lambda_w)}{R_{rs}(\lambda_a)}(a(\lambda) - b) - b \tag{8}$$

where λ_a is the wavelength for the phytoplankton pigment (i.e., Chl-a or PC) and λ_w is the wavelength for water.

Li et al. [29,30] introduced the following definition of $a(\lambda)$ with a wavelength-dependent absorption coefficient:

$$a(\lambda_a) = \left(\frac{R_{rs}(\lambda_w)b(\lambda_a)(a_w(\lambda_w) + b(\lambda_w))}{R_{rs}(\lambda_a)b(\lambda_w)}\right) - b(\lambda_a) - a_w(\lambda_a) \tag{9}$$

where a_w is the absorption coefficient of water referred to by [73]. The expression for the PC and Chl-a concentrations uses both the absorption coefficient and the specific absorption coefficient as follows:

$$PC\ (\text{mg m}^{-3}) = \frac{a(\lambda_{pc})}{a^*(\lambda_{pc})} \tag{10}$$

$$Chl - a\ (\mathrm{mg\ m^{-3}}) = \frac{a(\lambda_{chl-a})}{a^*(\lambda_{Chl-a})} \quad (11)$$

where λ_{pc} is the PC wavelength, $a^*(\lambda_{pc})$ is the specific absorption coefficient of PC ($\mathrm{m^2\ mg^{-1}}$), λ_{chl-a} is the Chl-a wavelength, and $a^*(\lambda_{chl-a})$ is the specific absorption coefficient of Chl-a ($\mathrm{m^2\ mg^{-1}}$). A more detailed description of the IOP algorithm for the determination of PC and Chl-a can be found in [29–32].

The IOP algorithm requires many empirical parameters as well as wavelength bands that accurately reflect optical properties of the water body. This study optimized the Simis algorithm and Li algorithm using the multi-objective optimization, resulting in 622 nm for PC and 660 nm for Chl-a [38]. The band ratio (2), band ratio (3), Simis, and Li algorithms were applied to images which had atmospheric correction completed by MODTRAN 6, ATCOR 4, or ANN.

2.3. Performance Evaluation

Nash-Sutcliffe efficiency (NSE) and the root mean square error (RMSE) were used to evaluate the performances of the atmospheric correction, optimized bio-optical algorithm, and ANN simulation following Equations (12) and (13),

$$\mathrm{NSE_x} = 1 - \frac{\sum \left(X_{x,pre} - X_{x,\,obs}\right)^2}{\sum \left(X_{x,\,obs} - X_{x,\,obs}^{avg}\right)^2} \quad (12)$$

$$\mathrm{RMSE_x} = \sqrt{\frac{\sum \left(X_{x,pre} - X_{x,\,obs}\right)^2}{n}} \quad (13)$$

$$\mathrm{Bias_x} = \frac{\sum \left(X_{x,pre} - X_{x,\,obs}\right)}{n} \quad (14)$$

where $X_{x,pre}$ is the predicted value, $X_{x,\,obs}$ is the observed value, $X_{x,\,obs}^{avg}$ is the average observed value, and x represents either the reflectance ($\mathrm{sr^{-1}}$) value or PC and Chl-a concentrations ($\mathrm{mg\ m^{-3}}$).

3. Results

3.1. Algal Variation in the Baekje Weir

Figure 3 shows the temporal variation in PC, Chl-a, PC:Chl-a values, and TSS during the sampling period. The measured PC and Chl-a ranged from 0.19 to 150.90 mg $\mathrm{m^{-3}}$ and from 11.85 to 111.40 mg $\mathrm{m^{-3}}$, respectively (Table 1). Thus, the PC:Chl-a value ranged from 0.0062 to 2.72. The high PC concentration resulted in a high PC:Chl-a value (Figure 3). In addition, the observed concentration of TSS measured from 6.27 to 40.14 mg $\mathrm{L^{-1}}$.

On 12 and 24 August 2016, the PC, Chl-a, and TSS concentrations were relatively high compared to those of the other sampling events. On 14 September and 20 October 2016, the PC concentration was near 0, but Chl-a maintained a relatively high concentration ranging from 11.85 to 60.88 mg $\mathrm{m^{-3}}$. The TSS concentration was also maintained between 11.36 mg $\mathrm{L^{-1}}$ and 19.60 mg $\mathrm{L^{-1}}$.

3.2. Performance of Atmospheric Correction Techniques

Figure 4 and Figure S1 show the atmospheric correction results achieved using MODTRAN 6 and ATCOR 4. The averaged spectra showed good agreement with *in-situ* reflectance (Figure 4a–d) while the correlation between the corrected reflectance of the individual bands and the in-situ reflectance was concentrated along the 1:1 line (Figure 4e–h). In contrast, the reflectance spectra corrected using ATCOR 4, shown in Figure S1, are four orders of magnitude smaller than the in-situ reflectance and are less correlated with the observed reflectance. The overall atmospheric correction performance of both models is presented in Table S4. Most NSE values of MODTRAN 6 were higher than 0.8. In addition, the RMSE values for the MODTRAN 6 results were smaller than 0.0034 $\mathrm{sr^{-1}}$.

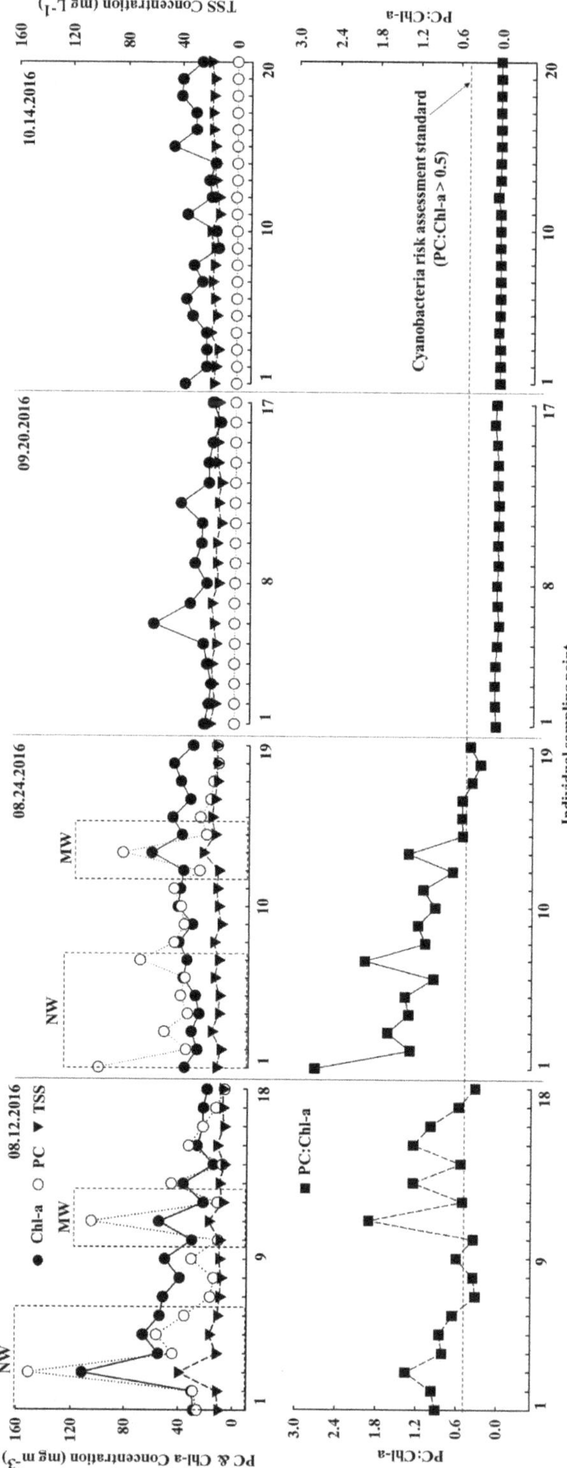

Figure 3. Variation in PC, Chl-a, TSS, and PC:Chl-a in the Baekje Reservoir. NW indicates sampling points near Baekje Weir and MW indicates sampling point in the middle of Baekje Weir.

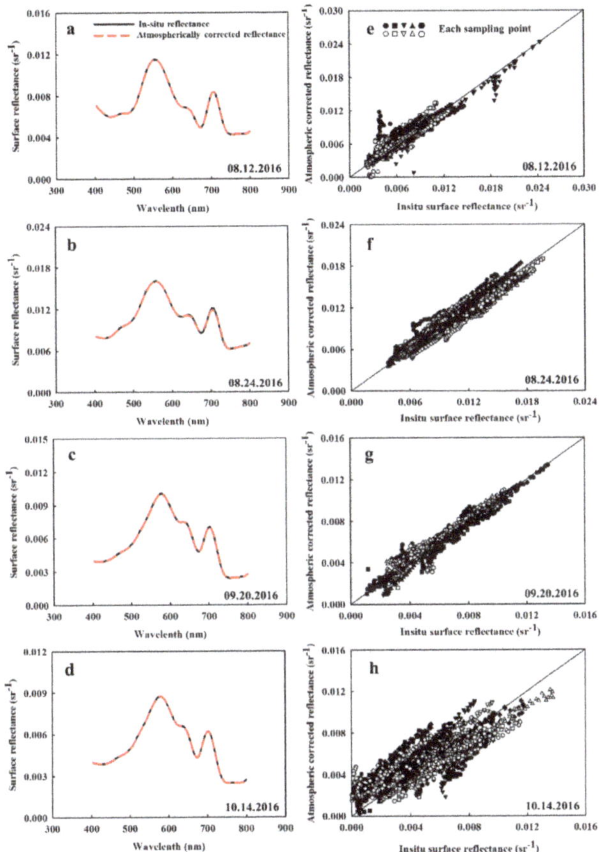

Figure 4. Atmospheric correction results using MODTRAN 6. Panels (**a**–**d**) show the average in-situ and corrected surface reflectance ρ_{surf}, respectively. Panels (**e**–**h**) show the correlation between the observed and corrected results at different wavelengths for each sampling point.

All reflectance corrected by the ATCOR 4 values had negative NSE values (Table S4) while their RMSE values were higher than those of the MODTRAN 6 results. The average error of the MODTRAN 6 and ATCOR 4 reflectance across all four sampling events is shown in Figure S3. The reflectance error of MODTRAN 6 was less than 30%. However, the errors in the wavelength ranges of $\lambda < 500$ nm and $\lambda > 700$ nm were higher than the errors in the other wavelength bands (Figure S3a–d, Table 2). In particular, the corrected reflectance at 439 nm, 445 nm, 755 nm, and 779 nm had higher errors than the other bands. Similar to MODTRAN 6, the errors of ATCOR 4 increased when the wavelength was less than 500 nm and greater than 700 nm (Figure S3e–h). The error of the ATCOR 4 correction was between 99% and 100%, which was three times higher than that of the MODTRAN 6 correction results. The simulated reflectance from the ANN model is shown in Figure S2. The simulation has an NSE value of 0.79 while the error was largely between 10% and 50% (Table 2). The reflectance error of the ANN simulation was greater than that of the MODTRAN 6 simulation but was less than that of the ATCOR 4. In addition, all three different methods did not show good performance for the imagery taken on 14 October 2016.

Table 2. Errors of atmospherically corrected reflectance.

	Reflectance Error (%)											
	12 August 2016			24 August 2016			20 September 2016			14 October 2016		
Band	MOD *	ATC **	ANN	MOD *	ATC **	ANN	MOD *	ATC **	ANN	MOD *	ATC **	ANN
439 nm	16.50	99.97	18.74	13.19	99.95	21.21	20.40	99.96	34.43	141.23	99.67	190.50
443 nm	15.54	99.97	18.20	13.09	99.95	21.54	15.51	99.96	31.41	115.49	99.66	151.80
534 nm	5.91	99.97	12.40	8.31	99.92	8.73	3.12	99.94	11.00	26.40	99.63	35.47
599 nm	8.03	99.97	11.78	9.64	99.94	10.85	3.17	99.92	9.22	22.27	99.72	25.72
618 nm	7.49	99.97	12.69	9.91	99.95	11.75	2.95	99.94	8.25	23.50	99.78	28.81
622 nm	7.26	99.97	13.47	9.65	99.95	12.74	2.58	99.94	8.62	23.76	99.79	30.19
627 nm	7.75	99.97	13.36	9.70	99.95	11.56	3.26	99.94	12.53	24.30	99.80	33.91
660 nm	9.16	99.97	14.73	10.39	99.96	11.15	2.96	99.96	12.00	29.70	99.86	40.24
674 nm	8.23	99.97	13.09	10.99	99.96	13.22	6.07	99.97	10.79	36.34	99.88	52.74
708 nm	7.25	99.97	13.86	7.15	99.94	10.16	3.04	99.95	10.23	26.65	99.87	28.21
755 nm	21.47	99.97	23.30	9.62	99.97	15.83	6.63	99.99	16.14	277.45	99.95	373.29
779 nm	22.10	99.96	23.15	9.23	99.97	19.76	5.48	99.98	16.96	373.08	99.95	508.48

* and ** indicate MODTRAN 6 and ATCOR 4, respectively.

3.3. Performance of the Bio-Optical Algorithm

Figures 5 and 6 show the results of the bio-optical algorithm for estimating PC and Chl-a, respectively. Figures S4 and S5 show the performance of the absorption coefficient with respect to PC and Chl-a estimation, respectively. Multi-objective optimization of the IOP algorithm was conducted using the observed reflectance data. The optimized parameters were applied to build the IOP algorithm using the reflectance data that had been atmospherically corrected using MODTRAN 6, ATCOR 4, or the ANN simulation (Table 3). The reflectance corrected by MODTRAN 6 showed good agreement with the observed PC concentration with R^2 values ranging from 0.68 to 0.77. The R^2 values of the Chl-a algorithms ranged from 0.49 to 0.53.

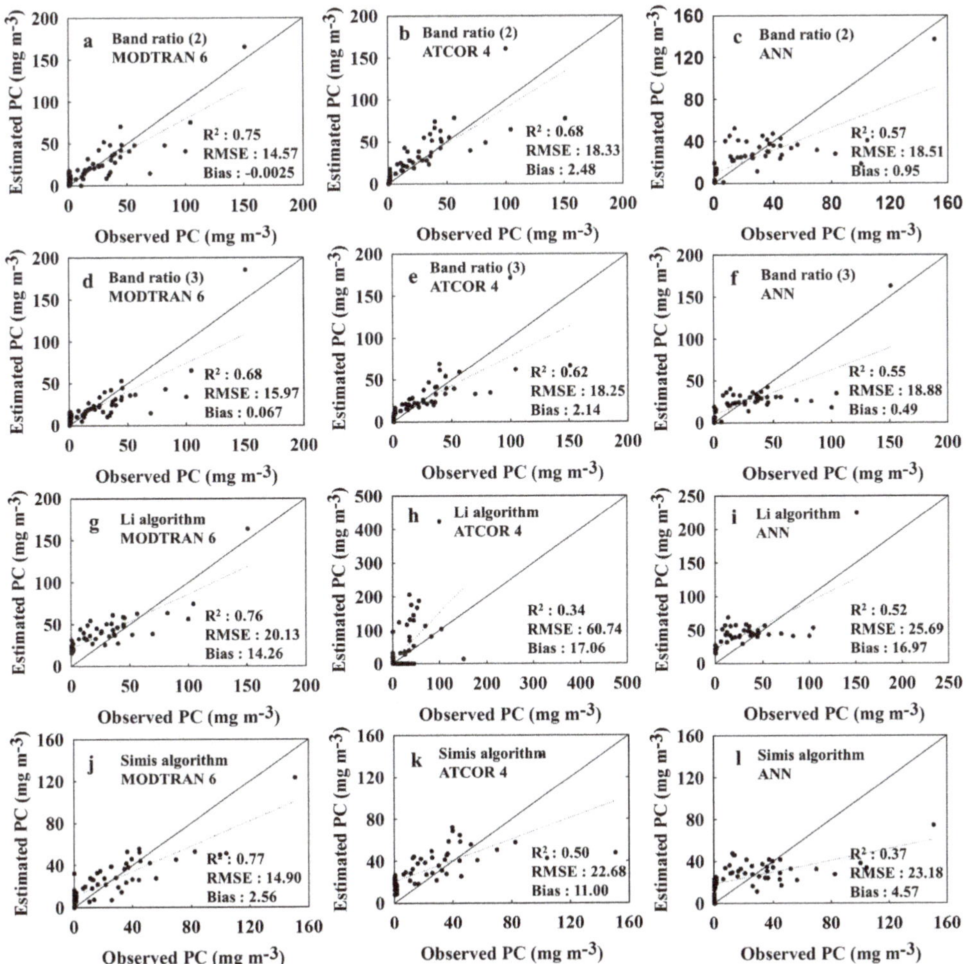

Figure 5. Optimized *PC* algorithm results with respect to in-situ and atmospherically corrected reflectance. Panels (**a–c**) show the band ratio (2) results. Panels (**d–f**) show the band ratio (3) results. Panels (**g–i**) show the Li algorithm results. Panels (**j–l**) show the Simis algorithm results.

Table 3. Optimized algorithm performance.

PC	MODTRAN 6			ATCOR 4			ANN		
	R^2	RMSE	Bias	R^2	RMSE	Bias	R^2	RMSE	Bias
Band (2)	0.75	14.57	−0.0025	0.68	18.33	2.48	0.57	18.51	0.95
Band (3)	0.68	15.97	0.0673	0.62	18.25	2.14	0.55	18.88	0.49
Li	0.76	20.13	14.26	0.34	60.74	17.06	0.56	25.69	16.97
Simis	0.77	14.90	2.56	0.50	22.68	11.00	0.37	23.18	4.57
Chl-a	R^2	RMSE	Bias	R^2	RMSE	Bias	R^2	RMSE	Bias
Band (2)	0.49	12.24	−5.44	0.29	13.91	−4.94	0.46	13.03	−6.54
Band (3)	0.51	11.03	−3.01	0.25	13.63	−2.90	0.56	11.09	−4.35
Li	0.53	10.62	−1.71	0.025	156.73	150.82	0.52	10.90	−2.30
Simis	0.53	10.88	−1.20	0.29	13.17	−2.08	0.53	11.19	−1.96

* Unit of root mean square error (RMSE) and bias of *PC* and *Chl-a* is mg m^{-3}.

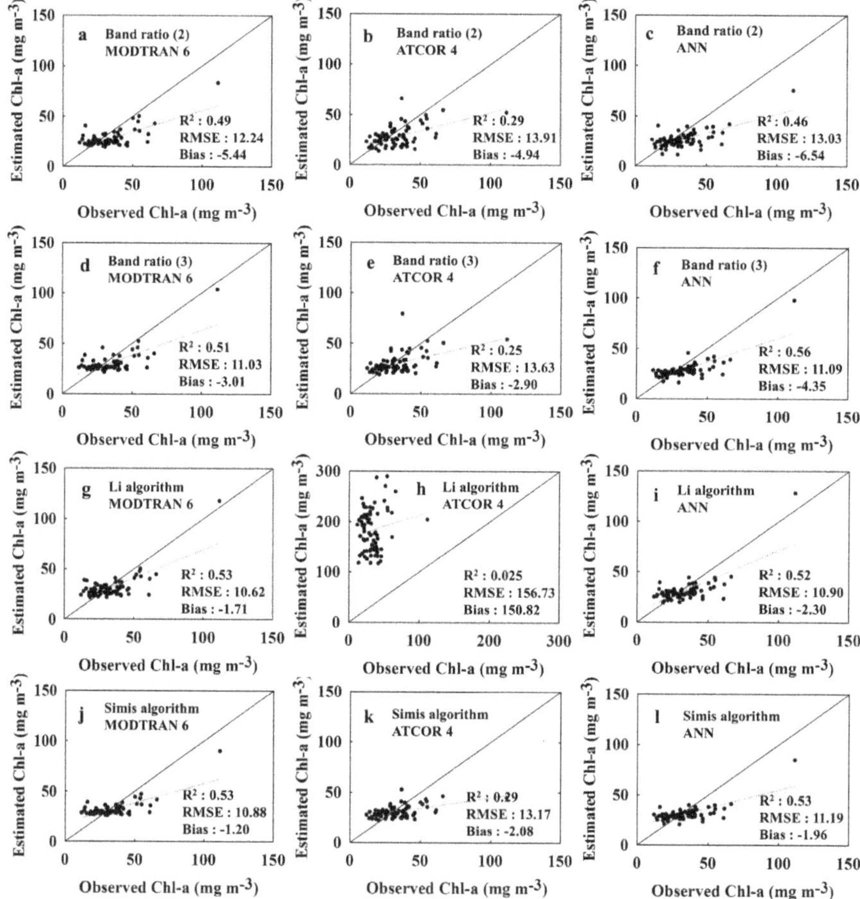

Figure 6. Optimized *Chl-a* algorithm results with respect to in-situ and atmospherically corrected reflectance. Panels (**a**–**c**) show the band ratio (2) results. Panels (**d**–**f**) show the band ratio (3) results. Panels (**g**–**i**) show the Li algorithm results. Panels (**j**–**l**) show the Simis algorithm results.

The PC and Chl-a algorithm results from ATCOR 4 showed a lower accuracy than MODTRAN 6-based results in terms of R^2 and RMSE (second column in Figures 5 and 6). Based on the ANN simulation (fourth column in Figures 5 and 6), the R^2 values of the PC and Chl-a algorithms ranged from 0.37 to 0.57 and 0.46 to 0.56, respectively.

Among the PC algorithms, the Simis algorithm with MODTRAN 6 correction showed the highest accuracy with an R^2 value of 0.77, an RMSE of 14.90 mg m^{-3}, and a bias of 2.56 mg m^{-3} (Figure 5j). This resulted from a good agreement of the estimated absorption coefficient with the observed coefficient (Figure S4a). For Chl-a estimation, the Simis algorithm with MODTRAN correction showed the highest performance with an R^2 value of 0.53, an RMSE of 10.88 mg m^{-3}, and a bias of −1.20 mg m^{-3} (Figure 6g). This proved the accurate estimation of the absorption coefficient (Figure S5d). Both the Li and Simis algorithms overestimated the PC and Chl-a concentrations when both concentrations were below 25 mg m^{-3} (third and fourth row in Figures 5 and 6). Under atmospheric correction by ATCOR 4, the Li algorithm was not responsive to PC and Chl-a concentration. This resulted in the lowest R^2 values of 0.34 and 0.025 and the highest RMSE values of 60.74 mg m^{-3} and 156.73 mg m^{-3}, as well as the highest biases of 17.06 mg m^{-3} and 150.82 mg m^{-3}, respectively.

3.4. PC and Chl-a Distribution Map

Following MODTRAN 6, ATCOR 4, and ANN correction, the spatial distribution of PC and Chl-a showed a similar pattern (Figures 7–14). However, the PC and Chl-a concentration obtained from the reflectance data corrected using MODTRAN 6 was relatively high compared to the concentration obtained from reflectance data corrected using ATCOR 4 and ANN. A distinctive spatial distribution of high PC and Chl-a concentration was observed in Section 1 on 12 August 2016. In addition, a high PC and Chl-a concentration level was distributed along the left edge of the river on 12 and 24 August 2016.

The IOP algorithms showed higher concentrations of PC and Chl-a than those of the AOP algorithm in terms of MODTRAN 6, ATCOR 4, and ANN correction. However, the Simis algorithm with ANN correction did not correctly estimate the PC distribution (Figures 7k and 8k). The PC concentrations of the Li algorithm with MODTRAN 6 correction were underestimated in the area highlighted by the dotted circle (i.e., region 1 in Figure 7c); however, the Li algorithm with ATCOR 4 correction showed the opposite Chl-a concentration pattern compared to the other results (Figures 11g and 12g). On the other sampling dates (20 September and 14 October 2016), the PC and Chl-a concentrations and spatial distribution were fairly constant (Figures S6–S13). The band ratio (2) algorithm produced a reasonable concentration range of spatial distributions for PC during these sampling events. In contrast, the IOP algorithms using reflectance data corrected by MODTRAN 6, ATCOR 4, and ANN showed higher PC and Chl-a concentrations than those of the AOP algorithms. Most algorithms showed concentrations ranging from 10 to 40 mg m^{-3} for the Chl-a estimations.

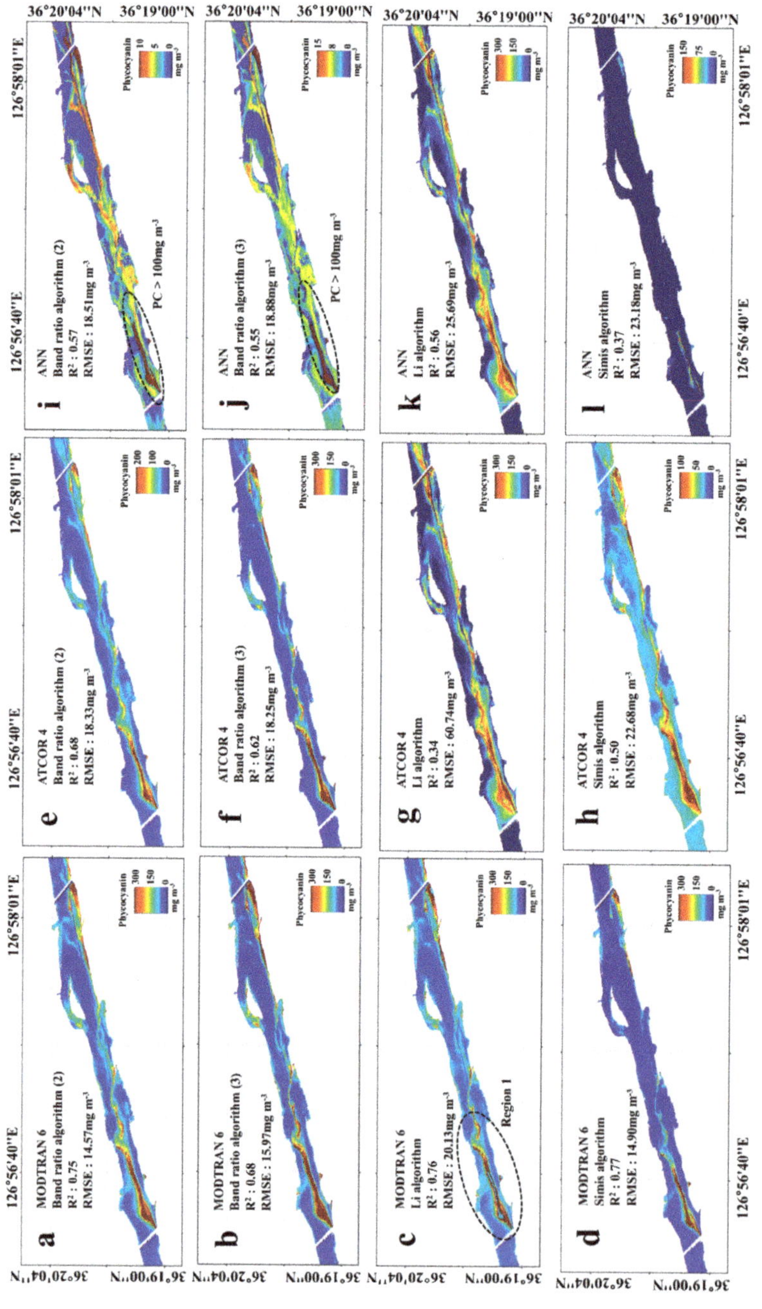

Figure 7. *Phycocyanin* (PC) concentration images on 12 August 2016 in Section 1. Panels (**a**–**d**) show the PC distribution driven by the MODTRAN 6 atmospheric correction. Panels (**e**–**h**) show the PC distribution driven by the ATCOR 4 atmospheric correction. Panels (**i**–**l**) show the PC distribution driven by the ANN atmospheric correction.

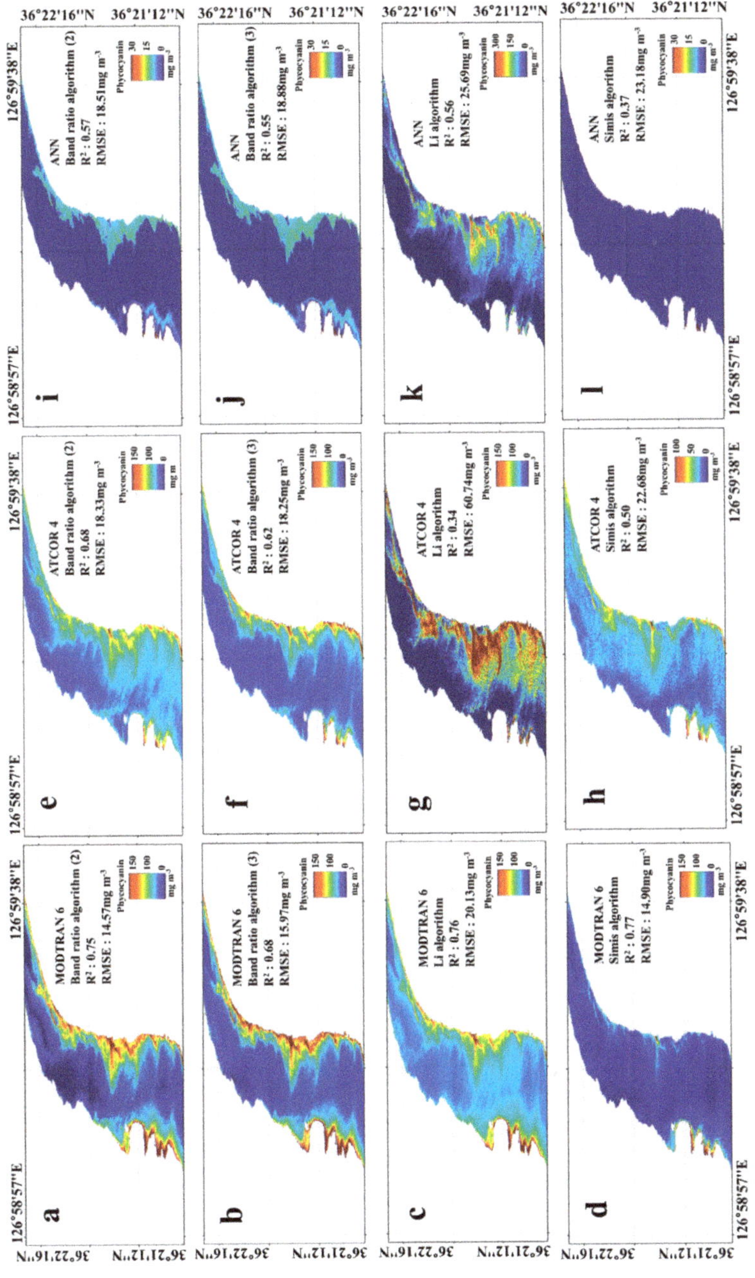

Figure 8. *Phycocyanin* (PC) concentration images on 12 August 2016 in Section 2. Panels (**a**–**d**) show the PC distribution driven by the MODTRAN 6 atmospheric correction. Panels (**e**–**h**) show the PC distribution driven by the ATCOR 4 atmospheric correction. Panels (**i**–**l**) show the PC distribution driven by the ANN atmospheric correction.

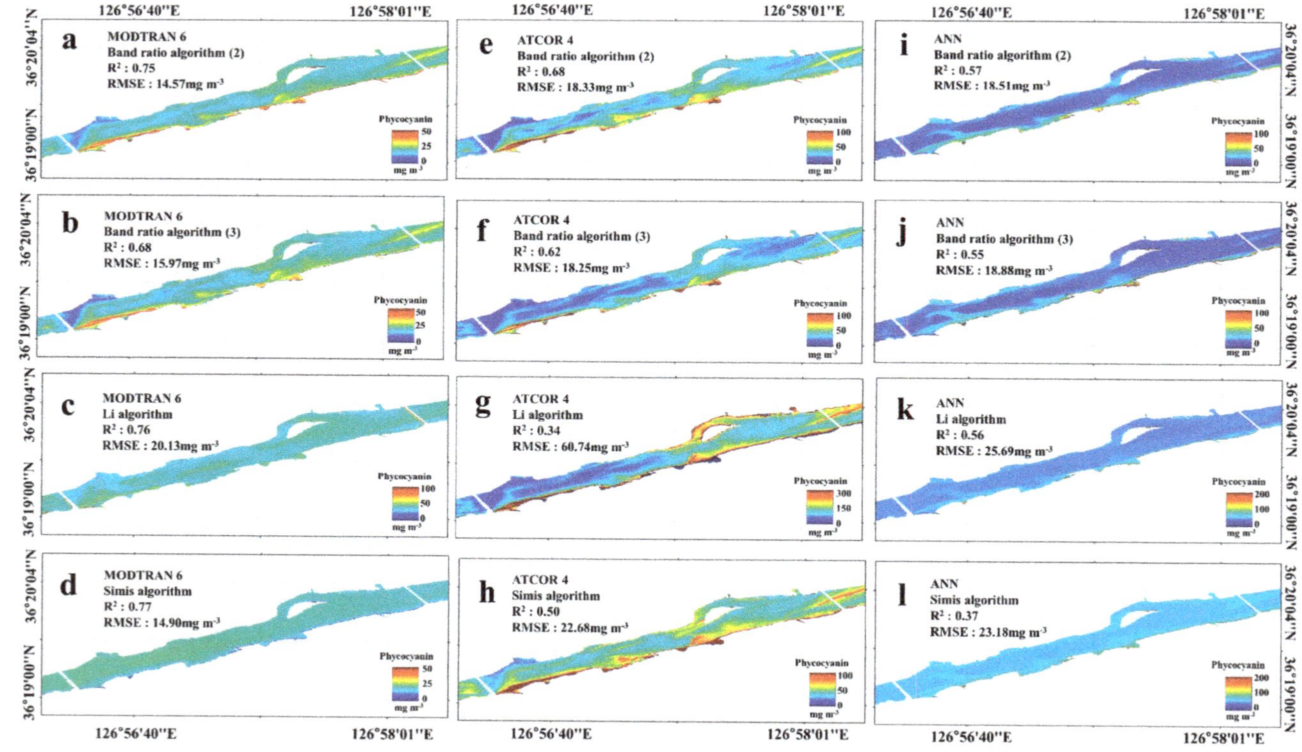

Figure 9. *Phycocyanin* (PC) concentration images on 24 August 2016 in Section 1. Panels (**a–d**) show the PC distribution driven by the MODTRAN 6 atmospheric correction. Panels (**e–h**) show the PC distribution driven by the ATCOR 4 atmospheric correction. Panels (**i–l**) show the PC distribution driven by the ANN atmospheric correction.

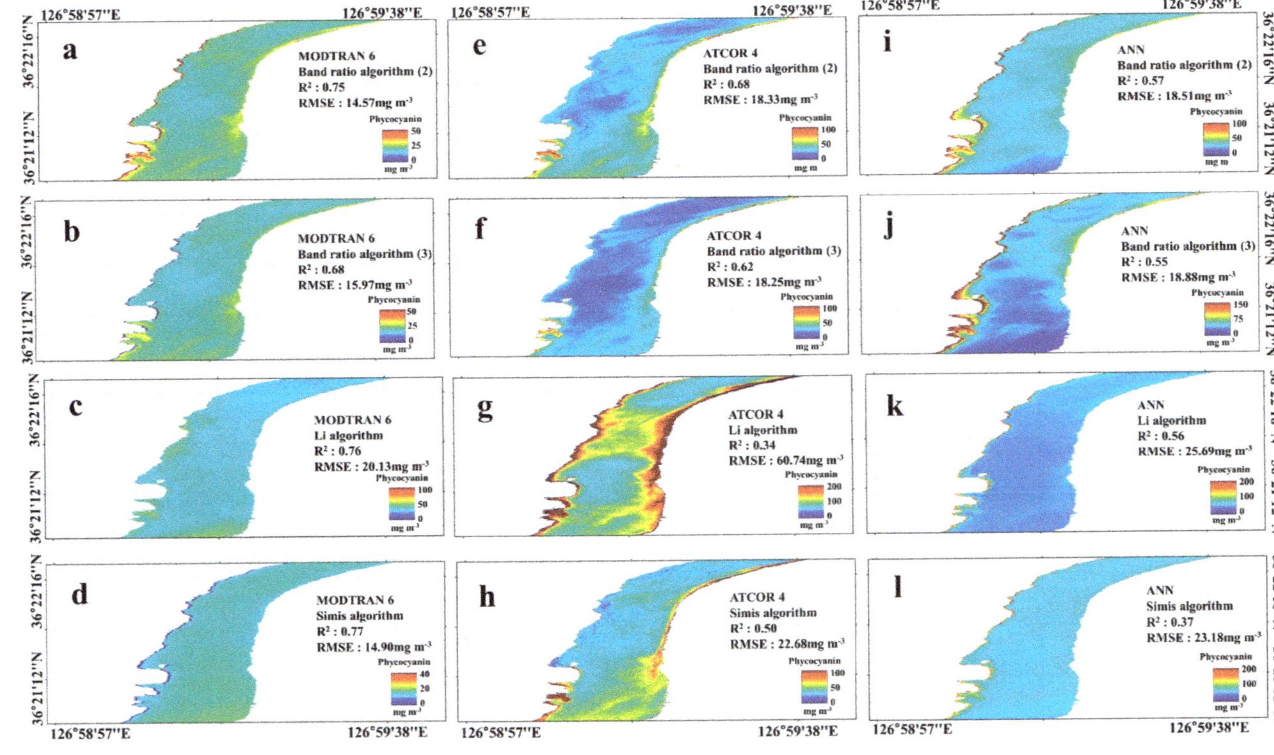

Figure 10. *Phycocyanin* (PC) concentration images on 24 August 2016 in Section 2. Panels (**a–d**) show the PC distribution driven by the MODTRAN 6 atmospheric correction. Panels (**e–h**) show the PC distribution driven by the ATCOR 4 atmospheric correction. Panels (**i–l**) show the PC distribution driven by the ANN atmospheric correction.

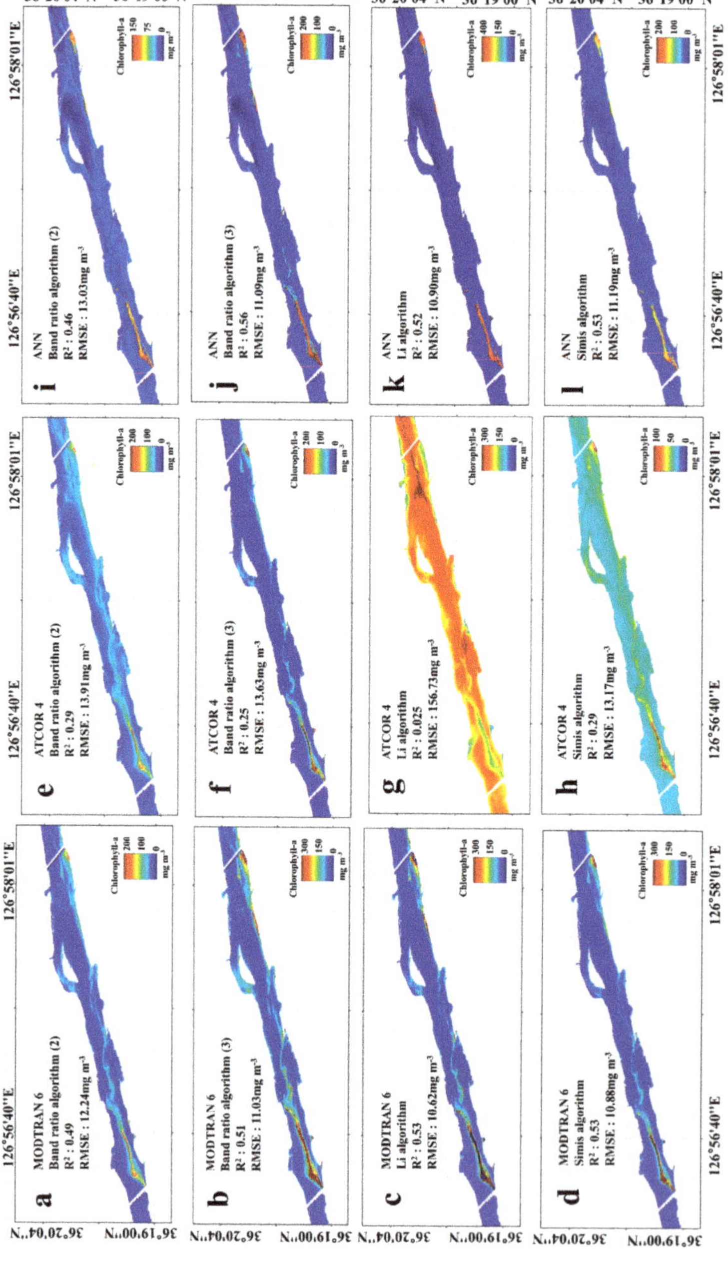

Figure 11. *Chlorophyll-a* (Chl-a) concentration images on 12 August 2016 in Section 1. Panels (**a–d**) show the Chl-a distribution driven by the MODTRAN 6 atmospheric correction. Panels (**e–h**) show the Chl-a distribution driven by the ATCOR 4 atmospheric correction. Panels (**i–l**) show the Chl-a distribution driven by the ANN atmospheric correction.

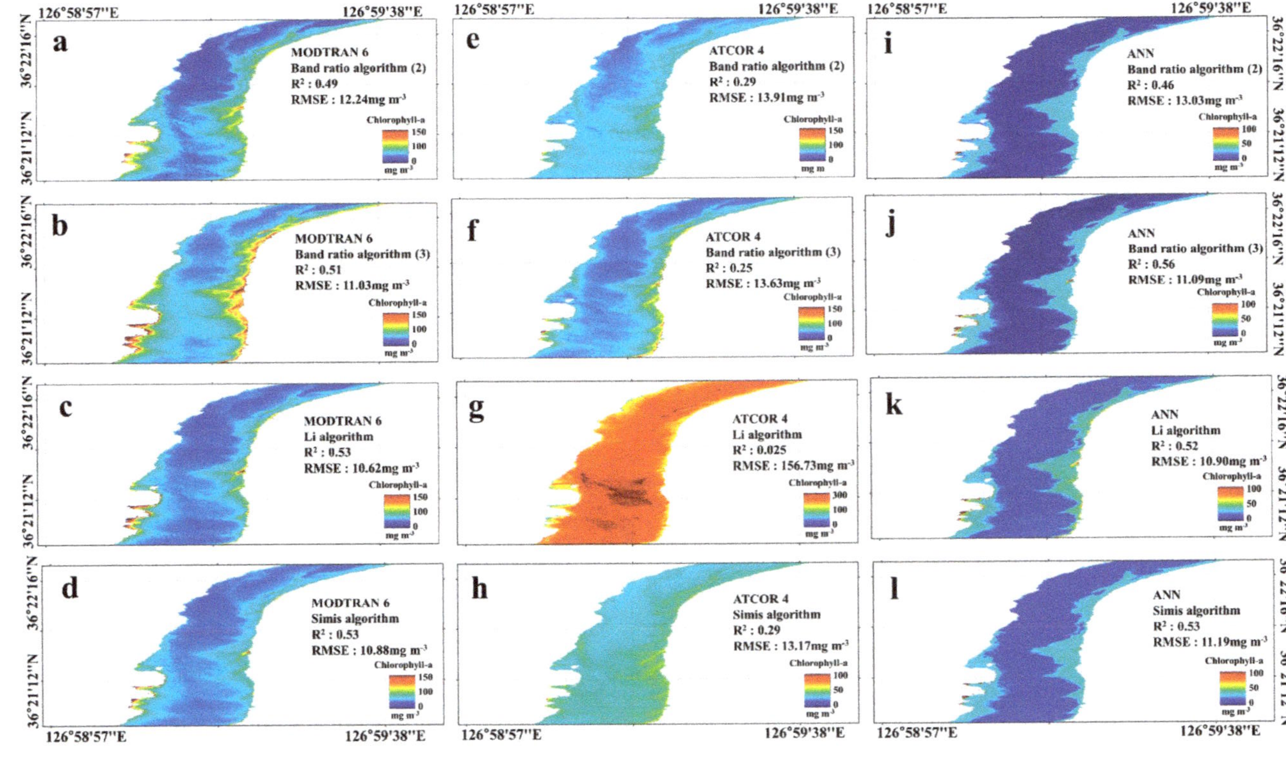

Figure 12. *Chlorophyll-a* (Chl-a) concentration images on 12 August 2016 in Section 2. Panels (**a–d**) show the Chl-a distribution driven by the MODTRAN 6 atmospheric correction. Panels (**e–h**) how the Chl-a distribution driven by the ATCOR 4 atmospheric correction. Panels (**i–l**) show the Chl-a distribution driven by the ANN atmospheric correction.

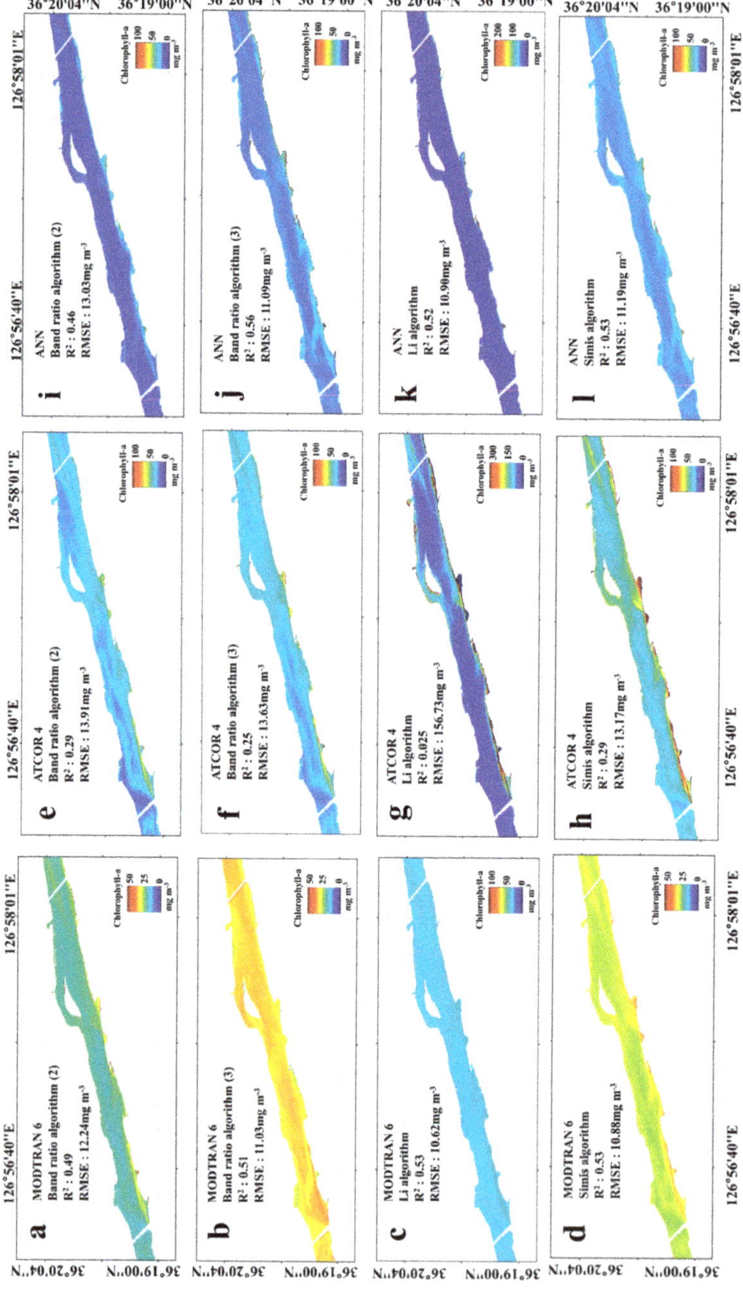

Figure 13. *Chlorophyll-a* (Chl-a) concentration images on 24 August 2016 in Section 1. Panels (**a**–**d**) show the Chl-a distribution driven by the MODTRAN 6 atmospheric correction. Panels (**e**–**h**) show the Chl-a distribution driven by the ATCOR 4 atmospheric correction. Panels (**i**–**l**) show the Chl-a distribution driven by the ANN atmospheric correction.

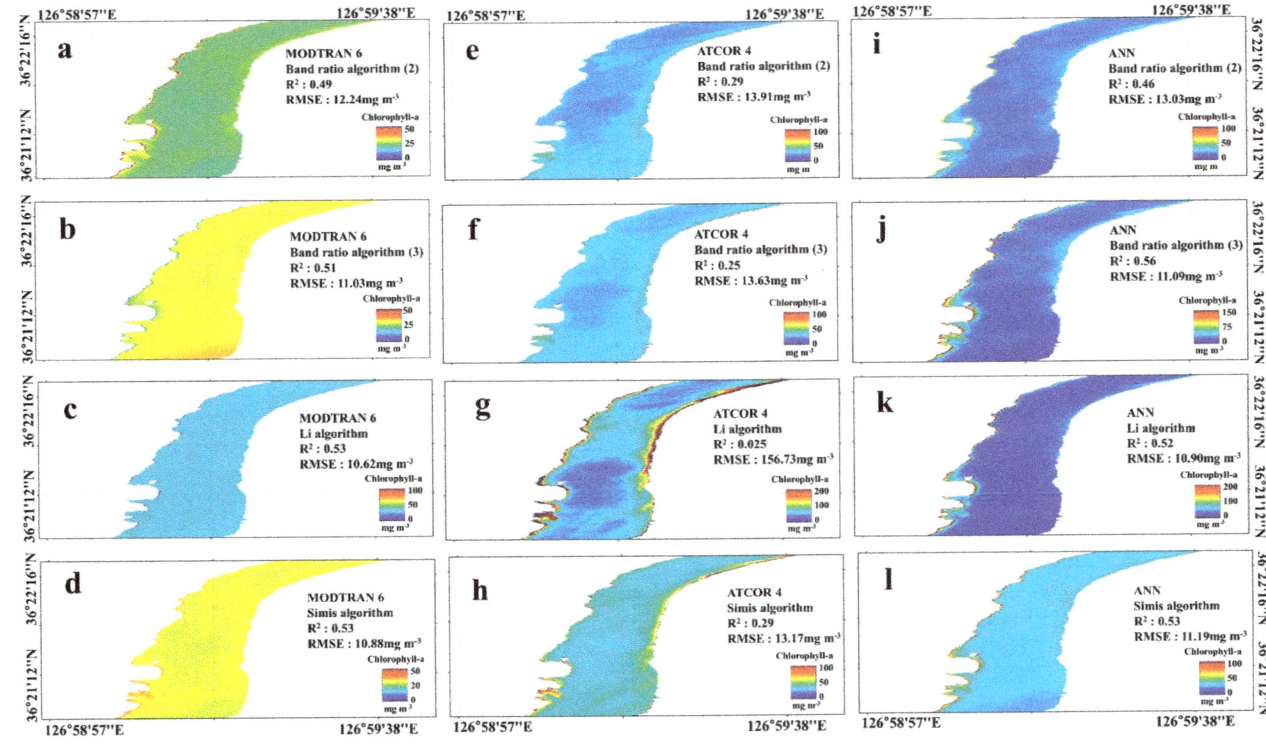

Figure 14. *Chlorophyll-a* (Chl-a) concentration images on 24 August 2016 in Section 2. Panels (**a**–**d**) show the Chl-a distribution driven by the MODTRAN 6 atmospheric correction. Panels (**e**–**h**) show the Chl-a distribution driven by the ATCOR 4 atmospheric correction. Panels (**i**–**l**) show the Chl-a distribution driven by the ANN atmospheric correction.

4. Discussion

4.1. Variation in Algae in the Baekje Reservoir

The HSI images taken on 12 and 24 August 2016 succinctly identified a cyanobacteria-dominant bloom in the reservoir. Most of the PC:Chl-a values on 12 and 24 August 2016 were observed to be greater than 0.5, which is a standard for assessing whether the cyanobacteria bloom is at risk. In particular, the PC:Chl-a of 12 and 24 August 2016 could be classified as medium risk because of the relatively high Chl-a concentrations [60].

The low PC concentration and relatively high Chl-a concentration on 20 September and 14 October 2016 signified an algal species succession from cyanobacteria to diatoms and green algae. This occurrence was mainly because of the observed water temperature between 18 °C and 22 °C, which is not a preferred growth condition for cyanobacteria [9]. The total number of cyanobacteria cells significantly decreased from 105,840 cells mL^{-1} to 23,840 cells mL^{-1} on 20 September and 14 October 2016, respectively, while the total cells of green algae and diatoms did not change substantially: 116,512 cells mL^{-1} (i.e., 30% green algae and 70% diatoms) on 20 September and 110,816 cells mL^{-1} (i.e., 30% green algae and 70% diatoms) on 14 October 2016.

4.2. Atmospheric Correction Performance

Overall, the atmospheric correction performance of MODTRAN 6 was acceptable, with an NSE value greater than 0.80. Although the averaged reflectance spectra of the MODTRAN 6 atmospheric correction was in good agreement with the in-situ spectra, the corrected reflectance result had 20–30% error in the blue and green bands (i.e., $\lambda < 500$ nm) and the near infrared bands (i.e., $\lambda > 700$ nm) (Figure S3). In addition to the uncertainty of the in-situ measurement, the reflectance errors caused by atmospheric correction have been documented by several authors. Bernstein et al. [52] reported that reflectance corrected using MODTRAN differed from the observations because of the lack of in-situ reflectance data as well as difficulties in the elimination of absorption and scattering properties in the atmospheric correction. Gao et al. [53] emphasized that retrieval of aerosol information allowed description of the absorption and scattering properties, which could be used to reduce the reflectance error in wavelength regions where $\lambda < 500$ nm and $\lambda > 700$ nm. Adler-Golden et al. [74] highlighted that the poor performance of the atmospheric correction method was driven by the high water column which increased the fractional error of the reflectance bands because of the combined effects of atmospheric absorption and scattering. Hunter et al. [33] noted that intensive scattering of aerosol and water vapor resulted in poor correction performance in the wavelength regions of $\lambda < 500$ nm and $\lambda > 750$ nm.

Similarly, the atmospheric correction performance of ATCOR 4 in this study may have been affected by the limited available data on the atmospheric conditions during the measurement campaigns. Hadjimitsis et al. [75] insisted that given sufficient data on the atmospheric conditions at the time of measurement, molecular absorption and scattering in the atmosphere could be described, resulting in accurate atmospheric correction using a physically based model such as ATCOR. In this study, the inaccuracy of the MODTRAN 6 atmospheric correction in certain reflectance bands and the poor performance of ATCOR 4 may have been caused by the lack of available data on important atmospheric conditions such as water vapor column and aerosol optical depth, as L_{path} and S depend on the water vapor column, which might not have been well defined in this study (see Appendix A in the Supplementary Material). Uncertainty in the atmospheric parameters would have led to uncertainty being distributed throughout the image [22]. Another possible cause of the correction error was suggested by Matthews et al. [22], who insisted that the absence of Lambertian bidirectional reflectance distribution functions could negatively affect the accuracy of the corrected reflectance.

In addition, imperfect time-matching between ground-based and airborne monitoring may have caused distortion of the corrected reflectance because of changes in water vapor over time [33]. The outliers of the corrected reflectance were observed on 12 August 2016, because of the phytoplankton influence on the in-situ reflectance. When the massive phytoplankton occurred on the water surface, the reflectance spectra had higher values greater than 700 nm because of the increased scattering of the phytoplankton and lower values less than 500 nm because of the increased absorption of the phytoplankton [76]. Then, averaged atmospheric parameters might not consider this abnormal circumstance to estimate the surface reflectance. The corrected reflectance on 14 October 2016 was less concentrated along the 1:1 line than the reflectance results of the other sampling periods. A haze effect on the water surface might be a cause of the uncertainty of the in-situ reflectance measurement [77,78]. This might increase the scattering, which results in distorted measurement of in-situ reflectance.

The atmospheric correction using the ANN model in this study showed satisfactory performance during both the training and validation steps, which had NSE values of 0.80 and 0.76, respectively. Compared to previous studies, the authors of [51] applied an ANN to the atmospheric correction of Medium Resolution Imaging Spectrometer (MERIS) imagery to retrieve remote-sensing reflectance under the water conditions in case 2. Their model showed a high correlation between the in-situ and corrected reflectance. Schroeder et al. [50] atmospherically corrected a MERIS image in the water in case 1 that had a low RMSE value for the at-sensor radiance by using top-of-atmosphere radiance, humidity, and angle data. Goyens et al. [49] corrected atmospheric effects in a MODIS-Aqua image using an ANN model which achieved an R^2 value greater than 0.8.

4.3. Bio-Optical Algorithm Application

The MODTRAN 6 correction led to the higher performance of the AOP and IOP algorithms compared to that of the ATCOR 4 correction in both PC and Chl-a estimation (Figure 15 and Figure S14). However, the AOP algorithm with ATCOR 4 correction resulted in an acceptable estimation (Figure 5b,e) as it can compensate for magnitude differences by using simple ratios [79]. The low performance of the ANN simulation might have been because of an insufficient number of reflectance input data points. Goyens et al. [49] and Schroeder et al. [51] used over 10,000 and 30,000 data points, respectively, to construct an ANN model. They showed an acceptable correction accuracy with R^2 and RMSE values of 0.8 and $\pm$1.1 W m^{-2} μm^{-1} sr^{-1}, respectively. The IOP algorithms were directly influenced by the correction performance, because these algorithms directly use the corrected reflectance in various wavelength bands. This could be caused by monitoring uncertainty in remote-sensing reflectance data [30,80]. Thus, the IOP algorithms were more significantly affected by the performance of the atmospheric correction than the AOP algorithms because of the intervention of the various reflectance bands (Figure S14). IOP algorithms commonly overestimate low PC concentrations because of the difficulty in accurately measuring the optical intensity at low PC concentrations. Li et al. [30] found degraded performance for the Li and Simis algorithms in low PC conditions because of optical interference, for example, from colored dissolved organic matter.

Although Li et al. [30] improved the algorithmic results at low PC concentrations by considering the interference effect in the algorithm, the Li algorithm still overestimated the PC concentration under those conditions. This implies that, in the case of IOP algorithms, it might be difficult to correctly estimate lower PC concentrations using the combination of various reflectance bands. Thus, AOP algorithms would be a straightforward means to describe low PC concentrations and their spatial distribution.

Even though Chl-a estimates showed lower precision than those of the PC estimates, the MODTRAN 6 correction showed more accurate results of the bio-optical algorithms than those using reflectance data corrected by ATCOR 4 and ANN (first column in Figure 6). The poor performance of the Chl-a algorithms was mainly from the relatively low level of Chl-a concentrations on 20 September and 14 October 2016. During these sampling events, the influence of interference such as TSSs might have increased as the TSS concentration was maintained from 12 mg L^{-1} to 20 mg L^{-1} (Figure 3). This was proven by the nonlinear relationship between the particulate matter and Chl-a as suggested by Bricaud et al. [81], Garver et al. [82], and Yentsch and Phinney [83]. They found that as Chl-a concentration decreased, the particulate optical properties increased. The particulate interference eventually resulted in poor performance of the Chl-a algorithms.

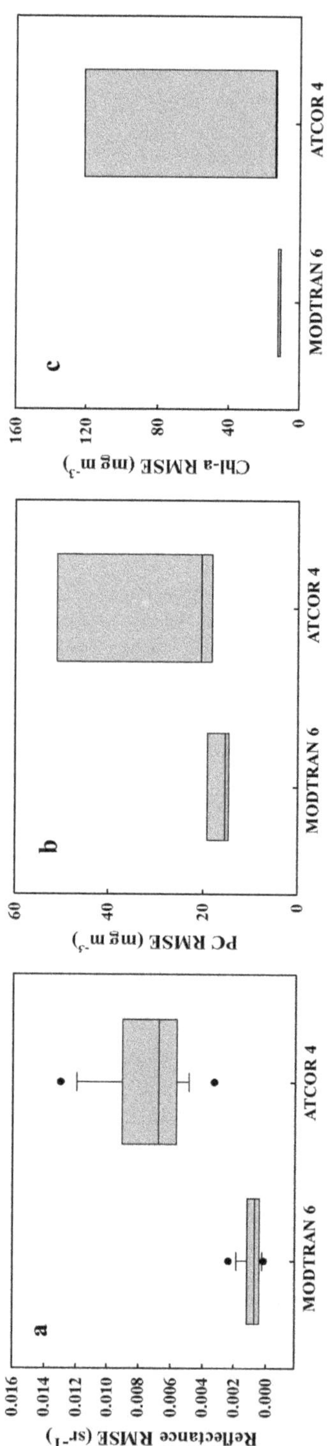

Figure 15. Influence of atmospheric correction using MODTRAN 6 and ATCOR 4 on the PC and Chl-a quantification. (**a–c**) are error of atmospheric correction, PC, and Chl-a, respectively, in terms of MODTRAN 6 and ATCOR 4.

4.4. Spatial Distribution Map of Algal Concentration

Under the MODTRAN 6 correction, the Li algorithm was able to describe the spatial distribution of PC appropriately on 12 August 2016. However, it still underestimated the PC concentration in the middle of region 1 because of the unstably corrected reflectance in bands greater than 700 nm (Figure 7c). In addition, the Chl-a distribution of the Li algorithm showed reverse concentration pattern because low reflectance values greater than 700 nm decreased $b(\lambda_w)$, which resulted in an abnormally high Chl-a estimation (Equation (9)).

In the middle of the river in Section 1 on 12 August (Figures 7 and 11), the gate operation for the hydropower plant caused a tailed shape of the PC and Chl-a distribution [9]. In this near-Baekje Weir region, high concentrations of PC and Chl-a were observed. This resulted in a high PC:Chl-a value, which resulted in a caution index for the cyanobacteria dominant bloom (first row in Figure S15). Likewise, the PC:Chl-a value on 24 August 2016 was high near the Baekje Weir region (second row in Figure S15). Thus, this near-Baekje Weir region could be classified as a medium risk zone on 12 and 24 August 2016 [33,60]. PC concentrations tended to be high along the edge of the river because the water flow was slower there than in the middle of the river, leading to longer water retention times. Figure 16 shows the negative relationship between flow velocity and PC:Chl-a on 12 and 24 August 2016 (p-value = 0.003). This proved that the cyanobacteria favor a long water retention time to form a dominant cyanobacteria bloom. Previous studies are in agreement with cyanobacteria blooms occurring when the water retention time is long [10,84–86]. Further, Park et al. [9] reported that a flow velocity less than 0.06 m s^{-1} was a suitable physical condition for cyanobacteria growth. In addition, a high water temperature and stable nutrient concentration were proven as dominant environmental factors for PC distribution on 12 and 24 August 2016 [9]. Overall, atmospheric correction using the ANN simulation resulted in a similar PC and Chl-a distribution as that using MODTRAN 6 or ATCOR 4. Sufficient input datasets are required to obtain a reasonable performance using the ANN model. If there is not sufficient input data, the results are often observed to have a single simulation value with various observed values on the dotted line as shown in Figure S2 [87]. Therefore, it is assumed that the scarcity of the input data for the ANN simulation resulted in relatively poor performance in developing the PC and Chl-a concentration map.

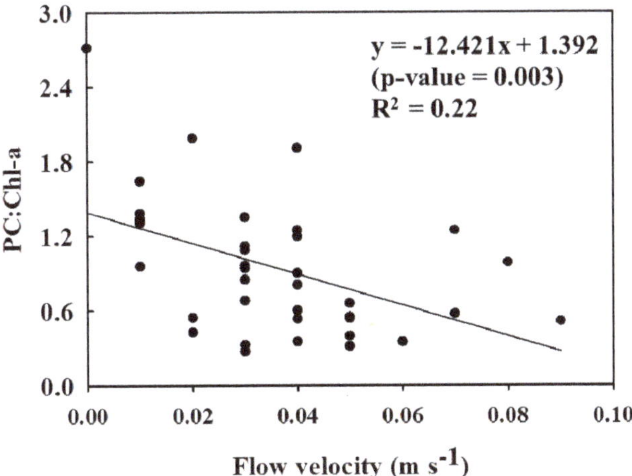

Figure 16. Relationship between flow velocity and the ratio of PC and Chl-a on 12 and 24 August 2016.

5. Conclusions

This study estimated the spatial distribution of PC and Chl-a concentrations using hyperspectral image data and identified how the performance of bio-optical algorithms depended on which atmospheric correction method was used. In addition, how the algae were distributed by influence factors such as flow velocity was also analyzed. The atmospheric correction methods investigated in this study were MODTRAN 6, ATCOR 4, and ANN. Field monitoring and experimental analysis were conducted, after which, bio-optical algorithms were built to quantify PC and Chl-a concentrations using hyperspectral image data. IOP algorithms were optimized using multi-objective optimization. MODTRAN 6, ATCOR 4, and ANN all succeeded in correcting for atmospheric effects on the hyperspectral image obtained from airborne monitoring. Both AOP and IOP algorithms generated maps of the spatial distribution of PC and Chl-a concentrations using the corrected images. The major findings of this study are as follows:

- The cyanobacteria bloom on 12 and 24 August 2016 occurred as the PC:Chl-a value was greater than 0.5. A succession of algal species from cyanobacteria to diatoms was then observed on 20 September and 14 October 2016.
- MODTRAN 6 provided reasonable atmospheric correction performance compared to that of ATCOR 4. However, the accuracy was low in certain regions of the reflectance spectra ($\lambda < 500$ nm and $\lambda > 700$ nm). This was mainly because of insufficient atmospheric observations during the campaigns.
- The most accurate atmospheric correction by MODTRAN 6, compared to ATCOR 4 and ANN, contributed to improving the performance of the bio-optical algorithms in terms of the estimation of PC and Chl-a concentration. The ANN model was found to require large quantities of input data to achieve accurate simulation results.
- The spatial distribution of a high PC:Chl-a value was derived using the flow velocity of less than 0.06 m s^{-1}. This study directly proved that the influence factor of the dominant PC bloom was a long water retention time.

This study identified the effect of the atmospheric correction method used in HSI on PC and Chl-a concentrations derived from images, and it evaluated the influence factor on the algal distribution. Thus, atmospheric correction performance has been shown to be critical in providing useful, informative, and precise maps of the spatial distribution of PC when employing airborne or satellite imagery.

Supplementary Materials: The following are available online at http://www.mdpi.com/2072-4292/10/8/1180/s1, Appendix A: Atmospheric correction of ATCOR 4, MODTRAN 6, and ANN, Table S1: MODTRAN input composition, Table S2: Solar angle for geometry specific input, Table S3: Input information for the ANN, Table S4. Atmospheric correction performances of MODTRAN 6 and ATCOR 4, Figure S1: Atmospheric correction results using ATCOR 4. Panels a–d show the average in-situ and corrected surface reflectance ρ_{surf}. Panels e–h show the correlation between the observed and corrected results at different wavelength for each sampling point, Figure S2: ANN simulation atmospheric correction results for overall wavelengths, Figure S3: Reflectance error (%) of the atmospheric correction. Panels a-d show the MODTRAN 6 correction error and panels e-h show the ATCOR 4 correction error, Figure S4: Optimized absorption coefficient results of *PC* algorithm with respect to in-situ and atmospheric corrected reflectance. Panels a–d show Li algorithm results. Panels d–f show Simis algorithm results. abs indicates absorption coefficient at 622 nm, Figure S5: Optimized absorption coefficient results of *Chl-a* algorithm with respect to in-situ and atmospheric corrected reflectance. Panels a-d show Li algorithm results. Panels d-f show Simis algorithm results. abs indicates absorption coefficient at 660 nm, Figure S6: Phycocyanin concentration images 20 September 2016 in Section 1. Panels a–d show the PC distribution driven by the MODTRAN 6 atmospheric correction. Panels e–h show the PC distribution driven by the ATCOR 4 atmospheric correction. Panels i–l show the PC distribution driven by the ANN atmospheric correction, Figure S7: Phycocyanin concentration images on 20 September 2016 in Section 2. Panels a–d show the PC distribution driven by the MODTRAN 6 atmospheric correction. Panels e–h show the PC distribution driven by the ATCOR 4 atmospheric correction. Panels i–l show the PC distribution driven by the ANN atmospheric correction, Figure S8: Phycocyanin concentration images on 14 October 2016 in Section 1. Panels a–d show the PC distribution driven by the MODTRAN 6 atmospheric correction. Panels e–h show the PC distribution driven by the ATCOR 4 atmospheric correction. Panels i–l show the PC distribution driven by the ANN atmospheric correction, Figure S9: Phycocyanin concentration images on 14 October 2016 in Section 2. Panels a–d show the

PC distribution driven by the MODTRAN 6 atmospheric correction. Panels e–h show the PC distribution driven by the ATCOR 4 atmospheric correction. Panels i–l show the PC distribution driven by the ANN atmospheric correction, Figure S10: Chlorophyll-a concentration images on 20 September 2016 in Section 1. Panels a–d show the Chl-a distribution driven by the MODTRAN 6 atmospheric correction. Panels e–h show the Chl-a distribution driven by the ATCOR 4 atmospheric correction. Panels i–l show the Chl-a distribution driven by the ANN atmospheric correction, Figure S11: Chlorophyll-a concentration images on 20 September 2016 in Section 2. Panels a–d show the Chl-a distribution driven by the MODTRAN 6 atmospheric correction. Panels e–h show the Chl-a distribution driven by the ATCOR 4 atmospheric correction. Panels i–l show the Chl-a distribution driven by the ANN atmospheric correction, Figure S12: Chlorophyll-a concentration images on 14 October 2016 in Section 1. Panels a–d show the Chl-a distribution driven by the MODTRAN 6 atmospheric correction. Panels e–h show the Chl-a distribution driven by the ATCOR 4 atmospheric correction. Panels i–l show the Chl-a distribution driven by the ANN atmospheric correction, Figure S13: Chlorophyll-a concentration images on 14 October 2016 in Section 2. Panels a–d show the Chl-a distribution driven by the MODTRAN 6 atmospheric correction. Panels e–h show the Chl-a distribution driven by the ATCOR 4 atmospheric correction. Panels i–l show the Chl-a distribution driven by the ANN atmospheric correction. Figure S14: Influence of atmospheric correction with MODTRAN 6 and ATCOR 4 on a: PC algorithm and b: Chl-a algorithm.* indicates the band ratio algorithm, ** indicates the Li algorithm, and *** indicates the Simis algorithm. Figure S15: PC:Chl-a map estimated by Li algorithm from reflectance data corrected by MODTRAN 6 in 12 and 24 August 2016.

Author Contributions: Conceptualization, J.C.P, K.H.C., and Y.P.; Field Data Collection and Experiment, J.C.P., M.L., Y.S.K, K.K., H.L., and T.K.; Data analysis, J.C.P; Atmospheric correction, J.C.P., S.B.C., and M.-H.A.; Writing-Original Draft Preparation, J.C.P.; Writing-Review and Editing, K.H.C, Y.P., and M.L.; Supervision, K.H.C and Y.P.

Acknowledgments: This research was supported by the ICT R&D program of MSIP/IITP. [1711070420, Space-time complex artificial intelligence blue-green algae prediction technology based on direct-readable water quality complex sensor and hyperspectral image] and in part by the Basic Core Technology Development Program for the Oceans and the Polar Regions of the National Research Foundation (NRF) funded by the Ministry of Science, ICT & Future Planning [grant number NRF-2016M1A5A1027457].

Conflicts of Interest: The authors declare no conflict of interest.

References

1. Chiswell, R.K.; Shaw, G.R.; Eaglesham, G.; Smith, M.J.; Norris, R.L.; Seawright, A.A.; Moore, M.R. Stability of cylindrospermopsin, the toxin from the cyanobacterium, cylindrospermopsis raciborskii: Effect of ph, temperature, and sunlight on decomposition. *Environ. Toxicol. Int. J.* **1999**, *14*, 155–161. [CrossRef]
2. Gerard, C.; Poullain, V. Variation in the response of the invasive species potamopyrgus antipodarum (smith) to natural (cyanobacterial toxin) and anthropogenic (herbicide atrazine) stressors. *Environ. Pollut.* **2005**, *138*, 28–33. [CrossRef] [PubMed]
3. Paerl, H.W.; Hall, N.S.; Calandrino, E.S. Controlling harmful cyanobacterial blooms in a world experiencing anthropogenic and climatic-induced change. *Sci. Total Environ.* **2011**, *409*, 1739–1745. [CrossRef] [PubMed]
4. Shurin, J.B.; Dodson, S.I. Sublethal toxic effects of cyanobacteria and nonylphenol on environmental sex determination and development in daphnia. *Environ. Toxicol. Chem.* **1997**, *16*, 1269–1276. [CrossRef]
5. Wagner, C.; Adrian, R. Cyanobacteria dominance: Quantifying the effects of climate change. *Limnol. Oceanogr.* **2009**, *54*, 2460–2468. [CrossRef]
6. Zanchett, G.; Oliveira, E.C. Cyanobacteria and cyanotoxins: From impacts on aquatic ecosystems and human health to anticarcinogenic effects. *Toxins* **2013**, *5*, 1896–1917. [CrossRef] [PubMed]
7. Feng, T.; Wang, C.; Wang, P.; Qian, J.; Wang, X. How physiological and physical processes contribute to the phenology of cyanobacterial blooms in large shallow lakes: A new euler-lagrangian coupled model. *Water Res.* **2018**, *140*, 34–43. [CrossRef] [PubMed]
8. Kim, D.M.; Park, H.S.; Chung, S.W. Relationship of the thermal stratification and critical flow velocity near the baekje weir in geum river. *J. Korean Soc. Water Envrion.* **2017**, *33*, 449–459.
9. Park, Y.; Pyo, J.; Kwon, Y.S.; Cha, Y.; Lee, H.; Kang, T.; Cho, K.H. Evaluating physico-chemical influences on cyanobacterial blooms using hyperspectral images in inland water, korea. *Water Res.* **2017**, *126*, 319–328. [CrossRef] [PubMed]
10. Romo, S.; Soria, J.; Fernandez, F.; Ouahid, Y.; Barón-Solá, Á. Water residence time and the dynamics of toxic cyanobacteria. *Freshw. Biol.* **2013**, *58*, 513–522. [CrossRef]
11. Ju, H.; Choi, I.; Yoon, J.; Lee, J.; Lim, B.; Lee, S. Analysis of cyanobacterial growth pattern in bekjae weir during recent 3 years. *Korean Soc. Water Enviorn.* **2016**, *2016*, 562–563.

12. Yoon, J.; Ju, H.; Yoon, H.; Lee, J.; Lee, J.; Lee, S. Effect of hydrological conditions on distribution characteristics of algal concentration in the geum river main stream. *Korean Soc. Water Environ.* **2015**, *2015*, 631–632.
13. Cheung, M.Y.; Liang, S.; Lee, J. Toxin-producing cyanobacteria in freshwater: A review of the problems, impact on drinking water safety, and efforts for protecting public health. *J. Microbiol.* **2013**, *51*, 1–10. [CrossRef] [PubMed]
14. Davies, J.-M.; Mazumder, A. Health and environmental policy issues in Canada: The role of watershed management in sustaining clean drinking water quality at surface sources. *J. Environ. Manag.* **2003**, *68*, 273–286. [CrossRef]
15. Brando, V.E.; Dekker, A.G. Satellite hyperspectral remote sensing for estimating estuarine and coastal water quality. *IEEE Trans. Geosci. Remote Sens.* **2003**, *41*, 1378–1387. [CrossRef]
16. Brezonik, P.; Menken, K.D.; Bauer, M. Landsat-based remote sensing of lake water quality characteristics, including chlorophyll and colored dissolved organic matter (CDOM). *Lake Reserv. Manag.* **2005**, *21*, 373–382. [CrossRef]
17. Dekker, A.G. Detection of Optical Water Quality Parameters for Eutrophic Waters by High Resolution Remote Sensing. Ph.D. Thesis, University of Amsterdam, Amsterdam, The Netherlands, 1993.
18. Hilton, J. Airborne remote sensing for freshwater and estuarine monitoring. *Water Res.* **1984**, *18*, 1195–1223. [CrossRef]
19. Kallio, K.; Kutser, T.; Hannonen, T.; Koponen, S.; Pulliainen, J.; Vepsäläinen, J.; Pyhälahti, T. Retrieval of water quality from airborne imaging spectrometry of various lake types in different seasons. *Sci. Total Environ.* **2001**, *268*, 59–77. [CrossRef]
20. Kiefer, I.; Odermatt, D.; Anneville, O.; Wüest, A.; Bouffard, D. Application of remote sensing for the optimization of in-situ sampling for monitoring of phytoplankton abundance in a large lake. *Sci. Total Environ.* **2015**, *527*, 493–506. [CrossRef] [PubMed]
21. Kwon, Y.S.; Jang, E.; Im, J.; Baek, S.H.; Park, Y.; Cho, K.H. Developing data-driven models for quantifying cochlodinium polykrikoides using the geostationary ocean color imager (GOCI). *Int. J. Remote Sens.* **2018**, *39*, 68–83. [CrossRef]
22. Matthews, M.W.; Bernard, S.; Winter, K. Remote sensing of cyanobacteria-dominant algal blooms and water quality parameters in zeekoevlei, a small hypertrophic lake, using meris. *Remote Sens. Environ.* **2010**, *114*, 2070–2087. [CrossRef]
23. Page, B.P.; Kumar, A.; Mishra, D.R. A novel cross-satellite based assessment of the spatio-temporal development of a cyanobacterial harmful algal bloom. *Int. J. Appl. Earth Obs. Geoinf.* **2018**, *66*, 69–81. [CrossRef]
24. Ritchie, J.C.; Zimba, P.V.; Everitt, J.H. Remote sensing techniques to assess water quality. *Photogramm. Eng. Remote Sens.* **2003**, *69*, 695–704. [CrossRef]
25. Song, K.; Li, L.; Tedesco, L.P.; Li, S.; Hall, B.E.; Du, J. Remote quantification of phycocyanin in potable water sources through an adaptive model. *ISPRS J. Photogramm. Remote Sens.* **2014**, *95*, 68–80. [CrossRef]
26. Tyler, A.; Svab, E.; Preston, T.; Présing, M.; Kovács, W. Remote sensing of the water quality of shallow lakes: A mixture modelling approach to quantifying phytoplankton in water characterized by high-suspended sediment. *Int. J. Remote Sens.* **2006**, *27*, 1521–1537. [CrossRef]
27. Zhou, L.G.; Roberts, D.A.; Ma, W.C.; Zhang, H.; Tang, L. Estimation of higher chlorophylla concentrations using field spectral measurement and hj-1a hyperspectral satellite data in dianshan lake, china. *ISPRS J. Photogramm. Remote Sens.* **2014**, *88*, 41–47. [CrossRef]
28. Gitelson, A.A.; Dall'Olmo, G.; Moses, W.; Rundquist, D.C.; Barrow, T.; Fisher, T.R.; Gurlin, D.; Holz, J. A simple semi-analytical model for remote estimation of chlorophyll-a in turbid waters: Validation. *Remote Sens. Environ.* **2008**, *112*, 3582–3593. [CrossRef]
29. Li, L.; Li, L.; Song, K.; Li, Y.; Tedesco, L.P.; Shi, K.; Li, Z. An inversion model for deriving inherent optical properties of inland waters: Establishment, validation and application. *Remote Sens. Environ.* **2013**, *135*, 150–166. [CrossRef]
30. Li, L.; Li, L.; Song, K. Remote sensing of freshwater cyanobacteria: An extended iop inversion model of inland waters (IIMIW) for partitioning absorption coefficient and estimating phycocyanin. *Remote Sens. Environ.* **2015**, *157*, 9–23. [CrossRef]
31. Simis, S.G.; Peters, S.W.; Gons, H.J. Remote sensing of the cyanobacterial pigment phycocyanin in turbid inland water. *Limnol. Oceanogr.* **2005**, *50*, 237–245. [CrossRef]

32. Simis, S.G.; Ruiz-Verdú, A.; Domínguez-Gómez, J.A.; Peña-Martinez, R.; Peters, S.W.; Gons, H.J. Influence of phytoplankton pigment composition on remote sensing of cyanobacterial biomass. *Remote Sens. Environ.* **2007**, *106*, 414–427. [CrossRef]
33. Hunter, P.D.; Tyler, A.N.; Carvalho, L.; Codd, G.A.; Maberly, S.C. Hyperspectral remote sensing of cyanobacterial pigments as indicators for cell populations and toxins in eutrophic lakes. *Remote Sens. Environ.* **2010**, *114*, 2705–2718. [CrossRef]
34. Matsushita, B.; Yang, W.; Yu, G.; Oyama, Y.; Yoshimura, K.; Fukushima, T. A hybrid algorithm for estimating the chlorophyll-a concentration across different trophic states in asian inland waters. *ISPRS J. Photogramm. Remote Sens.* **2015**, *102*, 28–37. [CrossRef]
35. Mishra, S.; Mishra, D.R.; Schluchter, W.M. A novel algorithm for predicting phycocyanin concentrations in cyanobacteria: A proximal hyperspectral remote sensing approach. *Remote Sens.* **2009**, *1*, 758–775. [CrossRef]
36. Vincent, R.K.; Qin, X.; McKay, R.M.L.; Miner, J.; Czajkowski, K.; Savino, J.; Bridgeman, T. Phycocyanin detection from landsat tm data for mapping cyanobacterial blooms in Lake Erie. *Remote Sens. Environ.* **2004**, *89*, 381–392. [CrossRef]
37. Duan, H.; Ma, R.; Hu, C. Evaluation of remote sensing algorithms for cyanobacterial pigment retrievals during spring bloom formation in several lakes of East China. *Remote Sens. Environ.* **2012**, *126*, 126–135. [CrossRef]
38. Pyo, J.; Pachepsky, Y.; Baek, S.-S.; Kwon, Y.; Kim, M.; Lee, H.; Park, S.; Cha, Y.; Ha, R.; Nam, G. Optimizing semi-analytical algorithms for estimating chlorophyll-a and phycocyanin concentrations in inland waters in Korea. *Remote Sens.* **2017**, *9*, 542. [CrossRef]
39. Watanabe, F.; Mishra, D.R.; Astuti, I.; Rodrigues, T.; Alcântara, E.; Imai, N.N.; Barbosa, C. Parametrization and calibration of a quasi-analytical algorithm for tropical eutrophic waters. *ISPRS J. Photogramm. Remote Sens.* **2016**, *121*, 28–47. [CrossRef]
40. Roessner, S.; Segl, K.; Heiden, U.; Kaufmann, H. Automated differentiation of urban surfaces based on airborne hyperspectral imagery. *IEEE Trans. Geosci. Remote Sens.* **2001**, *39*, 1525–1532. [CrossRef]
41. Segl, K.; Roessner, S.; Heiden, U.; Kaufmann, H. Fusion of spectral and shape features for identification of urban surface cover types using reflective and thermal hyperspectral data. *ISPRS J. Photogramm. Remote Sens.* **2003**, *58*, 99–112. [CrossRef]
42. Selige, T.; Böhner, J.; Schmidhalter, U. High resolution topsoil mapping using hyperspectral image and field data in multivariate regression modeling procedures. *Geoderma* **2006**, *136*, 235–244. [CrossRef]
43. Shippert, P. Introduction to hyperspectral image analysis. *Online J. Space Commun.* **2003**, *28*, 1–13.
44. Wang, F.; Gao, J.; Zha, Y. Hyperspectral sensing of heavy metals in soil and vegetation: Feasibility and challenges. *ISPRS J. Photogramm. Remote Sens.* **2018**, *136*, 73–84. [CrossRef]
45. Richter, R. Atmospheric correction of dais hyperspectral image data. *Comput. Geosci.* **1996**, *22*, 785–793. [CrossRef]
46. Thiemann, S.; Kaufmann, H. Lake water quality monitoring using hyperspectral airborne data—A semiempirical multisensor and multitemporal approach for the mecklenburg lake district, germany. *Remote Sens. Environ.* **2002**, *81*, 228–237. [CrossRef]
47. Berk, A.; Conforti, P.; Kennett, R.; Perkins, T.; Hawes, F.; van den Bosch, J. Modtran® 6: A major upgrade of the modtran® radiative transfer code. In Proceedings of the 2014 6th Workshop on Hyperspectral Image and Signal Processing: Evolution in Remote Sensing (WHISPERS), Lausanne, Switzerland, 24–27 June 2014; pp. 1–4.
48. Giardino, C.; Brando, V.E.; Dekker, A.G.; Strömbeck, N.; Candiani, G. Assessment of water quality in Lake Garda (Italy) using hyperion. *Remote Sens. Environ.* **2007**, *109*, 183–195. [CrossRef]
49. Goyens, C.; Jamet, C.; Schroeder, T. Evaluation of four atmospheric correction algorithms for modis-aqua images over contrasted coastal waters. *Remote Sens. Environ.* **2013**, *131*, 63–75. [CrossRef]
50. Schroeder, T.; Fischer, J.; Schaale, M.; Fell, F. Artificial-Neural-Network-Based Atmospheric Correction Algorithm: Application to Meris data. In *Ocean Remote Sensing and Applications*; International Society for Optics and Photonics: Bellingham, WA, USA, 2003; pp. 124–133.
51. Schroeder, T.; Behnert, I.; Schaale, M.; Fischer, J.; Doerffer, R. Atmospheric correction algorithm for meris above case-2 waters. *Int. J. Remote Sens.* **2007**, *28*, 1469–1486. [CrossRef]

52. Bernstein, L.S.; Adler-Golden, S.M.; Sundberg, R.L.; Levine, R.Y.; Perkins, T.C.; Berk, A.; Ratkowski, A.J.; Felde, G.; Hoke, M.L. Validation of the quick atmospheric correction (QUAC) algorithm for vnir-swir multi-and hyperspectral imagery. In *Algorithms and Technologies for Multispectral, Hyperspectral, and Ultraspectral Imagery XI*; International Society for Optics and Photonics: Bellingham, WA, USA, 2005; pp. 668–679.
53. Gao, B.-C.; Montes, M.J.; Davis, C.O.; Goetz, A.F. Atmospheric correction algorithms for hyperspectral remote sensing data of land and ocean. *Remote Sens. Environ.* **2009**, *113*, S17–S24. [CrossRef]
54. Jaelani, L.M.; Matsushita, B.; Yang, W.; Fukushima, T. An improved atmospheric correction algorithm for applying meris data to very turbid inland waters. *Int. J. Appl. Earth Obs. Geoinf.* **2015**, *39*, 128–141. [CrossRef]
55. Pons, X.; Pesquer, L.; Cristóbal, J.; González-Guerrero, O. Automatic and improved radiometric correction of landsat imagery using reference values from modis surface reflectance images. *Int. J. Appl. Earth Obs. Geoinf.* **2014**, *33*, 243–254. [CrossRef]
56. Rodríguez-Caballero, E.; Paul, M.; Tamm, A.; Caesar, J.; Büdel, B.; Escribano, P.; Hill, J.; Weber, B. Biomass assessment of microbial surface communities by means of hyperspectral remote sensing data. *Sci. Total Environ.* **2017**, *586*, 1287–1297. [CrossRef] [PubMed]
57. Igamberdiev, R.M.; Grenzdoerffer, G.; Bill, R.; Schubert, H.; Bachmann, M.; Lennartz, B. Determination of chlorophyll content of small water bodies (kettle holes) using hyperspectral airborne data. *Int. J. Appl. Earth Obs. Geoinf.* **2011**, *13*, 912–921. [CrossRef]
58. Jupp, D.L.; Kirk, J.T.; Harris, G.P. Detection, identification and mapping of cyanobacteria—Using remote sensing to measure the optical quality of turbid inland waters. *Mar. Freshw. Res.* **1994**, *45*, 801–828. [CrossRef]
59. Navarro-Cerrillo, R.M.; Trujillo, J.; de la Orden, M.S.; Hernández-Clemente, R. Hyperspectral and multispectral satellite sensors for mapping chlorophyll content in a mediterranean pinus sylvestris l. Plantation. *Int. J. Appl. Earth Obs. Geoinf.* **2014**, *26*, 88–96. [CrossRef]
60. Shi, K.; Zhang, Y.; Li, Y.; Li, L.; Lv, H.; Liu, X. Remote estimation of cyanobacteria-dominance in inland waters. *Water Res.* **2015**, *68*, 217–226. [CrossRef] [PubMed]
61. Shanmugam, P. Caas: An atmospheric correction algorithm for the remote sensing of complex waters. In *Annales Geophysicae*; Copernicus GmbH: Göttingen, Germany, 2012; p. 203.
62. Duan, H.; Tao, M.; Loiselle, S.A.; Zhao, W.; Cao, Z.; Ma, R.; Tang, X. Modis observations of cyanobacterial risks in a eutrophic lake: Implications for long-term safety evaluation in drinking-water source. *Water Res.* **2017**, *122*, 455–470. [CrossRef] [PubMed]
63. Song, Y.; Zhang, L.-L.; Li, J.; Chen, M.; Zhang, Y.-W. Mechanism of the influence of hydrodynamics on microcystis aeruginosa, a dominant bloom species in reservoirs. *Sci. Total Environ.* **2018**, *636*, 230–239. [CrossRef] [PubMed]
64. Bhaskar, S.U.; Gopalaswamy, G.; Raghu, R. A simple method for efficient extraction and purification of c-phycocyanin from spirulina platensis geitler. *Indian J. Exp. Biol.* **2005**, *43*, 277–279. [PubMed]
65. Arnon, D.I.; McSwain, B.D.; Tsujimoto, H.Y.; Wada, K. Photochemical activity and components of membrane preparations from blue-green algae. I. Coexistence of two photosystems in relation to chlorophyll a and removal of phycocyanin. *Biochim. Biophys. Acta Bioenerg.* **1974**, *357*, 231–245. [CrossRef]
66. Bennett, A.; Bogorad, L. Complementary chromatic adaptation in a filamentous blue-green alga. *J. Cell Biol.* **1973**, *58*, 419–435. [CrossRef] [PubMed]
67. Association, A.P.H.; Association, A.W.W.; Federation, W.P.C.; Federation, W.E. *Standard Methods for the Examination of Water and Wastewater*; American Public Health Association: Washington, DC, USA, 1915; Volume 2.
68. Richter, R.; Schläpfer, D. Geo-atmospheric processing of airborne imaging spectrometry data. Part 2: Atmospheric/topographic correction. *Int. J. Remote Sens.* **2002**, *23*, 2631–2649. [CrossRef]
69. Tuominen, J.; Lipping, T. Atmospheric correction of hyperspectral data using combined empirical and model based method. In Proceedings of the 7th European association of remote sensing laboratories SIG-imaging spectroscopy workshop, Edinburgh, UK, 11–13 April 2011.
70. Ruddick, K.G.; Gons, H.J.; Rijkeboer, M.; Tilstone, G. Optical remote sensing of chlorophyll a in case 2 waters by use of an adaptive two-band algorithm with optimal error properties. *Appl. Opt.* **2001**, *40*, 3575–3585. [CrossRef] [PubMed]
71. Tan, J.; Cherkauer, K.A.; Chaubey, I. Using hyperspectral data to quantify water-quality parameters in the wabash river and its tributaries, indiana. *Int. J. Remote Sens.* **2015**, *36*, 5466–5484. [CrossRef]

72. Olmanson, L.G.; Brezonik, P.L.; Bauer, M.E. Airborne hyperspectral remote sensing to assess spatial distribution of water quality characteristics in large rivers: The mississippi river and its tributaries in minnesota. *Remote Sens. Environ.* **2013**, *130*, 254–265. [CrossRef]
73. Buiteveld, H.; Hakvoort, J.; Donze, M. Optical properties of pure water. In *Ocean Optics XII*; International Society for Optics and Photonics: Bellingham, WA, USA, 1994; pp. 174–184.
74. Adler-Golden, S.M.; Matthew, M.W.; Bernstein, L.S.; Levine, R.Y.; Berk, A.; Richtsmeier, S.C.; Acharya, P.K.; Anderson, G.P.; Felde, J.W.; Gardner, J. Atmospheric correction for shortwave spectral imagery based on modtran4. In *Imaging Spectrometry V*; International Society for Optics and Photonics: Bellingham, WA, USA, 1999; pp. 61–70.
75. Hadjimitsis, D.G.; Clayton, C.; Hope, V. An assessment of the effectiveness of atmospheric correction algorithms through the remote sensing of some reservoirs. *Int. J. Remote Sens.* **2004**, *25*, 3651–3674. [CrossRef]
76. Gitelson, A.; Garbuzov, G.; Szilagyi, F.; Mittenzwey, K.; Karnieli, A.; Kaiser, A. Quantitative remote sensing methods for real-time monitoring of inland waters quality. *Int. J. Remote Sens.* **1993**, *14*, 1269–1295. [CrossRef]
77. Chen, W.-T.; Zhang, Z.; Wang, Y.-X.; Wen, X.-P. Atmospheric Correction of spot5 Land Surface Imagery. In Proceedings of the 2009 2nd International Congress on Image and Signal Processing, Tianjin, China, 17–19 October 2009; pp. 1–5.
78. Kaufman, Y.J. Atmospheric effects on remote sensing of surface reflectance. In *Remote Sensing: Critical Review of Technology*; International Society for Optics and Photonics: Bellingham, WA, USA, 1984; pp. 20–34.
79. Kaufman, Y.; Wald, A.; Remer, L.; Gao, B.-C.; Li, R.-R.; Flynn, L. *The Modis 2.1-μm Channel-Correlation with Visible Reflectancefor Use in Remote Sensing of Aerosol Geoscience and Remote Sensing*; IEEE: Piscataway, NJ, USA, 1997; Volume 35.
80. Kutser, T.; Metsamaa, L.; Dekker, A.G. Influence of the vertical distribution of cyanobacteria in the water column on the remote sensing signal. *Estuar. Coast. Shelf Sci.* **2008**, *78*, 649–654. [CrossRef]
81. Bricaud, A.; Babin, M.; Morel, A.; Claustre, H. Variability in the chlorophyll-specific absorption coefficients of natural phytoplankton: Analysis and parameterization. *J. Geophys. Res. Oceans* **1995**, *100*, 13321–13332. [CrossRef]
82. Garver, S.A.; Siegel, D.A.; Greg, M.B. Variability in near-surface particulate absorption spectra: What can a satellite ocean color imager see? *Limnol. Oceanogr.* **1994**, *39*, 1349–1367. [CrossRef]
83. Yentsch, C.S.; Phinney, D.A. A bridge between ocean optics and microbial ecology. *Limnol. Oceanogr.* **1989**, *34*, 1694–1705. [CrossRef]
84. Cromar, N.J.; Fallowfield, H.J. Effect of nutrient loading and retention time on performance of high rate algal ponds. *J. Appl. Phycol.* **1997**, *9*, 301–309. [CrossRef]
85. Soares, M.C.S.; Marinho, M.M.; Huszar, V.L.; Branco, C.W.; Azevedo, S.M. The effects of water retention time and watershed features on the limnology of two tropical reservoirs in Brazil. *Lakes Reserv. Res. Manag.* **2008**, *13*, 257–269. [CrossRef]
86. Soares, M.C.S.; Marinho, M.M.; Azevedo, S.M.; Branco, C.W.; Huszar, V.L. Eutrophication and retention time affecting spatial heterogeneity in a tropical reservoir. *Limnol.-Ecol. Manag. Inland Waters* **2012**, *42*, 197–203. [CrossRef]
87. Anctil, F.; Michel, C.; Perrin, C.; Andreassian, V. A soil moisture index as an auxiliary ann input for stream flow forecasting. *J. Hydrol.* **2004**, *286*, 155–167. [CrossRef]

Article

Underwater Use of a Hyperspectral Camera to Estimate Optically Active Substances in the Water Column of Freshwater Lakes

Michael Seidel [1,*], Christopher Hutengs [1], Felix Oertel [1], Daniel Schwefel [2], András Jung [3] and Michael Vohland [1,4]

1 Geoinformatics and Remote Sensing, Institute for Geography, Leipzig University, Johannisallee 19a, 04103 Leipzig, Germany; christopher.hutengs@uni-leipzig.de (C.H.); mail@felix-oertel.de (F.O.); michael.vohland@uni-leipzig.de (M.V.)
2 SphereOptics GmbH, Gewerbestraße 13, 82211 Herrsching, Germany; dschwefel@sphereoptics.de
3 Department of Remote Sensing and Geomatics, Szent István University, Villányi ut 29-43, 1118 Budapest, Hungary; jung.andras@kertk.szie.hu
4 Remote Sensing Centre for Earth System Research, Leipzig University, 04103 Leipzig, Germany
* Correspondence: michael.seidel@uni-leipzig.de

Received: 17 April 2020; Accepted: 26 May 2020; Published: 29 May 2020

Abstract: Freshwater lakes provide many important ecosystem functions and services to support biodiversity and human well-being. Proximal and remote sensing methods represent an efficient approach to derive water quality indicators such as optically active substances (OAS). Measurements of above-ground remote and in situ proximal sensors, however, are limited to observations of the uppermost water layer. We tested a hyperspectral imaging system, customized for underwater applications, with the aim to assess concentrations of chlorophyll a (CHLa) and colored dissolved organic matter (CDOM) in the water columns of four freshwater lakes with different trophic conditions in Central Germany. We established a measurement protocol that allowed consistent reflectance retrievals at multiple depths within the water column independent of ambient illumination conditions. Imaging information from the camera proved beneficial for an optimized extraction of spectral information since low signal areas in the sensor's field of view, e.g., due to non-uniform illumination, and other interfering elements, could be removed from the measured reflectance signal for each layer. Predictive hyperspectral models, based on the 470 nm–850 nm reflectance signal, yielded estimates of both water quality parameters (R^2 = 0.94, RMSE = 8.9 μg L^{-1} for CHLa; R^2 = 0.75, RMSE = 0.22 m^{-1} for CDOM) that were more accurate than commonly applied waveband indices (R^2 = 0.83, RMSE = 13.2 μg L^{-1} for CHLa; R^2 = 0.66, RMSE = 0.25 m^{-1} for CDOM). Underwater hyperspectral imaging could thus facilitate future water monitoring efforts through the acquisition of consistent spectral reflectance measurements or derived water quality parameters along the water column, which has the potential to improve the link between above-surface proximal and remote sensing observations and in situ point-based water probe measurements for ground truthing or to resolve the vertical distribution of OAS.

Keywords: chlorophyll a; colored dissolved organic matter; in situ measurements; vertical distribution; water column; snapshot hyperspectral imaging

---

1. Introduction

Lake ecosystems provide essential functions and services, including contributions to biodiversity, hydrologic regulation and water supply, and human well-being through their recreational benefits [1,2]. At the same time, they are subject to various threats from climate change, alterations of catchment land use, anthropogenic pollutants, aquatic invasive species, or human harvest including aquaculture [3]. Hence, appropriate monitoring adapted to relevant temporal and spatial scales is necessary for an improved understanding of lake ecosystems and their feedbacks.

Remote sensing in the visible and near-infrared (VNIR) range (400–1000 nm) allows for the spatio-temporal monitoring of various water quality parameters in freshwater lakes [4,5]. The most important indicators of water quality, in general, are phytoplankton, colored dissolved organic matter (CDOM) and total suspended matter (TSM), which represent optically active substances (OAS) [4,6]. Changes in the quantity of the OAS have a direct effect on the spectral signature detected by remote (or proximal) sensors, which, in turn, enables the estimation of OAS contents from measured spectra through physically-based analytical or empirical models [4,7,8]. Nevertheless, in the case of optically complex inland water bodies, the variety of OAS concentrations and their specific inherent optical properties is wide and independent from each other [6,9]. This complexity limits the use of simple band ratio approaches and might affect the accuracy of analytical models due to partly unknown optical properties of contributing OAS [7,10].

Beyond this, the general application of remote sensing methods may be limited, e.g., by cloud cover during overflight. Accurate atmospheric correction is another critical issue for retrieving surface reflectances from remotely sensed data; inaccuracies might affect the OAS retrieval, especially in the case of optically complex inland waters [9,11–14].

While remote sensing observations can per se provide consistent, spatially-distributed measurements of water quality parameters at large scales, such spectral measurements can be limited by the lake specific penetration depth of light, which might be shallower than the actual constituent layer; otherwise the constituent layer might just form a thin layer within the remotely sensed water layer [15–18]. Consequently, remotely-sensed measurements cannot resolve the distribution of constituents in the water column, which may impede the correct retrieval of column OAS contents when strong density gradients occur in the remotely-sensed water layer and below.

Water-quality probes, on the other hand, can acquire information from the entire water column, which is relevant, for example, for a series of ecological issues including the detection and analysis of the deep chlorophyll maximum as a hot spot of primary production and nutrient cycling [19]. Besides OAS such as CHLa and CDOM, these sensors can also retrieve additional water parameters including, e.g., temperature, dissolved oxygen, conductivity and pH (e.g., [20–22]). Remote sensing and in situ methods can therefore complement each other, for example by allowing ground truthing of satellite-derived biochemical data products or, conversely, the extension of point information across larger spatial scales [23].

In this context, the in situ hyperspectral measurement of water columns, from the uppermost layer observable by remote sensing through deeper layers that are limited to point sensor observations, offers the potential to improve the link between in situ water monitoring networks and Earth observation systems through consistent radiometric measurements along a water profile. Various studies have used hyperspectral point or imaging sensors to provide ground truth data for overflight campaigns and to validate satellite imagery products (e.g., [22,24,25]), but also for the direct derivation of OAS products for water monitoring purposes [21,26,27]. Recently, Keller et al. (2018) [21] deployed a hyperspectral snapshot camera mounted on a boat to collect hyperspectral imagery (450–950 nm) along the Elbe river in Germany with the goal to quantify multiple OAS such as CDOM and CHLa. While they could successfully estimate OAS quantities with surface measurements, it could also be advantageous to transfer this technology into the water column to measure OAS at multiple depths with the same device.

In this study, we evaluated the capabilities of a hyperspectral snapshot camera system to resolve the vertical distribution of CHLa and CDOM in pre-defined segments in the water column. The camera's spectral imaging quality and capabilities for underwater sensing were first tested in a laboratory experiment against a well-established point spectrometer. Afterward, we conducted a field campaign with multiple water column measurements in four freshwater lakes in Central Germany with the aims to (i) develop an approach to measure water column reflectance without distortions through variations in ambient illumination and (ii) to estimate CHLa and CDOM concentrations through multivariate calibrations with partial least squares regression (PLSR) in comparison to commonly applied CHLa and CDOM indices.

2. Materials and Methods

2.1. Hyperspectral Snapshot Camera

We collected hyperspectral measurements with a snapshot camera system (UHD 285; Cubert GmbH, Ulm, Germany) incorporated in a waterproof casing. The camera used a silicon CCD chip, which enabled the simultaneous acquisition of an entire three-dimensional hyperspectral image cube with one trigger pull. Its built-in sensor covered the spectral range of 450–998 nm with 8 nm spectral resolution at a 4 nm sampling interval. The acquired hyperspectral image cubes had a resolution of 50 × 50 pixels, resulting in 2500 spectra at 138 wavelengths. Due to spectral artifacts in the first few spectral bands and a Si-induced sensitivity loss at the end of the spectrum [28], we reduced the final spectral range to 470–850 nm with 96 spectral bands.

2.2. Laboratory Experiment

To test the camera's ability to capture small changes in OAS contents, we compared its performance with parallel measurements using an ASD FieldSpec 4 (Malvern Panalytical Ltd., Almelo, The Netherlands) point spectrometer in a laboratory experiment.

The laboratory setup included a small water tank with a 20 × 20 cm Zenith Polymer® (white panel with an average absolute reflectance of 0.95) placed at the bottom. Both spectrometers were installed with a nadir viewing geometry, and the scene was illuminated with an ASD ProLamp (14.5 V, 50 W) at a 45° zenith angle. After calibrating the instruments, we filled the tank with distilled water up to a column height of 20 cm. The ASD measurements were carried out directly above the water surface as the instruments' fiber optics cable could not be immersed in the water. To exclude contributions to the measured radiance due to specular reflection at the air-water interface, the fibre optics cable was encased with a non-reflective material. The measured reflectance thus referred to the water-leaving radiance after passage through the interface:

$$\rho_{ASD} = \frac{L_w}{L_{0,lamp}} \tag{1}$$

where L_w is the water-leaving radiance and $L_{0,lamp}$ is the radiance of the light source at the water surface, measured through the reference panel.

Measurements with the hyperspectral camera were carried out with the camera opening placed slightly below the water surface. The measured reflectance thus refers to the upwelling underwater radiance:

$$\rho_{Cam} = \frac{L_u}{L_{0,lamp}} \tag{2}$$

where L_u is the upwelling radiance before transmission through the surface.

The two reflectances can be related through a dimensionless proportionality factor that accounts for the transmission through the water–air interface [29]. Since we were mainly interested in the quality of the camera's data acquisition, i.e., shape of reflectance spectra, resolution of peaks, signal-to-noise ratio, and also carried out the field measurements (Section 2.5) below the water surface, we decided not to convert the ASD spectra to camera-equivalent reflectances.

We carried out two separate series of measurements to test the spectral response of CDOM and CHLa. Humic acid-sodium salt and a commercial *Chlorella* sp. powder were used as surrogates for CDOM and CHLa, respectively. Both substances were each mixed into the tank's water at increasing concentrations (CDOM with absorption coefficients at 440 nm: 0.0–0.5 m^{-1} in 0.1 increments, and 0.5–3.0 m^{-1} in 0.5 increments; CHLa: 0.0–12.5 μg L^{-1} in 2.5 increments, and 12.5–112.5 μg L^{-1} in 12.5 increments) and reflectance spectra were recorded at each stage.

2.3. Field Campaign

For the field campaign, we investigated four artificial freshwater lakes in Central Germany (Figure 1), which were characterized by significant differences in size, trophic state index, and management practices (Table 1). The studied lakes were selected to cover a broad range in terms of depths of visibility, OAS concentrations and trophy with the aim to test the camera's image acquisition and CHLa and CDOM modelling capabilities in different environments (Section 2.6). The shallow, hypereutrophic lake Auensee is a flooded, groundwater-fed former gravel pit with an average depth of 3.5 m, located in an inner city hardwood floodplain forest [30]. The groundwater-fed Cospuden Lake, with a maximum depth of ~54 m, represents a former open cast lignite mine [31], currently used as a recreational area. The Mulde and Kriebstein sites are both reservoirs, fed by the Mulde and the Zschopau river, respectively.

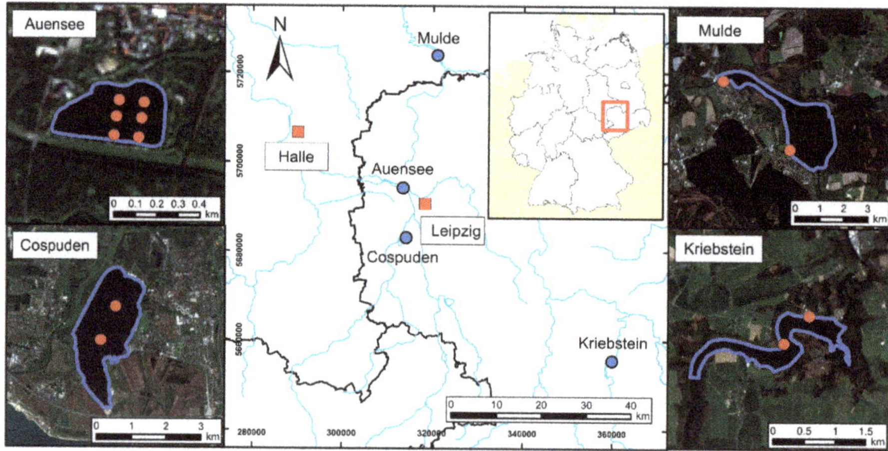

Figure 1. Location of the lake sites in Central Germany and sampling points (red dots) for water column measurements (satellite imagery: Sentinel 2A - RGB – 432 (17/04/2019); coordinate system of the map: ETRS89/UTM zone 33N).

Table 1. Characteristics of the investigated lake sites in Central Germany.

Site	Area (ha)	Trophic State Index	Type	Secchi Depth * (m)	Number of Sampling Points	Number of Sampling Units
Auensee	12	Eutrophic/hypereutrophic	Former gravel pit	0.40–0.65	6	27
Cospuden	436	Oligotrophic	Former open cast mine	6.00–6.05	2	10
Kriebstein	132	Oligotrophic	Reservoir	2.35–2.45	2	10
Mulde	630	Mesoeutrophic	Reservoir	1.25–1.30	2	9

* Refers to the viewing depth at the time of measurement using a 20 cm Secchi disk.

At each lake, spectral measurements and reference samples were collected at near-shore sampling points accessible by footbridges and at fixed markings within the lakes (Figure 1). Additionally, we determined the viewing depth using a 20 cm Secchi depth at each sampling point. For the Mulde, Kriebstein and Cospuden sites, measurements were carried out at two sampling points in each lake; whereas for the more variable Auensee site, samples were collected at six sampling points. Measurements were carried out for up to five continuous 0.5 m segments from the water surface down to a depth of 2.5 m, if possible, and for four segments in shallower waters. In total, 56 samples were taken.

2.4. CHLa and CDOM Reference Analysis

In parallel with the spectral measurements of the water column, we collected bulk water samples for each segment with a Ruttner water sampler (1.7 l, height: 24.5 cm). The samples were stored in cooling boxes and transported to the laboratory on the same day for the analysis of chlorophyll a (CHLa) and colored dissolved organic matter (CDOM). CHLa absorption was determined photometrically after hot ethanol extraction by using a SPECORD double-beam photometer with pure water (Milli-Q) as reference; CHLa concentrations were calculated afterwards according to ISO 10620 [32]. CDOM contents were also determined photometrically after filtering subsamples through Whatman GF/F-filters (pore size of 0.45 µm). The remaining filtrate was used to measure the absorbance of CDOM at 440 nm in 1 cm quartz cuvettes by using a SPECORD double-beam photometer with pure water (Milli-Q) as reference. Absorption coefficients were calculated according to the following expression [33]:

$$a_{CDOM}(440\ nm) = 2.303 \cdot \frac{A(440\ nm)}{l} \tag{3}$$

where aCDOM(440 nm) is the CDOM absorption coefficient at 440 nm, A(440 nm) is the measured absorbance at 440 nm, and l is the path length of the cuvette in m.

2.5. Hyperspectral Image Acquisition and Processing

For water column measurements, the hyperspectral camera system was mounted on a customized rack equipped with a portable halogen lamp (100 W). A Zenith Polymer® reference panel (average absolute reflectance of 0.95; 25 × 25 cm) was attached in front of the camera at a fixed distance of 40 cm so that it covered the entire field of view of the camera (Figure 2).

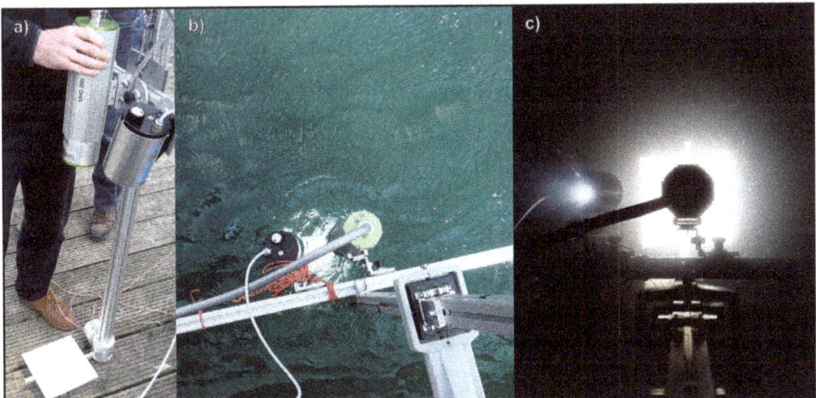

Figure 2. Hyperspectral camera system for underwater measurements: (**a**) camera system mounted on a rack with halogen light source and reference panel; (**b**) in situ measurement of uppermost water layer (0–0.5 m); (**c**) top-down view of night-time measurement in the water column (0.5–1.0 m).

Images of the reference panel above the water surface and of the individual water column segments were recorded in the raw digital number format (DN). Under optimal illumination conditions, the conversion from DN to radiance is a linear function and reflectance values can be calculated as:

$$\rho_{sample} = \frac{DN_{sample}}{DN_{ref}} \cdot \rho_{ref} \cdot \left(\frac{t_{ref}}{t_{sample}}\right) \tag{4}$$

where ρ_{sample} = reflectance of sample, DN = digital number, ρ_{ref} = reflectance factor of the reference panel and t = integration time during the measurement.

At each sampling point, we measured the reference panel above the water surface to acquire a calibration file for the entire water column. Since ambient illumination varies for each measured segment due to non-linear sunlight attenuation through the water column, we compensated for this effect through parallel measurements of the reflectance target with and without artificial illumination to retrieve the final reflectance curves. That is, two separate images were taken for each measurement, both for the reference panel above the water surface and for the individual segments within the water column. The first image was taken with the external lamp switched on, the second image with the lamp switched off. The difference between the two respective images then represents the signal of the measured water column segment without the impact of ambient stray light (Figure 3). Before calculating the final mean reflectance spectrum, we performed two processing steps for each image (Figure 4) to define an optimally illuminated region within the image while minimizing the impact of interfering objects. First, we applied a binary mask by thresholding pixel values at 710 nm, the wavelength of maximum signal intensity of the halogen lamp. This wavelength yielded a high discrimination accuracy for the applied threshold due to an optimal signal-to-noise ratio and was also less influenced by absorption processes of our target variables (see Section 3.1).

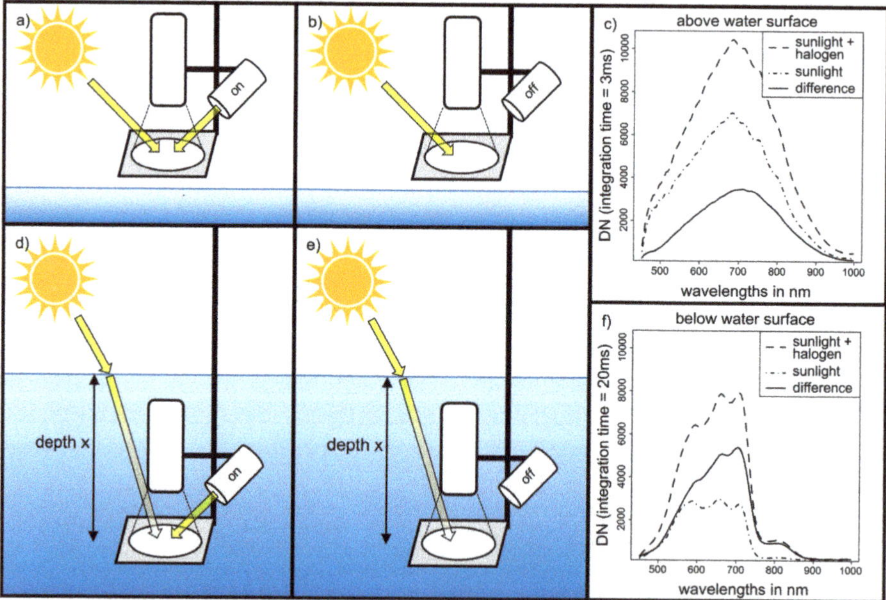

Figure 3. Field measurement setup for underwater reflectance retrieval: the upper row shows radiance measurements on the reference panel for calibration with light source turned on (**a**) and turned off (**b**). The difference of (**a**) and (**b**) represents the signal contributed by the light source only, which was used for instrument calibration (**c**). The lower row shows measurements below the water surface with the light source turned on (**d**) and turned off (**e**). The difference between measured spectra in configuration of (**d**) and (**e**) represents the reflected signal that only refers to the illumination from the artificial light source (**f**).

Since the illumination conditions and OAS contents varied between the images, each image-specific threshold was defined as the mean DN value at 710 nm. All pixels with a DN value less than the image mean were discarded to remove poorly illuminated pixels and interfering image objects (e.g., shadowing effects of surface waves, bubbles due to gaseous emissions from the seafloor, or floating plant residues in the water column). In the second step, we defined a square region of interest (ROI) with a maximum of 121 pixels centered at the pixel with the largest DN value (710 nm) to calculate the mean DN spectrum for each image (Figure 4).

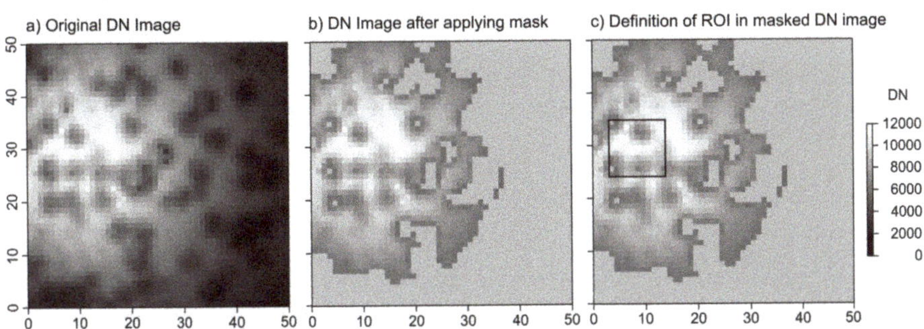

Figure 4. Image processing steps to extract optimal mean spectra for multivariate calibrations: (**a**) example of 2D image (at 710 nm) recorded near the bottom of the Auensee site where suspended particles and bubbles appeared during image acquisition; (**b**) masking of poorly illuminated image areas and interfering elements; (**c**) definition of a region of interest (ROI) around the 'brightest' pixel (at 710 nm).

Finally, the reflectance of each water column segment was calculated from the extracted mean DN spectra as:

$$\rho_{sample} = \frac{\overline{DN}(x)_{sun+lamp} - \overline{DN}(x)_{sun}}{\overline{DN}(x=0)_{sun+lamp} - \overline{DN}(x=0)_{sun}} \cdot \rho_{ref} \cdot \left(\frac{t_{x=0}}{t_x}\right) \tag{5}$$

where the numerator represents the averaged sunlight-compensated DN of the measurement at depth x, the denominator represents the averaged sunlight-compensated DN of the reference panel measured above the water surface (x = 0), and t_x and $t_{x=0}$ are the corresponding integration times.

Accordingly, the calculated reflectance curves were only dependent on the energy input of the external halogen light source and were thus comparable across all investigated water bodies.

To validate the compensation algorithm, we compared the calculated reflectance curves of daytime and nighttime measurements at the Cospuden site. During nighttime measurements, ambient light does not interfere with the measurements, so that these measurements only reflect the contributions of the halogen light source. As the Cospuden site was oligotrophic, no additional OAS variability was expected to contribute to the spectral information. The shape of the spectral signatures was therefore expected to remain constant throughout the entire vertical profile, regardless of any daytime ambient light effects.

2.6. Predictive Modeling of CHLa and CDOM

Based on the aggregated field dataset, we tested two different empirical approaches to estimate CHLa and CDOM including two waveband indices for each target variable and multivariate regression based on the full spectrum. For CHLa, we used the following three-band ratio, which is widespread in remote sensing applications [34], see in [8]:

$$CHLa = a + b\left(\frac{1}{R_{670}} - \frac{1}{R_{710}}\right) \cdot R_{750} \tag{6}$$

where a and b are model coefficients that were empirically re-optimized in the cross-validation loop, R is reflectance, and the subscript indicates the wavelength in nm. Additionally, we applied a single waveband model based on the first derivative of the reflectance signal at 690 nm, shown to work well for CHLa estimation in reservoirs by [35]:

$$CHLa = a + b(R'_{690}) \quad (7)$$

where R' is the first derivative of the reflectance curve.

For CDOM, we used a published two-band ratio model [36], which was tested successfully at various lakes by Zhu et al. (2014) [37]:

$$CDOM = a\left(\frac{R_{570}}{R_{654}}\right)^b \quad (8)$$

The choice between existing CDOM algorithms was limited to the spectral range we used for our dataset (470–850 nm) since many of the empirical algorithms are based on wavelengths <470 nm [37]. In parallel to the approach of [35] for CHLa, we empirically defined the wavelength at 602 nm as the one with the strongest correlation between CDOM and the first derivative of reflectance, resulting in the following model:

$$CDOM = a + b(R'_{602}) \quad (9)$$

To compare the performance of target variable-specific waveband indices with the use of hyperspectral data, we calibrated partial least squares regression (PLSR) models [38] with reflectance (PLS_{ref}) and first derivative spectra (PLS_{fda}) using the full spectral range. PLSR is widely used in chemometrics to develop multivariate calibrations with hyperspectral data. The method can cope with multicollinear and noisy datasets and has been applied in hyperspectral water spectroscopy of optically complex waters where OAS specific empirical band ratios might produce inaccurate results (e.g., [39–42]).

All models were evaluated with a 'leave-one-profile-out' cross-validation (CV). Therefore, we split the entire dataset iteratively up into eleven water column profiles for calibration, applying either the above-mentioned waveband models or full range PLSR, and the remaining water column profile for validation. Subsequently, we pooled the estimates of the individually cross-validated profiles for each method and calculated the following performance measures to evaluate the models: coefficient of determination (R^2), root mean square error (RMSE):

$$RMSE = \sqrt{\frac{\sum(\hat{y} - y)^2}{n}} \quad (10)$$

where y = measured value, $\hat{y}$ = estimated value and n = number of samples, and the ratio of performance to interquartile range (RPIQ):

$$RPIQ = \frac{IQR}{RMSE} \quad (11)$$

where IQR is the interquartile range of the reference data.

3. Results and Discussion

3.1. Laboratory Experiment

The results of the performed laboratory experiments are summarized in Figure 5. For reasons of comparison with the hyperspectral camera, we reduced the ASD spectra to a range between 470 and 850 nm. Additionally, we normalized all spectra to the measured reflectance at 810 nm to minimize potential scattering effects due to particulate characteristics of the added substances. The spectra recorded with the UHD 285 hyperspectral camera and the ASD FieldSpec point spectrometer were very similar overall. Reflectance patterns and absorption features at various levels of CDOM and CHLa were clearly defined and did not show any significant deviations between the instruments. The minor systematic offset between the spectra of the point spectrometer and the hyperspectral camera was presumably a result of the measurement setup, as the front of the camera was positioned slightly below the water surface, in contrast to the fiber optic cable of the ASD spectrometer with a position just above the water surface. The addition of CDOM caused a gradual increase in absorption in the 'blue-to-green' spectral range (<550 nm) that leveled off at higher wavelengths as already described in previous studies (e.g., [43–45]). CDOM did not show any distinct absorption features in the VNIR range. The high absorption in the shorter wavelengths presumably reflected large absorption features of dissolved organic matter (DOM) in the ultraviolet (UV) range that tailed off in the VIS [43]. Increasing the concentrations of algae showed a more differentiated effect on the spectra with a characteristic CHLa absorption feature around 670 nm, but induced also less pronounced peaks around 620 nm and 540 nm that might have originated from accessory pigments of *Chlorella* sp.

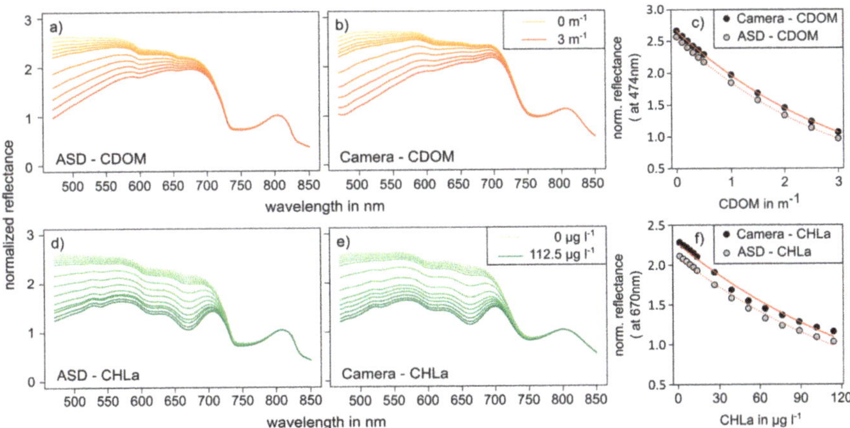

Figure 5. Comparison of experimental laboratory measurements between ASD FieldSpec 4 and UHD 285 hyperspectral camera at various levels of colored dissolved organic matter (CDOM) (**a**,**b**), and chlorophyll a (CHLa) (**d**,**e**) concentrations. Panels (**c**,**f**) show a direct comparison of the measured reflectance at specific key wavelengths sensitive to changes in CDOM and CHLa, respectively.

The normalization of the reflectance spectra to the wavelength at 810 nm resulted in an almost perfect match of the spectral signatures at wavelengths longer than 710 nm. This documented that both substances were not spectrally active in this range. On the other hand, both substances showed overlaps in the entire range below 710 nm. Hence, the presence of one substance might impair the spectral retrieval of the other substance, leading to a non-unique solution referred to as the ill-posed problem of spectra analysis [46]. Besides the absorption coefficient at a certain wavelength in the 400–460 nm range, CDOM could also be characterized by the spectral slope that describes the exponential decay of CDOM absorbance with increasing wavelength and which strongly depends on the molecular composition of DOM (see [43]). The reference analysis in the performed experiment with dissolved humic acid in the given concentration range indeed revealed a spectral slope of 0.008 in the 400–500 nm range. In freshwater lakes, however, the spectral slope of CDOM typically varies in a range between 0.014 and 0.020 (see [44,45,47]), and therefore has a smaller impact on the 'red' spectral range at each given absorption coefficient. Nevertheless, high amounts of CDOM with typical values of the described spectral slope might also affect empirical algorithms for CHLa retrieval in optically complex waters if based on wavelengths around the CHLa feature at 670 nm.

In summary, the hyperspectral camera was able to capture small OAS variabilities with an accuracy comparable to the ASD point spectrometer under laboratory conditions. The signal quality of the image mean was comparable to the point measurement of the ASD instrument and the minor divergence in total reflectance resulted from differences in the instrumental setup.

3.2. Water Quality Characteristics of the Investigated Lakes

The two target variables showed a high degree of variability between the studied lakes (see Table 2). The CHLa contents of the complete dataset varied between 0 and 96 μg L^{-1}, with an overall mean concentration of 37.2 μg L^{-1}. The Cospuden and Kriebstein sites generally had low CHLa contents throughout the complete measured water column with site-specific mean values of 0.6 μg L^{-1} and 2.3 μg L^{-1}, respectively, and a standard deviation of 0.5 μg L^{-1}. In the other two lakes mean concentrations were significantly higher with 63.9 μg L^{-1} at the Auensee site and 36.9 μg L^{-1} at the Mulde site.

Table 2. Descriptive Statistics of reference values of chlorophyll a—contents (CHLa, in μg L^{-1}) and CDOM absorbance at 440 nm (in m^{-1}). OAS = optically active substance, n = number of samples, min = minimum, Q1 = first quartile, Q2 = median, Q3 = third quartile, max = maximum, mean = arithmetic mean, sd = standard deviation.

OAS	Site	n	Min	Q1	Q2	Q3	Max	Mean	sd
CHLa	all	56	0	2	37	64	96	37.2	32.2
	Auensee	27	27	46	64	84	96	63.9	20.9
	Cospuden	10	0	0	1	1	1	0.6	0.5
	Kriebstein	10	2	2	2	3	3	2.3	0.5
	Mulde	9	15	35	36	41	53	36.9	10.5
CDOM	all	56	0.1	0.9	1.0	1.3	1.6	0.97	0.43
	Auensee	27	0.9	1.0	1.0	1.2	1.3	1.10	0.12
	Cospuden	10	0.1	0.1	0.2	0.2	0.2	0.16	0.03
	Kriebstein	10	1.4	1.4	1.4	1.6	1.6	1.45	0.11
	Mulde	9	0.7	0.9	0.9	1.2	1.2	0.95	0.18

A vertical CHLa stratification occurred in those lakes with relatively high CHLa contents (Figure 6). At the Auensee site, the layer-specific CHLa means ranged between 70 and 76 μg L^{-1} within the upper 1.5 m, and the upper layers were also characterized by a high variability (indicated by wide ranges within each layer). By contrast, the average concentration dropped to almost half (40 μg L^{-1}) at a depth of 2.5 m. A similar pattern was observed at the Mulde site with mean values between 38 and 45 μg L^{-1} in the upper 1.5 m and 15 μg L^{-1} in the lowest layer. Due to the overall low CHLa contents, no substantial stratification was observed at the Cospuden and Kriebstein sites.

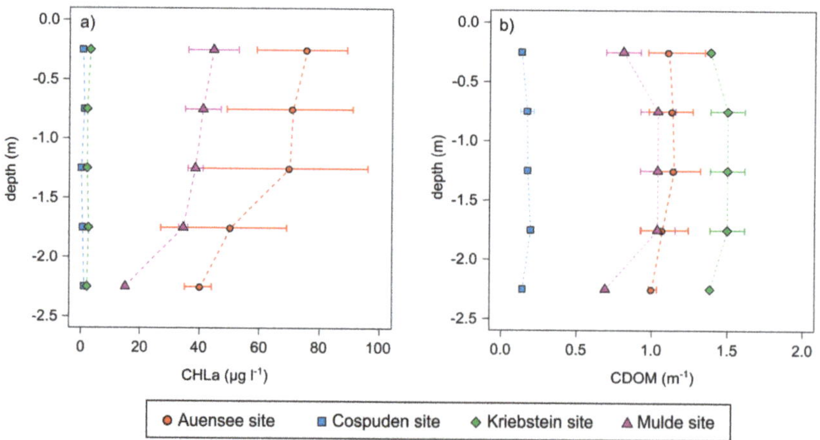

Figure 6. Vertical distribution of CHLa (**a**) and CDOM (**b**) for the investigated lake sites. Points mark the layer-specific mean, solid lines mark the layer-specific range of measured reference values.

The results of the CDOM analysis showed less variability across the lakes with CDOM values between 0.1 and 1.6 m^{-1}. We found the lowest CDOM levels at the Cospuden site with a maximum absorption coefficient of 0.2 m^{-1}, whereas the remaining lakes showed mean levels between 0.95 at the Mulde site and 1.45 m^{-1} at the Kriebstein site. In all four lakes, the vertical profile of CDOM showed an approximately uniform distribution.

In line with the relatively high CHLa concentrations in the surface layer, the Auensee site and the Mulde site showed the lowest Secchi disk depths measured during the field campaigns with ~0.5 m and ~1.3 m, respectively. Given that the Secchi depth approximates the water depth suitable for above-ground remote or proximal sensing, this documents the need for underwater in situ measurements to assess the complete vertical distribution of OAS.

3.3. Validation of Ambient Light Compensation

To validate the applicability of Equation (5) under realistic conditions, we compared the measured reflectance values of nighttime measurements at the Cospuden site with the calculated reflectance values of daytime measurements carried out only a few hours later at the same position (Figure 7).

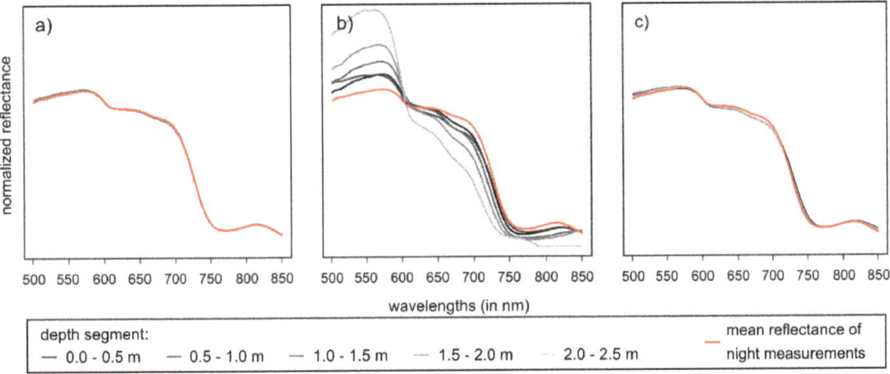

Figure 7. Comparison of reflectance spectra acquired at Cospuden site during (**a**) nighttime measurements, (**b**) daytime measurements without compensation for sunlight attenuation, and (**c**) daytime measurements with compensation for sunlight attenuation.

The spectra of the nighttime measurements were almost identical for all five increments. The reflectance curves showed no variability in terms of OAS absorption features (Figure 7a), which was in line with the results of the laboratory reference analysis, as the entire water column showed negligible contents of CHLa and CDOM. The calculation of reflectance based on daytime measurements without sunlight compensation (Equation (4)) resulted in an increased spectral variability throughout the entire spectral range (Figure 7b). This was solely caused by varying conditions of the ambient light field with increasing water depth. Accordingly, the application of the sunlight compensation algorithm (Equation (5)) removed these differences almost entirely (Figure 7c) and implied an optimization for the retrieval of CDOM and CHLa at different depths, as both absorb in the affected wavelength ranges (as shown in Section 3.1).

At the Auensee site, with a Secchi depth of ~0.5 m, the measured signal below a depth of ~2 m was very low, even with greatly increased integration times, and the retrievable spectral information was therefore limited to the range between 500–700 nm (data not shown), corresponding to the energy maximum of sunlight in the visible range. In the remaining parts of the spectrum, the signal was overlaid by dark current. The authors of [22] also reported a signal loss of >78 % for wavelengths longer than 620 nm within the first meter of a freshwater lake with noticeable algae and cyanobacteria contents. The use of a portable lamp, as shown in the present study, therefore allowed to compensate for the effects of sunlight attenuation and the associated signal loss with increasing depth. The combination of a constant light source and a reference panel at a fixed distance in front of the camera simulated shallow water with a standardized bottom and resulted in almost constant measurement conditions throughout the entire water column. Consequently, all acquired reflectance spectra were comparable between the investigated lakes and across different depths.

3.4. Predictive Modeling of CHLa and CDOM

The lake specific mean reflectance spectra (Figure 8) mirrored the measured OAS reference values. Starting with the mean spectrum of the Cospuden site, which represented low contents of CHLa and CDOM, a clear decrease of reflectance mainly at the shorter wavelengths was observed for the Kriebstein site, mainly attributable to high CDOM levels. Since CHLa was low, reduced reflectance values in the 'red' range might also be caused by CDOM at this site. Although CDOM absorption decays exponentially with increasing wavelengths, this finding suggests that high and variable CDOM contents may also affect CHLa retrieval based on the absorption feature at around 670 nm. The Auensee site showed the lowest overall reflectance curves due to both, high CDOM and CHLa contents, with a marked CHLa feature at ~670 nm. The application of the first derivative on the spectra resulted in the removal of the baseline and a narrowing of the reflective range especially in the region below 670 nm. Values at zero indicated pronounced peaks and troughs of the reflectance spectra, whereas slope differences were highlighted by the first derivative.

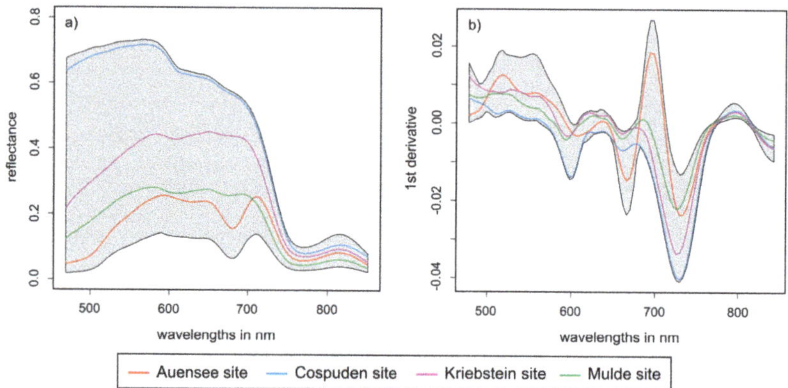

Figure 8. Mean visible and near-infrared (VNIR) reflectance spectra (**a**) and first derivative of mean VNIR reflectance spectra (**b**) of the investigated lake sites in the range 470–850 nm. The colored spectral curves represent the mean spectra of each lake. The shaded region displays the range between the minimum and maximum values at each wavelength for the entire data set.

Based on the found spectral sensitivities, the three-band ratio model worked well for CHLa retrieval in the case of the eutrophic and turbid Auensee waterbody (Table 3, Figure 9), which confirmed the applicability of this index for turbid and productive waters (see [8]).

Table 3. Root mean square error (RMSE) (in µg L^{-1}) for CHLa estimation per lake.

Site	Three-Band Ratio	Single Wavelength	PLS_{ref}	PLS_{fds}
Auensee (63.9) *	9.52	19.86	11.24	10.29
Cospuden (0.6) *	11.04	1.72	2.69	3.46
Kriebstein (2.3) *	12.07	13.78	5.17	5.98
Mulde (36.9) *	22.78	15.23	8.72	7.13

* Values in parentheses represent measured CHLa mean concentrations (in µg L^{-1}) of the lakes.

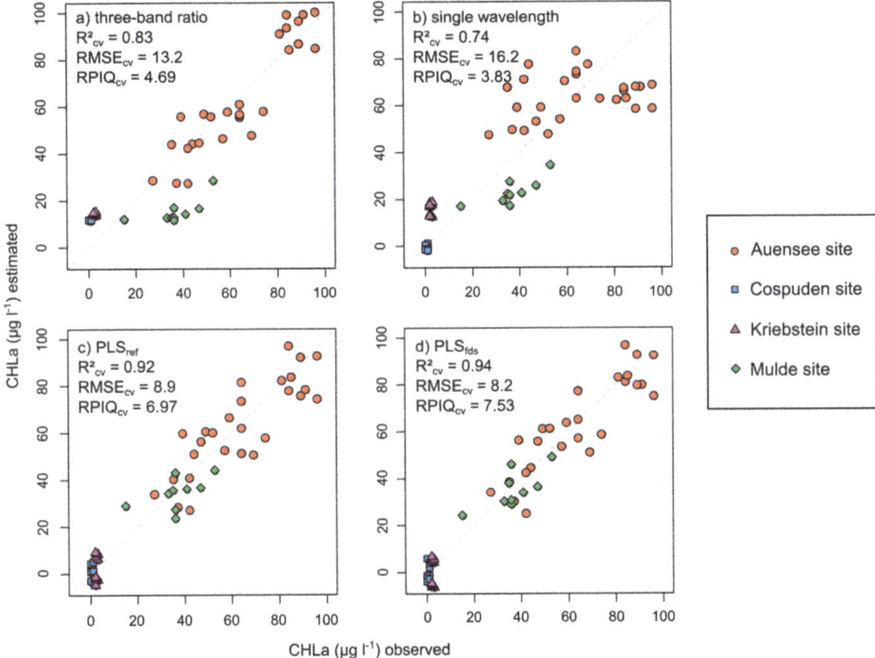

Figure 9. Measured and cross-validated prediction values of CHLa with error metrics based on the entire dataset: (**a**) three-band ratio, (**b**) single wavelength at 690 nm of first derivative, (**c**) partial least squares $(PLS)_{ref}$ = PLS regression based on reflectance spectra, and (**d**) PLS_{fds} = PLS regression based on first derivative spectra. The dashed line represents the 1:1 line.

For the Mulde site, however, values of nearly all samples were underestimated; conversely, we found a general overestimation of values for the remaining two sites (Cospuden, Kriebstein) with low CHLa contents. The single waveband approach based on spectral values of the first derivative at 690 nm showed similar results for the Mulde and Kriebstein sites compared to the three-band ratio. However, more accurate and precise estimates were achieved for the oligotrophic Cospuden site, whereas values of Auensee site samples with high CHLa levels being greater than 70 µg L^{-1} were all underestimated (Figure 9). For lakes with very low CHLa (and CDOM) concentrations, however, relative estimation errors, compared to the lake-specific mean values, were very large due to the small dynamic range in CHLa and CDOM. In the case of the Cospuden and Kriebstein sites, the water bodies were essentially transparent and the range of observed values was consequently lower than the sensitivity of the reflectance spectra given the uncertainties included in field measurements.

For all studied samples, the overall RMSE of cross-validation was 13.2 µg L^{-1} (three-band ratio) and 16 µg L^{-1} (single waveband), respectively. Similar results were found by Duan et al. (2010) [48] who investigated a single eutrophic lake and achieved slightly better results with the three-band ratio compared to the use of the single waveband of the first derivative at 680 nm. Nevertheless, Cheng et al. (2013) [49] showed that first derivative models using one waveband in the 690–700 nm range might be more robust when transferred to independent datasets compared to two-, three-, and four-band ratios of reflectance spectra.

With regard to the entire dataset, the PLSR models based on the full spectral information provided higher accuracies with RMSE values at 8.9 µg L^{-1}, and 8.2 µg L^{-1}, the latter for the first derivative (Figure 9). In addition, there was no systematic over- or under-estimation of a lake-specific sample set or a certain CHLa range. These results suggest that the use of full range reflectance in combination with an empirical multivariate model produces potentially more accurate and robust results than spectral indices, which is in line with other studies (e.g., [40–42]). The authors of [41], who combined PLSR with a genetic algorithm to identify most suitable CHLa sensitive wavelengths, emphasized a better transferability of models calibrated in that way to new sites compared to empirical models based on three-band indices. Nevertheless, the data in Table 3 indicate that CHLa related band indices may provide similarly good or even more accurate results at specific lakes compared to full spectrum approaches. However, prior expert knowledge of the lake under consideration seems to be necessary for the selection of a suitable index, as their estimation accuracies showed a higher variability between the lakes compared to the use of full spectrum models.

These findings suggest that the use of continuous hyperspectral data in the range between ~400 and ~1000 nm for CHLa retrieval is generally of advance compared to the use of band ratio models. Benefits relate to accuracy and transferability, especially for highly diverse water bodies or multiple water bodies with variable conditions.

For CDOM, the results were different (Table 4, Figure 10), which may be traced back to strong influences of e.g., algal biomass on the main region of CDOM absorption in the visible domain (see Section 3.1). Regarding the lake-specific error metrics (Table 4), the indices again show a variable pattern of estimation accuracies. While the single wavelength approach based on the first derivative value at 602 nm achieved the most consistent result of all models for the Cospuden site, the most accurate estimations for the Mulde site were yielded by the two-band ratio. Full spectrum models provided their best results for the Auensee and Cospuden sites.

Table 4. RMSE (in m^{-1}) for CDOM estimation per lake.

Site	Two-Band Ratio	Single Wavelength	PLS_{ref}	PLS_{fds}
Auensee (1.10) *	0.42	0.18	0.11	0.15
Cospuden (0.16) *	0.32	0.12	0.17	0.16
Kriebstein (1.45) *	0.22	0.35	0.27	0.27
Mulde (0.95) *	0.23	0.36	0.38	0.36

* Values in parentheses represent measured CDOM mean levels (in m^{-1}) of the lakes.

Overall estimation accuracies, as indicated by RPIQ values (Figure 10), were thus significantly lower than those for CHLa. The two-band ratio approach produced the poorest results. This contrasts to Zhu et al. (2014) [37], who achieved—with this index—RMSE values at 0.28 m^{-1} for lakes with CDOM levels between 0.9 and 2.1 m^{-1} and at 0.05 m^{-1} for CDOM levels beyond 3.4 m^{-1}. Nevertheless, they also stated that the algorithm might overestimate low CDOM levels. In our study, the single waveband index derived from the first derivative spectra outperformed the two-band ratio index. At this point, the physical relevance of the wavelength region at around 602 nm for the retrieval of CDOM is not obvious, but Brezonik et al. (2015) [47] also summarized several studies that included (at least with moderate success) wavelength regions beyond 500 nm for the retrieval of CDOM. Shao et al. (2016) [50] also tested different index approaches and found a ratio index calculated from reflectance values at 584 nm and 646 nm to outperform a single band index based on values of the first derivative at 406 nm (which showed the highest correlation with CDOM in their dataset). Additionally, they applied PLS with a back-propagation artificial neural network, but this provided less accurate results compared to the two-band ratio approach. Our results showed, different from that, that PLSR with both reflectance and first derivative spectra produced overall more accurate results with an RMSE at 0.22 m^{-1}. The plotted CDOM values revealed two clusters with markedly different levels of CDOM (which qualifies the applicability of one common linear approach and the retrieved statistical measures). Regardless of achieved estimation accuracies, the first derivative waveband approach and the PLS models were both able to separate these two classes, as evident from Figure 10.

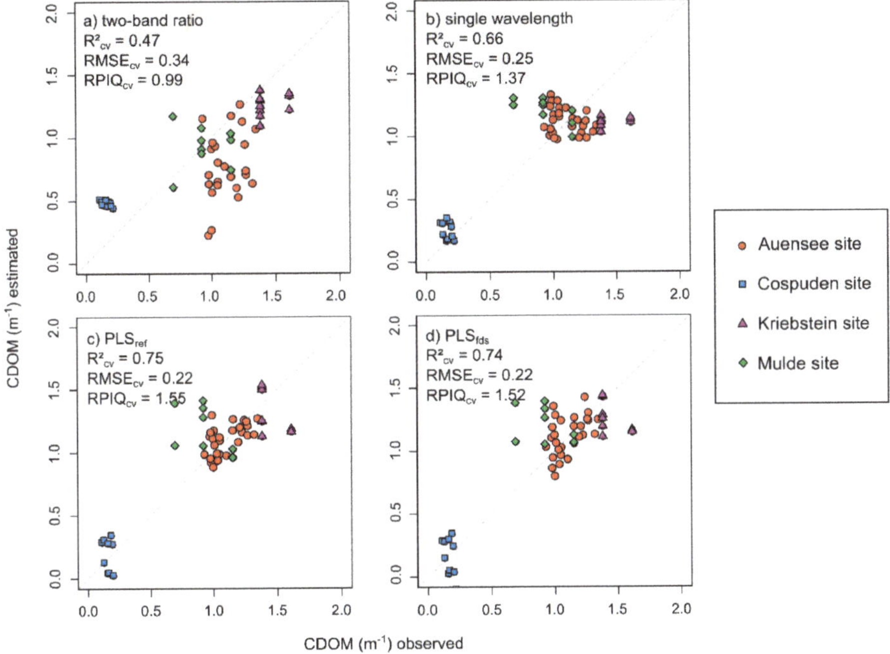

Figure 10. Measured and cross-validated prediction values of CDOM with error metrics based on the entire dataset: (**a**) two-band ratio, (**b**) single wavelength at 602 nm of first derivative, (**c**) PLS_{ref} = PLS regression based on reflectance spectra, and (**d**) PLS_{fds} = PLS regression based on first derivative spectra. The dashed line represents the 1:1 line.

Results for CDOM were generally poorer compared to CHLa, which might be due to a missing diagnostic absorption feature within the investigated spectral range of 470–850 nm and due to a large overlap between CDOM and CHLa absorption in the 'blue' spectral range. Additionally, further OAS such as detritus (non-living organic suspended matter), which is known to absorb in a similar pattern as CDOM does [51], may also affect CDOM retrieval. Therefore, accurate CDOM retrieval based on empirical methods seems to be still challenging, especially in optically complex waters.

Similar to our study, Abd-Elrahman et al. (2011) [26] also studied the retrieval of CHLa in fishery ponds by using a combination of hyperspectral measurements and submergible targets. To this end, they installed a hyperspectral scanner (400 to 1000 nm) above the water surface of 14 aquaculture ponds, where CHLa concentrations ranged from 0.8 to 494.4 μg L^{-1}. Additionally, they developed a three-level design of vertically arranged reflective targets. The first one was positioned above the water surface for calibration purposes, the second one 10 cm below, and the third one 30 cm below the water surface to test the effect of fixed viewing depths by using standardized bottoms. For CHLa retrieval, they used two-band and three-band indices. The best results were achieved with a three-band index and the target that was positioned 10 cm below the water surface (RMSE = 13.4 μg L^{-1}), whereas the lowest accuracy was obtained with the target 30 cm below the water surface (RMSE = 89.9 μg L^{-1}). They confirmed the advantage of using reflective targets in the upper water column to enhance the quality of the spectral signal. We successfully extended that approach by measuring multiple depths of 2.5 m water columns while being almost independent from ambient illumination conditions.

Generally, our results imply a strong potential for resolving the vertical water column at a fine scale for the provision of both, hyperspectral information and OAS products, which could be helpful for modelling approaches regarding the water leaving spectrum [15,16,52] and additionally provides insights into water layers below the viewing depth of above-ground remote or proximal sensors. Further, our measurement protocol proposes an experimental approach to cope with variable illumination conditions in order to obtain consistent reflectance spectra, which can be a critical aspect for in situ measurements [53]. The combination of its use for above-water surface measurements (e.g., [21,26]) and for underwater use (this study), hyperspectral cameras provide a potential link between point source underwater measurements and spectral imaging above the water surface.

4. Conclusions

We evaluated the performance of a submersible hyperspectral camera for underwater reflectance measurements and the estimation of CHLa and CDOM at various depths in four freshwater lakes with different trophic levels. The measurement configuration we developed allows a consistent retrieval of reflectance spectra throughout the water column with potential applications in OAS retrieval or radiometric ground truthing for remote sensing observations of the uppermost water layer. The available image information allows a pixel-wise analysis of the sensor's field of view to improve data quality through the removal of poorly illuminated areas or interfering objects. For quantitative OAS retrieval, predictive models based on hyperspectral reflectance data can achieve more robust and accurate estimates for CHLa and CDOM than empirical algorithms based on specific wavebands, at least in complex datasets that include multiple lakes at different trophic levels. As our comparison included only two common waveband indices, however, a lake-specific selection of different band ratios might yield similar results to hyperspectral algorithms.

While the UHD 285 camera used in this study is a commercial-grade instrument with a mature data acquisition and processing chain, the customizations for underwater use, including the camera mount and the required instrument calibration procedures, were at a research level. Further refinements in the technology are necessary to allow a more rapid deployment, data acquisition and analysis, e.g., for near-real time water monitoring, the integration in sensor networks or operational use by environmental agencies.

Despite these present and future challenges, hyperspectral measurements throughout the water column may potentially bridge the gap between spatially continuous remote sensing observations of the surface water layer and point sensors that can provide continuous water monitoring data at and below the surface.

Author Contributions: Conceptualization, M.S., D.S., A.J. and M.V.; methodology, M.S., C.H., F.O., D.S.; validation, M.S., C.H.; formal analysis, M.S., C.H., F.O.; investigation, M.S., F.O., D.S.; resources, D.S.; data curation, M.S.; writing—original draft preparation, M.S., C.H., F.O., D.S., A.J. and M.V.; writing—review and editing, M.S., C.H., F.O., D.S., A.J. and M.V.; visualization, M.S., D.S.; supervision, A.J. and M.V.; project administration, A.J. and M.V.; funding acquisition, A.J. and M.V. All authors have read and agreed to the published version of the manuscript.

Funding: This research was funded by the German Federal Ministry for Economic Affairs and Energy (BMWi) in the framework of the Central Innovation Program for SMEs (ZIM, Germany, project number: KF3273801SA3).

Acknowledgments: We would like to thank T. Hillmann and especially Olaf Keitsch (University of Applied Sciences, Neubrandenburg) for the design, construction and provision of the camera rack. We would also like to thank Torsten Jakob (Institute for Biology, Leipzig University, Leipzig) for facilitating the reference analyses and for his support. We are grateful to Oliver Elle for his support in the laboratory experiments. We acknowledge support from Leipzig University for Open Access Publishing.

Conflicts of Interest: The authors declare no conflict of interest.

References

1. Aylward, B.; Bandyopadhyay, J.; Belausteguigotia, J.-C.; Börkey, P.; Cassar, A.; Meadors, L.; Saade, L.; Siebentritt, M.; Stein, R.; Tognetti, S.; et al. Freshwater ecosystem services. In *Millenium Ecosystem Assessment: Ecosystems and Human Well-Being: Policy Responses*; Chopra, K., Leemans, R., Kumar, P., Simons, H., Eds.; Island Press: Washington, DC, USA, 2005; pp. 213–255.
2. Schallenberg, M.; de Winton, M.D.; Verburg, P.; Kelly, D.J.; Hamill, K.D.; Hamilton, D.P. Ecosystem services of lakes. In *Ecosystem Services in New Zealand–Conditions and Trends*; Dymond, J.R., Ed.; Whenua Press: Lincoln, New Zealand, 2013; pp. 203–225.

3. Carpenter, S.R.; Stanley, E.H.; Zanden, M.J.V. State of the World's Freshwater Ecosystems: Physical, Chemical, and Biological Changes. *Annu. Rev. Environ. Resour.* **2011**, *36*, 75–99. [CrossRef]
4. Dörnhöfer, K.; Oppelt, N. Remote sensing for lake research and monitoring – Recent advances. *Ecol. Indic.* **2016**, *64*, 105–122. [CrossRef]
5. Gholizadeh, M.H.; Melesse, A.; Reddi, L.N. A Comprehensive Review on Water Quality Parameters Estimation Using Remote Sensing Techniques. *Sensors* **2016**, *16*, 1298. [CrossRef]
6. Odermatt, D.; Gitelson, A.; Brando, V.; Schaepman, M.E. Review of constituent retrieval in optically deep and complex waters from satellite imagery. *Remote Sens. Environ.* **2012**, *118*, 116–126. [CrossRef]
7. Giardino, C.; Brando, V.; Gege, P.; Pinnel, N.; Hochberg, E.; Knaeps, E.; Reusen, I.; Doerffer, R.; Bresciani, M.; Braga, F.; et al. Imaging Spectrometry of Inland and Coastal Waters: State of the Art, Achievements and Perspectives. *Surv. Geophys.* **2018**, *40*, 401–429. [CrossRef]
8. Matthews, M. A current review of empirical procedures of remote sensing in inland and near-coastal transitional waters. *Int. J. Remote Sens.* **2011**, *32*, 6855–6899. [CrossRef]
9. Palmer, S.; Kutser, T.; Hunter, P.D. Remote sensing of inland waters: Challenges, progress and future directions. *Remote Sens. Environ.* **2015**, *157*, 1–8. [CrossRef]
10. Matthews, M.; Bernard, S. Characterizing the Absorption Properties for Remote Sensing of Three Small Optically-Diverse South African Reservoirs. *Remote Sens.* **2013**, *5*, 4370–4404. [CrossRef]
11. Bernardo, N.; Watanabe, F.; Rodrigues, T.; Alcantara, E. Atmospheric correction issues for retrieving total suspended matter concentrations in inland waters using OLI/Landsat-8 image. *Adv. Space Res.* **2017**, *59*, 2335–2348. [CrossRef]
12. Moses, W.J.; Sterckx, S.; Montes, M.J.; Keukelaere, L.D.; Knaeps, E. Atmospheric correction for inland waters. In *Bio-Optical Modeling and Remote Sensing of Inland Waters*, 1st ed.; Deepak, M., Ogashawara, I., Gitelson, A., Eds.; Elsevier: Amsterdam, The Netherlands, 2017; pp. 69–100.
13. Mouw, C.B.; Greb, S.; Aurin, D.; Digiacomo, P.M.; Lee, Z.; Twardowski, M.; Binding, C.; Hu, C.; Ma, R.; Moore, T.; et al. Aquatic color radiometry remote sensing of coastal and inland waters: Challenges and recommendations for future satellite missions. *Remote Sens. Environ.* **2015**, *160*, 15–30. [CrossRef]
14. Sterckx, S.; Knaeps, E.; Ruddick, K. Detection and correction of adjacency effects in hyperspectral airborne data of coastal and inland waters: The use of the near infrared similarity spectrum. *Int. J. Remote Sens.* **2011**, *32*, 6479–6505. [CrossRef]
15. Kutser, T.; Metsamaa, L.; Dekker, A. Influence of the vertical distribution of cyanobacteria in the water column on the remote sensing signal. *Estuar. Coast. Shelf Sci.* **2008**, *78*, 649–654. [CrossRef]
16. Pitarch, J.; Odermatt, D.; Kawka, M.; Wüest, A. Retrieval of vertical particle concentration profiles by optical remote sensing: A model study. *Opt. Express* **2014**, *22*, A947–A959. [CrossRef] [PubMed]
17. Xue, K.; Zhang, Y.; Duan, H.; Ma, R.; Loiselle, S.A.; Zhang, M. A Remote Sensing Approach to Estimate Vertical Profile Classes of Phytoplankton in a Eutrophic Lake. *Remote Sens.* **2015**, *7*, 14403–14427. [CrossRef]
18. Xue, K.; Zhang, Y.; Ma, R.; Duan, H. An approach to correct the effects of phytoplankton vertical nonuniform distribution on remote sensing reflectance of cyanobacterial bloom waters. *Limnol. Oceanogr. Methods* **2017**, *15*, 302–319. [CrossRef]
19. Leach, T.H.; Beisner, B.; Carey, C.C.; Pernica, P.; Rose, K.C.; Huot, Y.; Brentrup, J.A.; Domaizon, I.; Grossart, H.-P.; Ibelings, B.W.; et al. Patterns and drivers of deep chlorophyll maxima structure in 100 lakes: The relative importance of light and thermal stratification. *Limnol. Oceanogr.* **2017**, *63*, 628–646. [CrossRef]
20. Duan, W.; He, B.; Chen, Y.; Zou, S.; Wang, Y.; Nover, D.; Chen, W.; Yang, G. Identification of long-term trends and seasonality in high-frequency water quality data from the Yangtze River basin, China. *PLoS ONE* **2018**, *13*. [CrossRef]
21. Keller, S.; Maier, P.M.; Riese, F.; Norra, S.; Holbach, A.; Börsig, N.; Wilhelms, A.; Moldaenke, C.; Zaake, A.; Hinz, S. Hyperspectral Data and Machine Learning for Estimating CDOM, Chlorophyll a, Diatoms, Green Algae and Turbidity. *Int. J. Environ. Res. Public Heal.* **2018**, *15*, 1881. [CrossRef]
22. Kwon, Y.S.; Pyo, J.; Duan, H.; Cho, K.H.; Park, Y.; Kwon, Y.-H. Drone-based hyperspectral remote sensing of cyanobacteria using vertical cumulative pigment concentration in a deep reservoir. *Remote Sens. Environ.* **2020**, *236*, 111517. [CrossRef]
23. Greb, S.; Odermatt, D.; Schaeffer, B.A.; Spyrakos, E.; Wang, M. Complementarity of in situ and satellite measurements. In *Earth Observations in Support of Global Water Quality Monitoring*; Greb, S., Dekker, A., Binding, C., Eds.; IOCCG: Dartmouth, NS, Canada, 2018; pp. 29–42.

24. Kutser, T.; Paavel, B.; Verpoorter, C.; Ligi, M.; Soomets, T.; Toming, K.; Casal, G. Remote Sensing of Black Lakes and Using 810 nm Reflectance Peak for Retrieving Water Quality Parameters of Optically Complex Waters. *Remote Sens.* **2016**, *8*, 497. [CrossRef]
25. Pyo, J.; Ligaray, M.V.; Kwon, Y.; Ahn, M.-H.; Kim, K.; Lee, H.; Kang, T.; Cho, S.B.; Park, Y.; Cho, K.H. High-Spatial Resolution Monitoring of Phycocyanin and Chlorophyll-a Using Airborne Hyperspectral Imagery. *Remote Sens.* **2018**, *10*, 1180. [CrossRef]
26. Abd-Elrahman, A.; Croxton, M.; Pande-Chettri, R.; Toor, G.S.; Smith, S.E.; Hill, J. In situ estimation of water quality parameters in freshwater aquaculture ponds using hyperspectral imaging system. *ISPRS J. Photogramm. Remote Sens.* **2011**, *66*, 463–472. [CrossRef]
27. Wagner, A.; Hilgert, S.; Kattenborn, T.; Fuchs, S. Proximal VIS-NIR spectrometry to retrieve substance concentrations in surface waters using partial least squares modelling. *Water Supply* **2018**, *19*, 1204–1211. [CrossRef]
28. Jung, A.; Vohland, M.; Thiele-Bruhn, S. Use of A Portable Camera for Proximal Soil Sensing with Hyperspectral Image Data. *Remote Sens.* **2015**, *7*, 11434–11448. [CrossRef]
29. Mobley, C.D. Estimation of the remote-sensing reflectance from above-surface measurements. *Appl. Opt.* **1999**, *38*, 7442–7455. [CrossRef]
30. Dunker, S.; Nadrowski, K.; Jakob, T.; Kasprzak, P.; Becker, A.; Langner, U.; Kunath, C.; Harpole, S.; Wilhelm, C. Assessing in situ dominance pattern of phytoplankton classes by dominance analysis as a proxy for realized niches. *Harmful Algae* **2016**, *58*, 74–84. [CrossRef]
31. Abel, A.; Michael, A.; Zartl, A.; Werner, F. Impact of erosion-transported overburden dump materials on water quality in Lake Cospuden evolved from a former open cast lignite mine south of Leipzig, Germany. *Environ. Earth Sci.* **2000**, *39*, 683–688. [CrossRef]
32. ISO 10260. *Water Quality—Measurement of Biochemical Parameters—Spectrometric Determination of the Chlorophyll-A Concentration*; International Organization for Standardization: Geneva, Switzerland, 1992.
33. Mannino, A.; Novak, M.G.; Nelson, N.B.; Belz, M.; Berthon, J.-F.; Blough, N.V.; Boss, E.; Bricaud, A.; Chaves, J.; del Castillo, C.; et al. Measurement protocol of absorption by chromophoric dissolved organic matter (CDOM) and other dissolved materials (DRAFT). In *Inherent Optical Property Measurements and Protocols: Absorption Coefficient*; Mannino, A., Novak, M.G., Eds.; IOCCG: Dartmouth, NS, Canada, 2019; pp. 22–27.
34. Gitelson, A.A.; Gritz, Y.; Merzlyak, M.N. Relationships between leaf chlorophyll content and spectral reflectance and algorithms for non-destructive chlorophyll assessment in higher plant leaves. *J. Plant Physiol.* **2003**, *160*, 271–282. [CrossRef]
35. Han, L.; Rundquist, D.C. Comparison of NIR/RED ratio and first derivative of reflectance in estimating algal-chlorophyll concentration: A case study in a turbid reservoir. *Remote Sens. Environ.* **1997**, *62*, 253–261. [CrossRef]
36. Ficek, D.; Zapadka, T.; Dera, J. Remote sensing reflectance of Pomeranian lakes and the Baltic. *Oceanologia* **2011**, *53*, 959–970. [CrossRef]
37. Zhu, W.; Yu, Q.; Tian, Y.; Becker, B.L.; Zheng, T.; Carrick, H.J. An assessment of remote sensing algorithms for colored dissolved organic matter in complex freshwater environments. *Remote Sens. Environ.* **2014**, *140*, 766–778. [CrossRef]
38. Wold, S.; Sjöström, M.; Eriksson, L. PLS-regression: A basic tool of chemometrics. *Chemom. Intell. Lab. Syst.* **2001**, *58*, 109–130. [CrossRef]
39. Ali, K.A.; Ortiz, J.D. Multivariate approach for chlorophyll-a and suspended matter retrievals in Case II type waters using hyperspectral data. *Hydrol. Sci. J.* **2015**, *61*, 200–213. [CrossRef]
40. Ryan, K.; Ali, K. Application of a partial least-squares regression model to retrieve chlorophyll-a concentrations in coastal waters using hyper-spectral data. *Ocean Sci. J.* **2016**, *51*, 209–221. [CrossRef]
41. Song, K.; Li, L.; Tedesco, L.; Li, S.; Duan, H.; Liu, D.; Hall, B.; Du, J.; Li, Z.; Shi, K.; et al. Remote estimation of chlorophyll-a in turbid inland waters: Three-band model versus GA-PLS model. *Remote Sens. Environ.* **2013**, *136*, 342–357. [CrossRef]
42. Wang, Z.; Kawamura, K.; Sakuno, Y.; Fan, X.; Gong, Z.; Lim, J. Retrieval of Chlorophyll-a and Total Suspended Solids Using Iterative Stepwise Elimination Partial Least Squares (ISE-PLS) Regression Based on Field Hyperspectral Measurements in Irrigation Ponds in Higashihiroshima, Japan. *Remote Sens.* **2017**, *9*, 264. [CrossRef]

43. Helms, J.; Stubbins, A.; Ritchie, J.D.; Minor, E.C.; Kieber, D.J.; Mopper, K. Absorption spectral slopes and slope ratios as indicators of molecular weight, source, and photobleaching of chromophoric dissolved organic matter. *Limnol. Oceanogr.* **2008**, *53*, 955–969. [CrossRef]
44. Strömbeck, N.; Pierson, D. The effects of variability in the inherent optical properties on estimations of chlorophyll a by remote sensing in Swedish freshwaters. *Sci. Total Environ.* **2001**, *268*, 123–137. [CrossRef]
45. Twardowski, M.S.; Boss, E.; Sullivan, J.M.; Donaghay, P.L. Modeling the spectral shape of absorption by chromophoric dissolved organic matter. *Mar. Chem.* **2004**, *89*, 69–88. [CrossRef]
46. Defoin-Platel, M.; Chami, M. How ambiguous is the inverse problem of ocean color in coastal waters? *J. Geophys. Res. Space Phys.* **2007**, *112*, 843. [CrossRef]
47. Brezonik, P.L.; Olmanson, L.G.; Finlay, J.C.; Bauer, M.E. Factors affecting the measurement of CDOM by remote sensing of optically complex inland waters. *Remote Sens. Environ.* **2015**, *157*, 199–215. [CrossRef]
48. Duan, H.; Ma, R.; Xu, J.; Zhang, Y.; Zhang, B. Comparison of different semi-empirical algorithms to estimate chlorophyll-a concentration in inland lake water. *Environ. Monit. Assess.* **2009**, *170*, 231–244. [CrossRef] [PubMed]
49. Cheng, C.; Wei, Y.; Sun, X.; Zhou, Y. Estimation of Chlorophyll-a Concentration in Turbid Lake Using Spectral Smoothing and Derivative Analysis. *Int. J. Environ. Res. Public Heal.* **2013**, *10*, 2979–2994. [CrossRef] [PubMed]
50. Shao, T.; Song, K.; Du, J.; Zhao, Y.; Liu, Z.; Zhang, B. Retrieval of CDOM and DOC Using In Situ Hyperspectral Data: A Case Study for Potable Waters in Northeast China. *J. Indian Soc. Remote Sens.* **2015**, *44*, 77–89. [CrossRef]
51. Babin, M.; Stramski, D.; Ferrari, G.M.; Claustre, H.; Bricaud, A.; Obolensky, G.; Hoepffner, N. Variations in the light absorption coefficients of phytoplankton, nonalgal particles, and dissolved organic matter in coastal waters around Europe. *J. Geophys. Res. Space Phys.* **2003**, *108*, 3211. [CrossRef]
52. Nouchi, V.; Odermatt, D.; Wüest, A.; Bouffard, D. Effects of non-uniform vertical constituent profiles on remote sensing reflectance of oligo- to mesotrophic lakes. *Eur. J. Remote Sens.* **2018**, *51*, 808–821. [CrossRef]
53. Göritz, A.; Berger, S.A.; Gege, P.; Grossart, H.-P.; Nejstgaard, J.C.; Riedel, S.; Röttgers, R.; Utschig, C. Retrieval of Water Constituents from Hyperspectral In-Situ Measurements under Variable Cloud Cover—A Case Study at Lake Stechlin (Germany). *Remote Sens.* **2018**, *10*, 181. [CrossRef]

Article

Automated Georectification and Mosaicking of UAV-Based Hyperspectral Imagery from Push-Broom Sensors

Yoseline Angel [1,*], Darren Turner [2], Stephen Parkes [1], Yoann Malbeteau [1,3], Arko Lucieer [2] and Matthew F. McCabe [1]

1 Hydrology, Agriculture and Land Observation Group (HALO), Division of Biological and Environmental Science and Engineering, King Abdullah University of Science and Technology, Thuwal 23955-6900, Saudi Arabia; stephen.parkes@kaust.edu.sa (S.P.); yoann.malbeteau@kaust.edu.sa (Y.M.); Matthew.McCabe@kaust.edu.sa (M.F.M.)

2 Discipline of Geography and Spatial Sciences, College of Sciences and Engineering, University of Tasmania, Hobart 7001, Australia; darren.turner@utas.edu.au (D.T.); arko.lucieer@utas.edu.au (A.L.)

3 Center for Remote Sensing Applications (CRSA), Mohammed VI Polytechnic University, Ben Guerir 43150, Morocco

* Correspondence: yoseline.angellopez@kaust.edu.sa

Received: 27 October 2019; Accepted: 30 November 2019; Published: 20 December 2019

Abstract: Hyperspectral systems integrated on unmanned aerial vehicles (UAV) provide unique opportunities to conduct high-resolution multitemporal spectral analysis for diverse applications. However, additional time-consuming rectification efforts in postprocessing are routinely required, since geometric distortions can be introduced due to UAV movements during flight, even if navigation/motion sensors are used to track the position of each scan. Part of the challenge in obtaining high-quality imagery relates to the lack of a fast processing workflow that can retrieve geometrically accurate mosaics while optimizing the ground data collection efforts. To address this problem, we explored a computationally robust automated georectification and mosaicking methodology. It operates effectively in a parallel computing environment and evaluates results against a number of high-spatial-resolution datasets (mm to cm resolution) collected using a push-broom sensor and an associated RGB frame-based camera. The methodology estimates the luminance of the hyperspectral swaths and coregisters these against a luminance RGB-based orthophoto. The procedure includes an improved coregistration strategy by integrating the Speeded-Up Robust Features (SURF) algorithm, with the Maximum Likelihood Estimator Sample Consensus (MLESAC) approach. SURF identifies common features between each swath and the RGB-orthomosaic, while MLESAC fits the best geometric transformation model to the retrieved matches. Individual scanlines are then geometrically transformed and merged into a single spatially continuous mosaic reaching high positional accuracies only with a few number of ground control points (GCPs). The capacity of the workflow to achieve high spatial accuracy was demonstrated by examining statistical metrics such as RMSE, MAE, and the relative positional accuracy at 95% confidence level. Comparison against a user-generated georectification demonstrates that the automated approach speeds up the coregistration process by 85%.

Keywords: georectification; mosaicking; push-broom; UAV; hyperspectral imaging

---

1. Introduction

Remote sensing has provided incredible advances in our capacity to observe and understand the earth system [1], with new and emerging technologies providing further opportunities for insights and understanding [2]. One of the key constraints in our observation capacity relates to the compromise between spatial and temporal resolution, i.e., space-based platforms tend to suffer from either spatial or temporal restrictions that affect the frequency and fidelity of retrievals. Within the last decade, developments in remote sensing using unmanned aerial vehicles (UAV) have provided a sensor-flexible platform that has lowered operational costs, while providing unprecedented spatial (<10 cm) and on-demand temporal resolution [3,4]. To leverage these technological advances, hyperspectral camera systems capturing radiances across the visible and near-infrared portions of the spectrum [5] have been developed, with diverse applications being presented in agriculture [6–8], forestry [9], and mining [10] studies. However, while UAV technology has progressed rapidly and there is a level of maturity in many sensing capabilities [11], routine application of hyperspectral imaging systems remains challenging and has been constrained by a lack of automation and processing options to streamline the image analysis. In particular, additional efforts are required in the integration of accurate positional sensors with spectral devices during image collection and subsequent automated geometric calibration frameworks based on image coregistration and ground control points. To realize this potential, UAV-based hyperspectral sensing systems need to provide radiometrically and geometrically accurate data that allow posterior quantitative analysis to be performed with confidence.

UAV-based image products are generally produced by stitching together hundreds of overlapping scanlines or frames captured on the fly [10]. However, when "matching" any two images, the transformation and reprojection undertaken by the fitting algorithms routinely introduce localized distortions. While the positional accuracy of an individual image may be on the order of a few centimeters, the accuracy of the completed mosaic may increase to the decimeter range as the positional errors accumulate through the merging process. In the case of scanning systems, image distortion can also result due to geometric noise induced by UAV movements. Likewise, accurately overlapping swaths requires a good number of matching and ground control points (GCPs) to avoid further distortion in the final mosaic [11]. A range of hyperspectral sensor configurations is available for UAV-based integration, including point [12] and push-broom spectrometers, as well as 2D spectral imagers [3]. In general, the traditional image georectification process for any such system relies on positional, orientation, rotation, and acceleration data collected by Global Navigation Satellite Systems (GNSS) and/or Inertial Navigation Systems (INS) [13]. In the case of point spectrometers [12,14–17], spectra are collected with no integrated spatial reference, requiring ancillary data (onboard and in situ) to georeference the imagery. Push-broom sensors [18–27] offer a high spectral and spatial resolution by sampling individual lines of spectra during flight. However, the spatial accuracy of each scanline is highly dependent on flying conditions, with the resulting error constrained by GNSS/INS sensors accuracy [28] and the stability provided by the gimbal setup. Conversely, 2D cameras collect band sequential spectra data in two spatial dimensions or by integrating multiple synchronized cameras [29–33]. Such is the case of snapshot systems [34–36], which record all the bands simultaneously, with the advantage of capturing spatial and spectral data with every scene. In both cases, the mosaicking process of UAV-based spectral imaging requires rectification approaches [37,38], which, when integrated with an optimal combination of complementary sensors, assure the spatial accuracy of the products.

Multiple applications using UAV-based hyperspectral systems have been proposed in the literature, exploring a range of prototypes and georectification methods. For instance, Zarco-Tejada et al. [18,20] investigated the early detection of plant diseases and the seasonal trends of narrow-band physiological and structural vegetation indices using a Headwall micro-Hyperspec [21] VNIR push-broom sensor onboard a fixed-wing platform. In their studies, the geometric rectification of the 30 cm [18] and 40 cm [20] pixel resolution imagery was conducted using the PARGE [39] software, which relies on GNSS/INS parameters and a digital elevation model (DEM) to perform the ortho-rectification of airborne optical scanner imagery. Lucieer et al. [22], Turner et al. [23], and Malenovsky et al. [24], used the same sensor mounted on a multirotor aircraft, collecting 2–4 cm ground sample distance (GSD) hypercubes to map the health and status of vegetation. In these cases, the geometrical rectification was based on a dense network of ground control points in addition to using PARGE [39] and achieved a root mean square error (RMSE) of around 5 cm. A different arrangement was employed by Sankey et al., 2017, who collected 12 cm pixel resolution data by integrating the Headwall Nano-Hyperspec [21], and a Light Detection and Ranging (LiDAR) system onboard an octocopter for forest monitoring. Sankey et al. preprocessed the individual hyperspectral tiles in the SpectralView software [40], and then manually tied the georectified swaths to produce a single mosaic, achieving an RMSE of 0.94 m and 1.1 m in the X and Y dimensions, respectively. Recently, a boresight calibration of GNSS/INS has been explored by Habib et al. [41], in an attempt to directly derive the scanner position and orientation by defining the optimal/minimal flight and control/tie point configuration.

With an aim of reducing the required ground sampling efforts and the payload onboard, some studies have explored photogrammetry-based computer vision approaches to determine sensor orientations. Suomalainen et al. [26] and Turner et al. [23] developed processing workflows that include RGB frame-based scenes captured simultaneously with the hyperspectral imagery, to produce a DSM by using Structure from Motion (SfM) algorithms, and then feeding PARGE [39] with this high-resolution model, with resulting imagery achieving accuracies below 10 cm RMSE. Ramirez-Paredes et al. [37] sought to exploit the homographies between RGB frames to align line-to-line the hyperspectral data in the frame camera image plane, by using a low-cost payload on a radio-controlled airplane. Habib et al. [38] proposed an alternative mosaicking approach relying on image coregistration algorithms to stitch together hyperspectral swaths, which were previously rectified by feeding SpectralView [40] with a base-frame DSM, reaching submetric accuracies. Further, computer vision coregistration approaches have even been explored as standard video stabilization techniques [42–44] by performing a robust feature detection using Scale-invariant Feature Transform (SIFT), Speeded Up Robust Features (SURF), Features from Accelerated Segment Test (FAST), and Binary Robust Independent Elementary Features (BRIEF) key points between adjacent frames, then smoothing the sensor path and finally rendering the stabilized frames of a video.

From the approaches presented above, the semiautomated [18,20,22,26,39] solutions require additional efforts, including collection of a high number of GCPs and manually detecting matching points to produce decimeter accurate georectified mosaics. In contrast, previous semiautomated [37,38] methods identify pairs of points based on image coregistration algorithms, with the limitations involving manually identifying geometrical features [28], being compute-intense, and not exceeding the accuracies achieved by manually-based approaches. In general, all of the described techniques highlight the necessity for further research towards highly accurate, fast and fully automated methods that provide a balance between sensors payload, ancillary field data needs, and computational efficiency. An additional challenge is the massive volume of UAV-based hyperspectral data cubes (on the order of terabytes) that are now being collected [45], particularly for those studies where millimeter scales may be required (i.e., phenotyping investigations) [46]. All of these factors highlight the need for speeding up geo-processing to achieve highly accurate hyperspectral imagery while optimizing the data collection demands.

In view of the above, the goal of this research was to conduct a fully automated workflow to produce highly accurate georectified UAV-based hyperspectral mosaics collected by push-broom scanners and to optimize the geoprocessing time by adopting an efficient computing coregistration strategy, requiring a small number of GCPs. UAV-based hyperspectral scans and RGB scenes from two different experiments were used to assess the applicability of the proposed workflow, which follows the five stages of image coregistration process between individual hyperspectral scans and an RGB frame-based orthophoto [47], including: (i) feature detection and description, (ii) feature matching, (iii) inlier selection, (iv) derivation of a transformation function, and (v) image mosaicking. RGB frame-based orthophotos were used as a reference to georectify preprocessed hyperspectral swaths by implementing a parallelized routine of feature detector functions that find the corresponding points between them. The Speeded Up Robust Features (SURF) [48] matching algorithm was integrated with the Maximum Likelihood Estimation SAmple and Consensus (MLSAC) [49], estimator method to automate the coregistration processing. Accordingly, individual geographical transformations per swath were estimated, and the georectified strips were mosaicked. The computational robustness of the approach was evaluated by timing each step-process, and the spatial accuracy was assessed by determining standard accuracy metrics such as root mean square error (RMSE), mean absolute error (MAE), with the relative positional accuracy determined at the 95% confidence level. The proposed methodology provides a novel solution to expedite one of the most costly postprocessing stages of UAV-based hyperspectral remote sensing for push-broom sensors, implementing a simplified coregistration strategy and achieving high positional accurate results.

2. Materials and Methods

2.1. Study Area and Experimental Design

Data were collected from two experimental facilities in Saudi Arabia (Figure 1). The first dataset supports a phenotyping study undertaken over a wild tomato crop at the King Abdulaziz University experimental farm, located at Hada Al-Sham, approximately 60 km east of Jeddah [50]. The site is characterized by a tropical arid climate [49] with an annual rainfall below 100 mm and is situated in a valley at an elevation of approximately 250 m above sea-level, with a predominantly sandy loam soil type. Four campaigns were conducted during the winter season from November 2017 to the end of January 2018, when the average air temperature (during UAV flight) was between 10 and 35 °C. The second study site was a commercial date palm plantation near Al-Kharj [51], a city approximately 200 km southeast of Riyadh. The site is located in a desert depression approximately 1300 m above sea level, it has an average annual rainfall of 51 mm and has sandy desert soils that are irrigated by a natural spring [52]. A single campaign was undertaken during May 2018, when average daytime temperatures reached highs of around 33 °C. Both sites present quite different crop types and geographic extents, which allows an assessment of the transferability of the proposed georectification approach. For instance, a square area of 80 m × 80 m was established for the tomato experiment, comprising four fields with rows aligned along the north-east direction at approximately 2 m spacing. For the date plantation, a total area of approximately 8.7 hectares (270 m × 320 m) was overflown following a north-east direction, with a total of 1300 individual palms (equally spaced at 8 m intervals) captured.

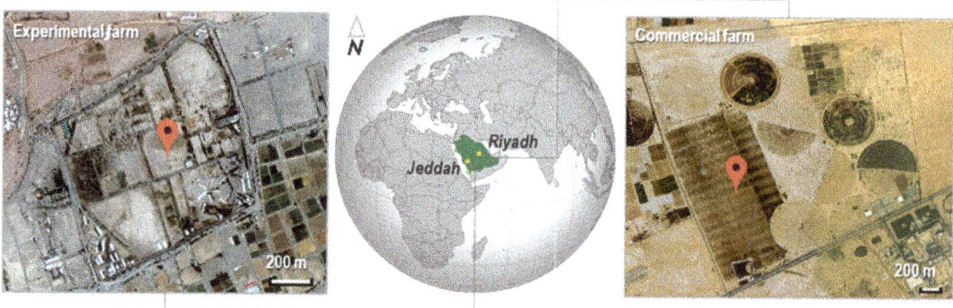

Figure 1. Study site locations, including (**left**) the tomato experiment at the Hada Al-Sham experimental facility (Lat. = 21.797°, Long. = 39.725°), approximately 60 km east of Jeddah, and (**right**) the commercial date farm near Al Kharj (Lat. = 24.231°, Long. = 47.633°), approximately 200 km southeast of Riyadh.

2.2. Unmanned Aerial Vehicles and Sensor Package

Two separate UAV-based remote sensing systems were used for data collection (Figure 2). Hyperspectral imagery was collected using a DJI Matrice 600 (M600) hexacopter [53] coupled with a Ronin-MX gimbal to reduce flight dynamic effects. The flight platform housed a Headwall Nano-Hyperspec [21] push-broom camera, with 12 mm lens and a horizontal field of view (FOV) of 21.1°, which gathered radiometric data in the 400–1000 nm range across 272 continuous bands and with 6 nm FWHM. Two GNSS antennas were mounted on the upper plate of the UAV, with one for the aircraft navigation and another for the hyperspectral camera. An Xsens inertial measurement unit (IMU) was paired with the camera and the GNSS antenna, to monitor the roll, pitch, and yaw motions. The total payload of the M600 was 3.65 kg, which constrains the flight time to approximately 20 min. Ancillary RGB imagery was captured using a DJI Matrice 100 (M100) quadcopter [54], which is paired with a 3-axis gimbal to keep the camera steady in the air, an IMU built in the main controller, and a single GNSS navigation antenna. An on-board Exmor CMOS Zenmuse X3 frame camera [55], with 20 mm optical lens and diagonal FOV of 94°, collected RGB data across a single spectral range (400–700 nm). The total payload of the M100 was 0.25 kg, constraining the flight time to approximately 20 min.

Figure 2. The DJI Matrice 600 and 100 unmanned aerial vehicle (UAV) systems, sensors, and payload used for data collection over the experimental sites. The Headwall Nano-Hyperspec collects surface radiance in the wavelength range from 400–1000 nm across 270 continuous bands. The Zenmuse X3 camera collects RGB radiance in the visible spectral range across a single 400–700 nm spectral range.

2.3. Flight Planning

Prior to each field campaign, a flight plan was designed depending on flight altitude, spatial resolution requirement, area to cover, overlap percentage between swaths, and lighting conditions (Figure 3). Additional preflight aspects to consider included planning for optimal atmospherical conditions. Morning hours close to solar noon under clear sky were preferred to avoid wind and thermals generated by environmental heating. The Universal Ground Control Station [56] desktop application was used to construct all UAV flight plans. For the tomatoes experiment, the hyperspectral swaths were collected using the M600, with 30% sidelap at a speed of 1 m/s and a height of 16 m, scanning at a frame rate of 100 fps to ensure square pixels. A total of four flights per campaign were required to collect 56 swaths, with a ground sampling distance (GSD) of 0.007 m. RGB data was captured with a 78% along-track overlap and 82% sidelap at a speed of 2 m/s and a height of 13 m, with a frame frequency of 0.33 fps. A total of 196 frames with a 0.005 m pixel size fully covered the area. For the date palms plantation, a total of 16 hyperspectral strips were scanned, reaching a GSD of 0.06 m with 40% sidelap, flying at a speed of 5 m/s and a height of 80 m above the ground, scanning at a frame rate of 100 fps. In addition, 184 RGB frames at a 0.04 m spatial resolution were captured with the M100, with an 82% along-track overlap and 87% sidelap, flying at a speed of 5 m/s and a height of 80 m. More detailed information on the specific flight configurations is provided in Table 1.

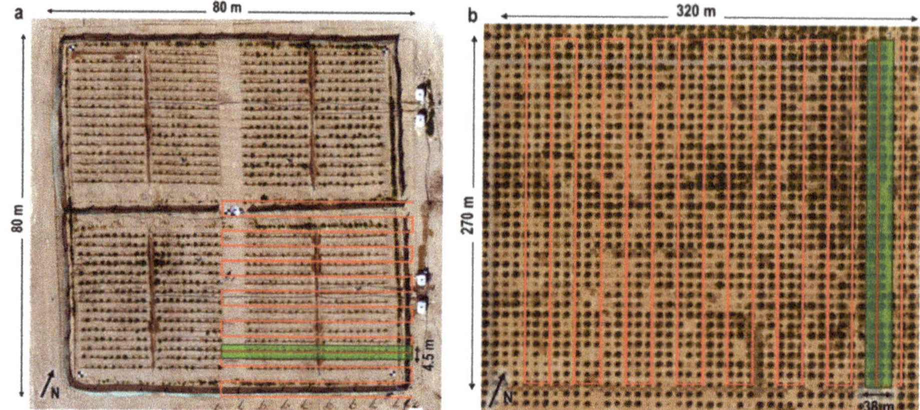

Figure 3. Flight plans and mission areas for (**a**) the experimental crop of tomatoes, where one flight per quarter of the field was required to cover the total area; and (**b**) the commercial plantation of date palms, which was covered by a single flight.

Table 1. Flights planning and collected data details per campaign.

Crop	Area (ha)	Year/DOY	RGB Frames	Hyperspectral Swaths Per Day	Hyperspectral Data Size (Gigabytes)	Ground Sampling Distance GSD (m)	GCPs
Tomato	0.64	2017/320 2017/334 2017/340 2018/007	196	56	232.8 220.4 202.2 274.7	0.007	5
Date Palms	8.70	2018/087	184	16	77	0.06	3

2.4. Ground Data Collection

The GNSS receivers fitted on the UAVs record the geographical location of the cameras with decimeter-level accuracy when an image is taken. However, this low geometric accuracy could affect the quality of the imagery and consequently the products derived from them. In order to assure the highest possible geometric accuracy, GCPs were spaced throughout each area of interest, surveying their center coordinates using a Leica Viva GS15 rover [57] and a RTK Leica AS10 GNSS base station [58]. All raw data from the base station and rover were postprocessed using Leica Geo Office package [59]. For the tomatoes field, five checkerboards of dimension 1 m × 1 m were used as GCPs, with four placed in each corner and one in the center of the field. For the date palms, three circular targets of 0.5 m diameter were spaced throughout the area of interest.

3. Methods

Raw remote sensing imagery is comprised of row and column coordinates pairs, i.e., pixels do not have preassociated geographic coordinates. Unprocessed images present geometric and location distortions that must be corrected through a process known as georectification. This process combines two key steps including rectification, whereby pixels are transformed to a common plane that corrects for geometric distortions; and georeferencing, where real-world coordinates are assigned to each pixel of the image. For an accurate georectification, an automated coregistration methodology between preprocessed hyperspectral scans and RGB orthorectified images is proposed herein (Figure 4). Under this approach, the pixel geometry and location in each data-cube is defined by its corresponding pixel in the RGB base image, which has been previously orthorectified using a digital elevation model reconstructed from a Structure from Motion technique (SfM) [60]. The automated georectification workflow was fully coded in Matlab and performed under a parallel computing scheme to speed up data processing. The desktop analysis employed an Intel Xeon E5-2680 v2 processor, 20 cores @2.8 GHz, and 200 GB RAM. The following sections describe the proposed methodological workflow to rectify, georeference, and ultimately to mosaic UAV-based hyperspectral imagery.

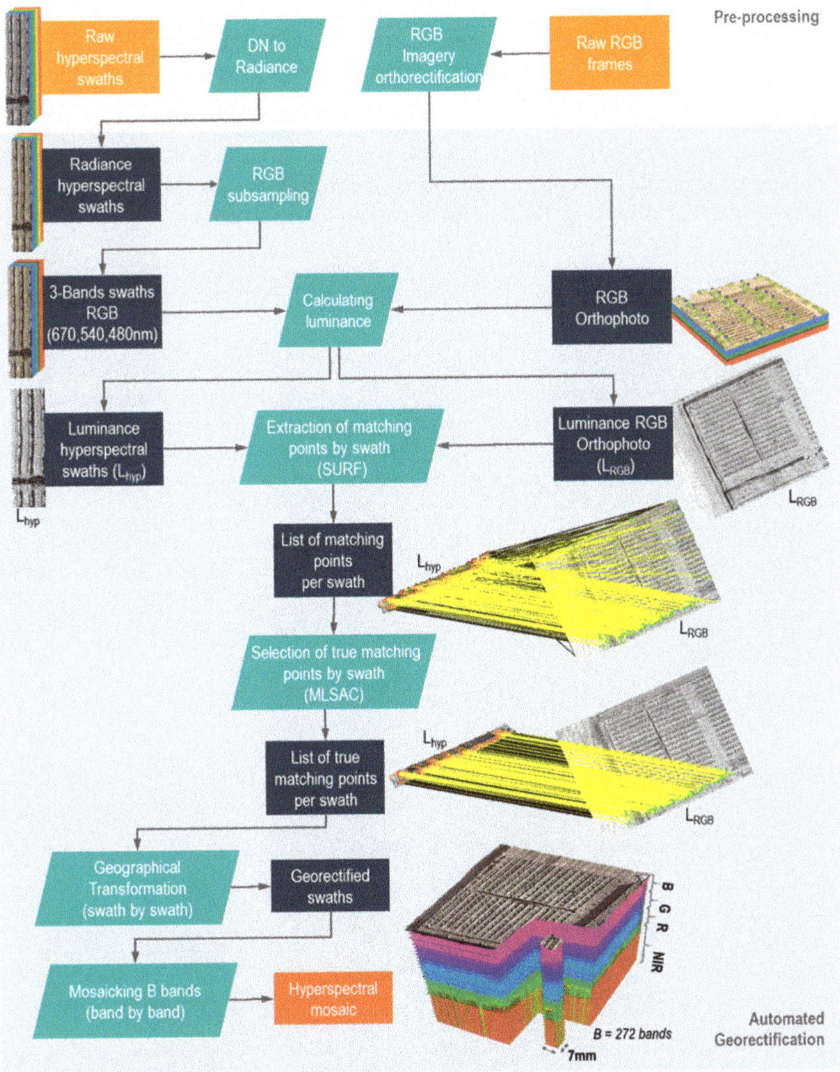

Figure 4. The workflow of the proposed methodology is divided into two main stages, preprocessing and automated processing. The preprocessing corresponds to raw data preparation before going through the georectification and mosaicking routine. The automated processing starts with an RGB subsampling from the hyperspectral swaths to calculate the illuminance image, followed by the coregistration strategy phase required to perform the geographical transformation by swath. Then, the set of georectified strips are merged together to retrieve the final hyperspectral mosaic.

3.1. RGB Imagery Orthorectification

The RGB data was processed in Agisoft PhotoScan Professional 1.3 [61] to produce a georeferenced orthomosaic for each experimental campaign. The digital photogrammetric routine implemented in Photoscan [62] includes several stages [63] based on SfM and computer vision algorithms. First, the frame camera positions measured by the GNSS/IMU sensors onboard the M100 aircraft and the set of matching points generated between overlapping images were used in a bundle adjustment to perform the imagery alignment. The default number of key and tie points, 40,000 and 4,000 pairs, respectively, were used to retrieve an initial cloud of matches. Then, the external and internal orientations of the frame camera were estimated. Based on the camera positions and a minimum of three GCPs manually identified, a dense cloud of georeferenced 3D points was generated and interpolated over the area to produce a digital elevation model (DEM) and an RGB orthomosaic. Ultra-high accuracy and moderate depth filtering options were set to discriminate most of the outlier points and retrieve the dense cloud. Because the GSD reached by the orthomosaics was smaller than the hyperspectral imagery GSD (Table 1), the orthomosaics were resampled to the hyperspectral pixel size by applying a bilinear interpolation, where the output pixel value is estimated by averaging the four surrounding pixels.

3.2. Raw Hyperspectral Data Preprocessing

Nonsystematic distortions are common in airborne sensing. For instance, turbulence and eddy-induced effects during the flight can cause scale and location errors, since the sensor direction and height above ground level varies while scanning. Initial preprocessing of the raw hyperspectral swaths was performed to correct for such distortions by using a parametric model developed by Headwall [40]. Under this approach, the difference (θ) between the effective view angle vector (V) and the theoretic view angle vector (V_t) is calculated by modeling the three-dimensional movements of the aircraft, i.e., roll (ω), pitch (φ), and yaw (κ), which are recorded by the onboard IMU (see Figure 5). This formulation considers adjustment features such as GPS coordinates, timestamps, IMU offsets, the field of view (FOV), lens parameters, and sensor orientation, to reconstruct the scanning geometry line by line and to compose each individual swath. However, this reconstruction approach is limited by the GPS/IMU accuracy leading to geometric errors in the preprocessed scans [28], hence requiring additional processing.

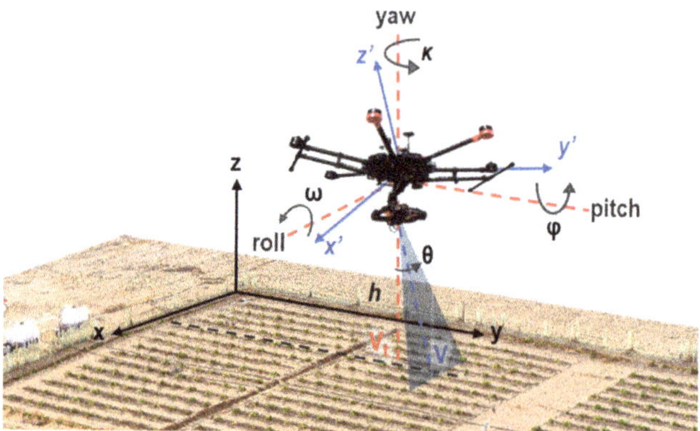

Figure 5. Three-dimensional range of motion of the UAV, where ω, φ, κ denote roll, pitch, and yaw angles, respectively. θ represents the difference between the theoretical view angle vector V_t and the effective look angle vector V.

3.3. Luminance Retrieval

Grayscale images are preferred over colored ones in order to simplify the image processing complexity, by transforming an RGB color image into a single channel image. Moreover, grayscale images contain the brightness, contrast, edges, shapes, contours, textures, perspective, and shadows of the original RGB data, easing the matching process between two scenes. From the variety of grayscale approaches, luminance images are considered as the best option to identify potential matching points in scenes composed of homogenous textures [64]. In this study, the RGB mosaic was converted to a grayscale luminance image (L_{rgb}) by eliminating the saturation and hue information, while retaining the original luminance, using the formulation defined in the international standard National Television System Committee (NTSC) [65] (1). Contrast enhancement of the luminance image was then performed using a histogram equalization process.

$$L = 0.299R + 0.587G + 0.114B, \tag{1}$$

A luminance image was also retrieved via an RGB composite from each preprocessed hyperspectral swath (L_{hyp}) by extracting the central wavelength red, green, and blue bands (670 nm, 540 nm, and 480 nm).

3.4. Extraction of Matching Points by SURF

An implementation of the Speed Up Robust Features (SURF) [48] computer vision technique was used to align the L_{hyp} based on corresponding points from the L_{rgb}. SURF is implemented since it is widely used as a scale-invariant feature detector method that is able to retrieve both matching points position and their correspondent descriptors. SURF performs the matching points (or features detection) by following three main stages: (i) extraction, (ii) description, and (iii) matching. Edges, corners, blobs, ridges, or any other specific pattern is considered as a feature, with the only condition to be unique, easily tracked, and comparable. First, the locations of key points that are likely to be found in both images are extracted by convolving two-dimensional box Gaussian smoothing filters, vertically and horizontally, with the integral images of L_{hyp} and L_{rgb}, which are an averaged version of the luminance L commonly used to speed up the convolution calculation. Thus, feature orientations are defined by the vector sum of vertical and horizontal responses for the neighborhood around each point. This process is done in parallel for different scales by using filters with different sizes, increasing the chances to detect both smaller and larger sized features, and identifying in this way, scale and rotation invariant key points such as corners, blobs, and T-junctions. The results of these convolutions are integrated into a Hessian matrix per each point. Then, a new neighborhood window is oriented along the dominant direction of each point, and by dividing each window into 4 × 4 sub-regions, horizontal (Σdx) and vertical (Σdy) Haar wavelet responses are again taken to form a vector descriptor V (2), which describes the luminance (L) distribution and polarity ($\Sigma|dx|$, $\Sigma|dy|$) of the surrounding pixels. Finally, the sign (-, +) of the Hessian matrix trace is used to classify bright features on dark backgrounds and dark features on a bright background. Only features from both images, L_{rgb} and L_{hyp}, with identical sign are compared, and the Euclidian distance between their descriptor vectors is calculated to select the set of matching points.

$$V = \left(\Sigma dx, \Sigma dy, \Sigma|dx|, \Sigma\left|dy\right|\right), \tag{2}$$

3.5. Selection of True Matching Points by MLSAC

The set of paired points obtained by SURF can contain both true and false feature matches, affecting the accuracy of the fitted geographical transformation. To address this, a parameter estimation approach is required, that adjusts the best transformation model from outlier-corrupted data. Here, we use the Maximum Likelihood Sample Consensus (MLSAC) [49] algorithm, which is an adaptation of the widely applied Random Sample Consensus (RANSAC) technique. RANSAC [66] is a hypothesis-verify iterative method used in coregistration applications to estimate model (projective, affine, etc.) parameters that best fit the set of paired points (true and false) retrieved by a feature detector (SIFT, SURF, etc.). It proceeds by repeatedly generating and testing solutions estimated from a minimal random set of matches gathered from the total paired points. The best solution relies on the highest number of true matches (inliers), with an error below a user-defined threshold. In contrast, MLSAC adopts the same iterative strategy to generate solutions from random samples of matches, but chooses the solution that minimizes the error, rather than just looking for the maximum number of inliers. The following three points motivate that use of MLSAC herein:

- MLSAC improves upon RANSAC by assuming the distance between paired points follows a Gaussian distribution, with a zero-mean error and a uniform standard deviation.
- A maximum likelihood cost function is evaluated in terms of finding the solution that minimizes the error.
- Since the optimal solution does not rely on a defined number of inliers, MLSAC is well suited to estimating complex geometric transformations that exist between images captured under different viewing geometries, where just a few true matches could be retrieved.

MLSAC workflow consists of five general stages. First, a randomly sampled set of matching points is considered to fit an initial transformation model, using the remaining points for testing. Then, each individual matching pair is evaluated by using the fitted model to estimate the distance error in pixels between the point in L_{rgb} and the projection of the corresponding point from L_{hyp}. The algorithm classifies as inliers those points whose distance error is below a threshold of N pixels and counts the total number of inlier candidates. The N limit depends on the aimed positional accuracy of the results, which in this case was set to a maximum of 1.5 pixels. Then, the likelihood of the probability distribution function of the errors is maximized, and the above process is repeated i times (3) to evaluate a statistically significant number of subsamples. These i iterations depend on the randomly sampled subset size (m), the percentage of outliers (w) allowed, and the probability of selecting a good subsample (q). Generally, a probability $q = 99\%$ is desired, considering $w = 50\%$ as the worst case scenario, and $m = 3$, or $m = 4$ when using an affine or projective transformation, respectively. After the loop is finished, the transformation model that maximizes the likelihood of the cost function with a 99% confidence of finding the maximum number of inliers is selected as the best solution.

$$i = \frac{\log(1-q)}{\log(1-w^m)} \tag{3}$$

3.6. Geographical Transformation and Mosaicking

Affine transformation, a special case of the projective approach, was used to convert the L_{hyp} units to real-world coordinates, based on the L_{rgb} mosaic, since it is one of the most flexible transformation methods (4) [67]. This transformation model requires a minimum of three pairs of matching points to translate, scale, shear, and rotate an image while preserving parallelism. Generally, the greater the number of true matching pairs, the higher the accuracy of the model.

$$\begin{bmatrix} x' \\ y' \\ 1 \end{bmatrix} = \begin{bmatrix} x \\ y \\ 1 \end{bmatrix} \begin{bmatrix} a_1 & a_2 & t_x \\ a_3 & a_4 & t_y \\ 0 & 0 & 1 \end{bmatrix} \tag{4}$$

where x' and y' are the coordinates of the transformed point, x and y are the original coordinates of the point, a_1, a_2, a_3, and a_4 define linear transformations composed by scale, shear, and rotation factors, and t_x and t_y specify the displacement or translation along the X and Y axis, respectively.

By running the routines previously described, individual geographic transformation solutions per swath were determined and operated band by band. Finally, the hyperspectral mosaic is produced by merging one by one these multiple georectified swaths into a single mosaic per band and stacking together these individual mosaics into a raster data cube, where the output pixel values for the overlapping areas are determined by the value from the last swath added into the mosaic.

3.7. Georectification Assessment

The relative positional accuracy between each georectified hyperspectral dataset and the correspondent RGB mosaic was determined by calculating the root mean square error (RMSE), the mean absolute error (MAE), and the accuracy at the 95% confidence limit. The RMSE (5) is determined by calculating the Euclidean distance between the rectified coordinates in the hyperspectral mosaic and the reference coordinates in the RGB mosaic. The closer the RMSE values are to zero, the more accurate the georectification. In this case, the reference coordinates were prespecified check points from each of the imagery. Check points are identifiable features in both the reference RGB image and the hyperspectral mosaic, whose locations are used to quantitatively assess the positional quality of the georectified data cube. To compute the RMSE, at least 20 well-defined checkpoints are used per mosaic, making sure that 25% are well distributed in each of the four quadrants of the image of interest (for the tomato experiment). A total of 52 (tomato) and 25 (date plantation) checkpoints were randomly spread over each dataset.

$$RMSE = \sqrt{\frac{1}{n}\sum_{i=1}^{n}\left(\left(x_{hyp}-x_{rgb}\right)^2+\left(y_{hyp}-y_{rgb}\right)^2\right)} \tag{5}$$

The MAE [68] (6) measures the average magnitude of the Euclidean distance in the set of checkpoints, where all individual differences have equal weight.

$$MAE = \frac{\sum_{i=1}^{n}\sqrt{\left((x_{hyp}-x_{rgb})^2+\left(y_{hyp}-y_{rgb}\right)^2\right)}}{n}, \tag{6}$$

According to the National Standard for Spatial Data Accuracy (NSSDA) [69,70], the relative horizontal positional accuracy is reported in meters at the 95% confidence level (9) and is determined in two separate components: x (7) and y (8). The value of 1.22385 in the accuracy expression in (9), is derived from the Chi-square statistical distribution for 2 degrees of freedom and a lower tail area of 0.05. In other words, 95% of the positions in the hyperspectral mosaic will have an error with respect to the RGB mosaic position that is equal to, or smaller than, the reported accuracy value.

$$RMSE_x = \sqrt{\frac{1}{n}\sum_{i=1}^{n}(x_{hyp}-x_{rgb})^2}, \tag{7}$$

$$RMSE_y = \sqrt{\frac{1}{n}\sum_{i=1}^{n}(y_{hyp}-y_{rgb})^2}, \tag{8}$$

$$Accuracy_{95\%} = 1.22385 \times \left(RMSE_x + RMSE_y\right), \tag{9}$$

In addition, the performance of the proposed automated method was evaluated with respect to a semiautomated approach by manually selecting matching points between the hyperspectral swaths and the RGB image. The number of good matches (or inliers) retrieved by the automated workflow was used as a reference to set the number of point pairs to be identified by hand and to fit an affine transformation per swath. The aligned strips were mosaicked together, and the above-mentioned positional accuracy metrics were estimated to compare the performance of both methods.

4. Experimental Results and Analysis

The UAV-based hyperspectral imagery for both field experiments was georectified and mosaicked using the methodology described above. In this section, the efficiency of the automated coregistration routine between hyperspectral data and RGB frame-based imagery is evaluated, together with a qualitative and quantitative assessment of the accuracy reached for the georectified high spatial and spectral resolution mosaics. An analysis of the computational cost of the automated process is also undertaken.

4.1. RGB Frame-Based Orthomosaic

As described in the previous section, the RGB orthomosaics derived from the collected frame images were processed using a SfM package and GCPs. All the mosaics over the tomatoes field (Figure 6a) were resampled from 0.005 to 0.007 m, with a rectification error of 0.002 m. Similarly, the native resolution of the date palms mosaic (Figure 6b) was resized from 0.034 to 0.060 m, with an RMSE error of 0.043 m. From visual inspection of these images, in general a good alignment was reached by the RGB mosaics, well preserving sizes and shapes. Figure 6a shows how some linear features are continuous, such as irrigation pipes, defined objects like individual plants are free of gaps or blur effects, and contrasting tones and textures are visible in the bare soil areas. From Figure 6b, road edges are continuous and well defined, date palms keep their characteristic shapes, and soil areas preserve smooth textures and contrasting tones.

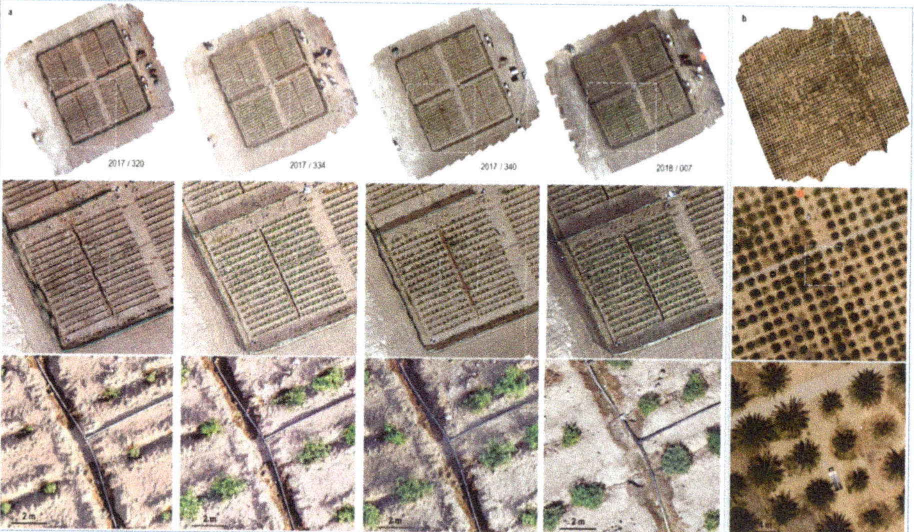

Figure 6. (**a**) Multitemporal RGB orthomosaics over the tomatoes field and close-ups of one of the quarters of the total area showing the good alignment and high spatial resolution achieved; (**b**) RGB orthomosaic and close-up over the date palms plantation.

4.2. Efficiency of the Automated Coregistration Routine

The most important steps in the coregistration processing are the extraction and selection of common features between the RGB reference image and each preprocessed hyperspectral swath. Under the proposed methodology, SURF was used to extract a set of matching points, which were purged of false positives or outlier pairs by using the MLSAC model. The efficiency of these combined routines relies on the number of inliers retrieved to fit the best affine transformation function, to align each swath to the RGB mosaic. The higher the number of inliers, the better the fitting of the transformation model. Table 2 presents the number of features detected in the RGB mosaics and the average detected per swath, together with the matches identified by SURF, and the total inliers selected by MLSAC. It is evident how the features retrieval varies from one flight to another, since this process is performed by using luminance images, which in turn vary with the illumination and surface conditions. Although a large number of features were extracted, from 10K in the hyperspectral data to 300K in the RGB approximately, only few matches were retrieved, between 505 and 951 pairs in the case of the tomato crop, and 103 pairs in the date palms dataset. This performance is explained not only by different illumination conditions but also when coregistering data from different sensors [71]. In the case of the tomato field experiments, the percentage of points pairs detected as inliers from the total of matches varies between 65% and 80%. An average of 26% of matches was selected as inliers for the date palms swaths. In both cases, the number of inliers was sufficient to fit the transformation models by swath and to ultimately stitch the hyperspectral mosaics.

Table 2. Features, matches, and inliers detected per flight.

Crop	Year/ DOY	RGB Features	Metrics Accounting all Swaths per Flight							
			Hyperspectral Features			Matching Points			Average Inliers	Inliers/ Matching (%)
			Min.	Aver.	Max.	Min.	Aver.	Max.		
Tomato	2017/320	309461	37135	38667	40199	818	951	1083	757	80
	2017/334	293013	32817	35161	37505	633	771	908	591	77
	2017/340	246575	29589	36487	43385	393	505	616	327	65
	2018/007	301210	36145	36798	37451	520	667	813	477	71
Date Palms	2018/087	448156	8963	9750	10537	80	103	125	27	26

Both the number of inliers and the distribution of the points along the swaths are determinant by the georectification quality. Inliers should be fairly uniform and located across the strips in order to avoid local distortions after performing the geometric transformation. Figure 7 shows some examples of the distribution and location of the matching points extracted by SURF, from which MLSAC selected the set of inliers. In the case of the tomato crop (Figure 7a), a dense cloud of matches was retrieved, including some outliers that are generated when the texture, color, or intensity of the surface are homogeneous, thus identifying similar patches between the hyperspectral strip and the RGB reference. After MLSAC prunes the false matches (or outliers), a good distribution of inliers is achieved. Figure 7a shows a close-up of an area where some calibration panels and GCPs were placed and where a good number of inliers were selected. However, the number of matches can decrease when repetitive forms are present within the images, i.e., the neighborhood around the features does not vary enough to allow for reliable comparison between both scenes. An example of this effect is shown in Figure 7b, where the crown of the palms represents a very homogeneous pattern. In this case, the density of matches is reduced, but the extracted inliers are still well distributed across the swath.

4.3. Qualitative Accuracy Assessment

As part of the accuracy assessment of the results, an evaluation of visual factors such as gaps, matches across boundaries, deformations, and patches was performed. Figure 8 shows a comparison between the preprocessed and the georectified multitemporal hyperspectral mosaics for the tomato experiment. As can be seen, the full dataset is free of gaps and patches, and the hyperspectral swath borders are dissembled. In the zoomed areas, the impact of the automated alignment can be seen on some linear features, such as irrigation pipelines, furrows, and fences, which are straight, parallel, or continuous across the stitched swaths. Likewise, the shapes and sizes of individual plants are well maintained. The high degree of visual consistency achieved indicates that the estimated affine transformations were well fitted with sufficient and well-distributed corresponding points.

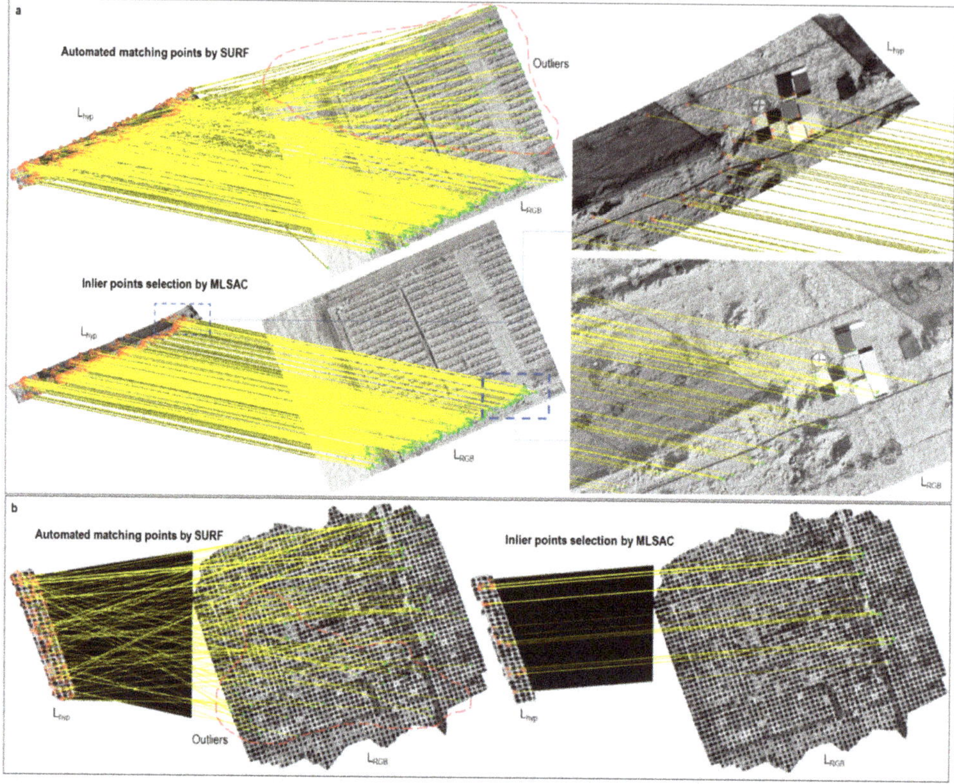

Figure 7. (**a**) Matches identified by Speeded-Up Robust Features (SURF) and inliers selected by Maximum Likelihood Estimator Sample Consensus (MLSAC) between both the hyperspectral strip (L_{hyp}) and the RGB reference (L_{RGB}) luminance images in the tomatoes field on 2017/320; (**b**) matching and inlier points identified between Lhyp and LRGB in the date palms field.

For the case of the date palms experiment, the misalignment between preprocessed passes can clearly be seen in Figure 9, with overlapping distortions of individual palms. After processing, a good fit between the RGB reference and the hyperspectral georectified mosaic was reached. The matching quality of linear geometries, such as the border of the roadway (Figure 9b), or the continuity of leaflets in the crown of the palms can be observed throughout the mosaic. As with the tomato experiment, the collinearity and equidistance between individual palms were recovered by the georectification process. Particularly noticeable is the good performance of the affine transformations at the extreme borders of the swaths, which are usually susceptible to deformation when insufficient or poorly-distributed stitching points are retrieved. While the automated routine produced a lower number of matches in this case than for the tomatoes experiment, the set of inliers was sufficient to fit a highly accurate transformation model.

4.4. Spatial Accuracy

Although the visual inspection of the hyperspectral mosaics provides an important qualitative indication of the spatial accuracy, quantifying statistical metrics such as the mean absolute error (MAE), root square mean error (RMSE), and relative positional accuracy (Table 3) is necessary to develop confidence in the approach.

Table 3. Relative positional accuracy assessment of the automated georectification.

Crop	Image (Year/DOY)	Check Points	Min. Error (m)	Max. Error (m)	MAE (m)	RMSE (m)	Accuracy 95% (m)	# Check Points Whose Error > MAE
Tomato	2017/320	52	0.005	0.151	0.044	0.054	0.092	3
	2017/334	52	0.001	0.214	0.046	0.063	0.107	2
	2017/340	52	0.003	0.289	0.060	0.083	0.137	1
	2018/007	52	0.003	0.224	0.056	0.074	0.126	3
Date Palms	Automated	25	0.001	0.222	0.095	0.113	0.188	1
	Semi-automated		0.032	0.275	0.096	0.102	0.167	3

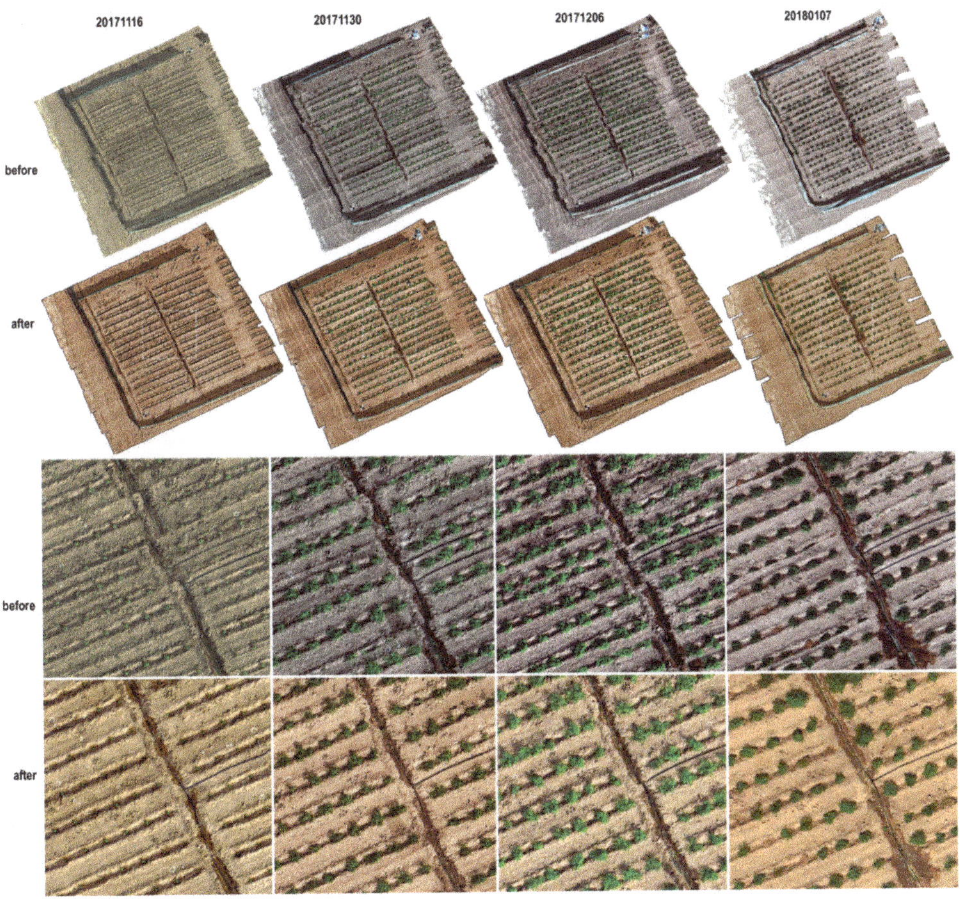

Figure 8. Comparison between preprocessed (**before**) and rectified multitemporal hyperspectral data (**after**). Close-ups of a central area show the good alignment achieved by continuous linear features, such as irrigation pipelines and furrows, which were originally shifted before performing the georectification.

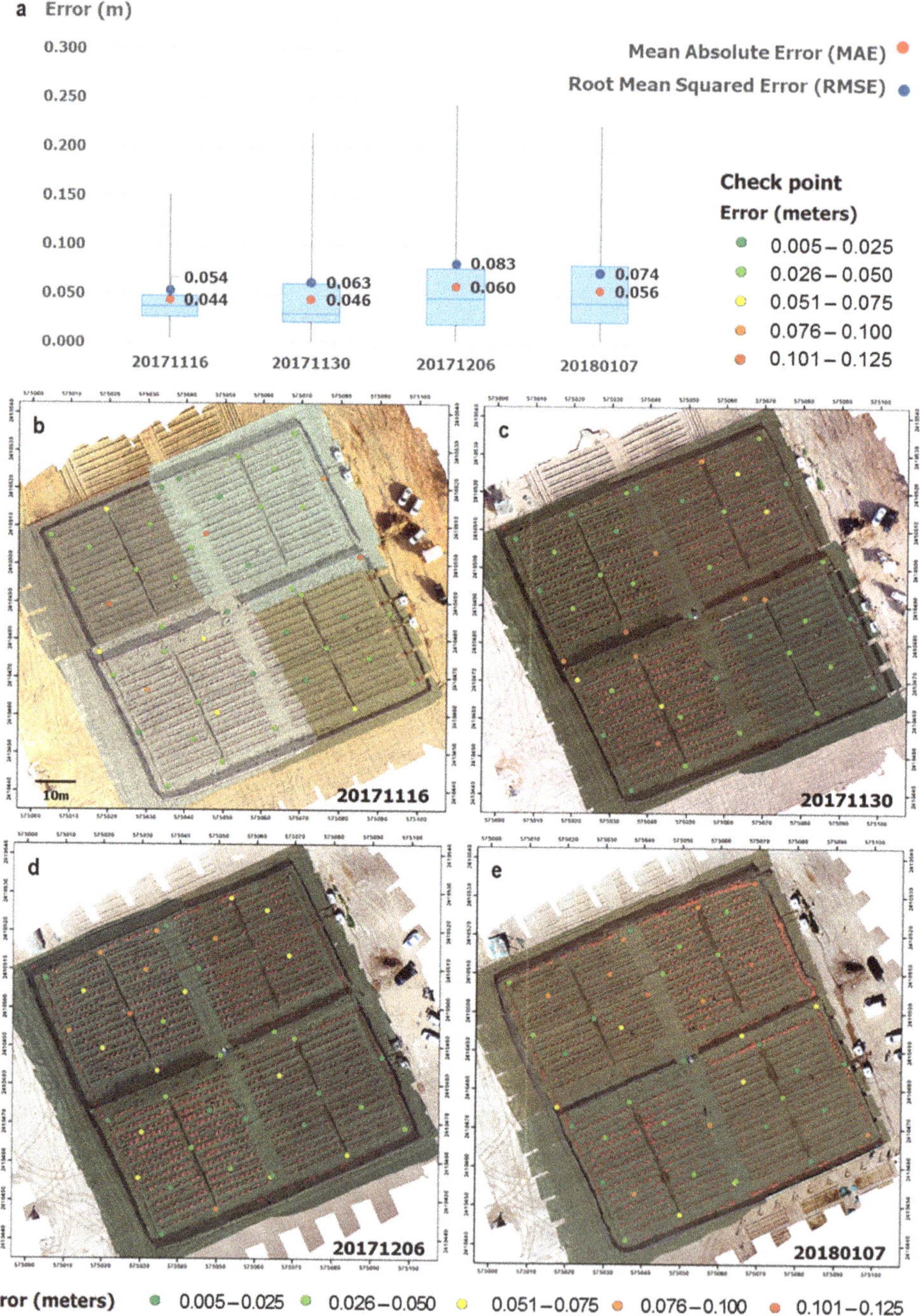

Figure 9. (**a**) Box-plot of the spatial error distribution for the tomatoes case study; (**b–e**) false infrared composition of the georectified hyperspectral mosaic of the tomato experimental field and positional error for 52 check points.

Figure 10 illustrates the checkpoints evaluation for the hyperspectral data series in the tomato experiment. The error is randomly distributed over the mosaics, reaching an overall MAE between 6 and 8 times the ground sampling distance (which represents around 5 cm) and an RMSE at the level of 7 to 11 times GSD (corresponding to approximately 6 cm). Figure 9 shows how the error is distributed throughout the checkpoints over the date palms crop. In this case, the MAE and RMSE were at the level of 1 and 1.5 times GSD, which equates to 6 and 9 cm, respectively. However, errors for some of the mosaics are more variable than others, which is the case for the last two datasets in the tomato experiment and single capture for the date palm experiment, showing a direct correlation between the achieved error and the percentage of inliers selected from the total matching points, i.e., the more inliers that are detected (Table 2), the lower the RMSE.

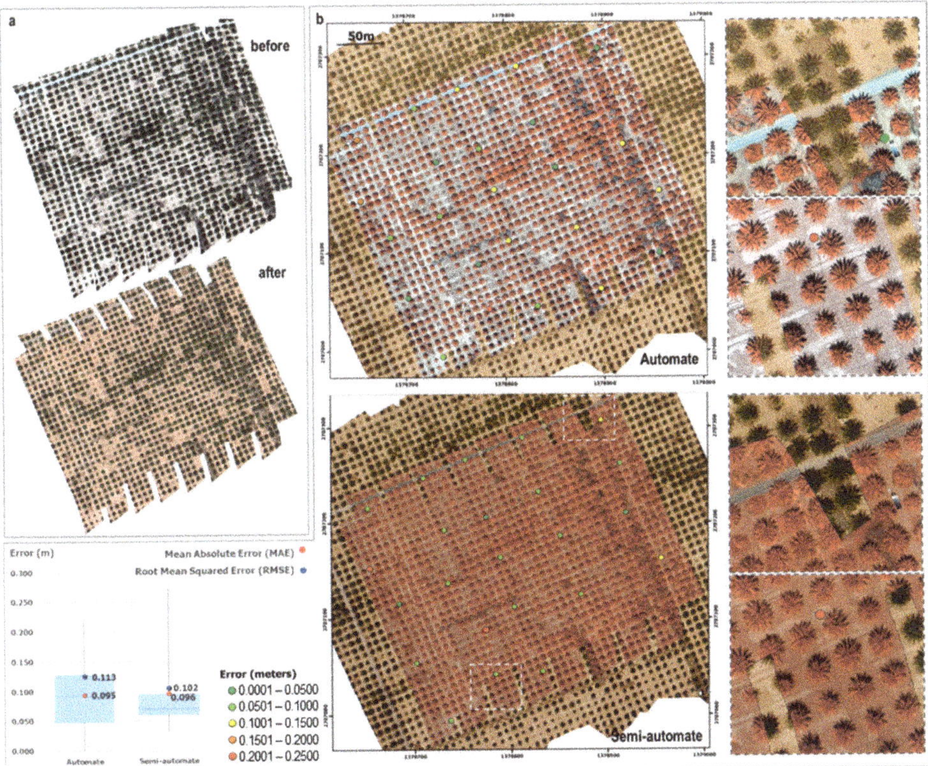

Figure 10. (**a**) Comparison of the hyperspectral mosaic before and after the automated georectification; (**b**) false infrared composition of the georectified hyperspectral mosaic overlapping the RGB base and positional error for 25 checkpoints. Some close-up (**right**) of two areas show the alignment of a road and the palm crowns.

The relative accuracy was tested by comparing the X (east) and Y (north) coordinates of the checkpoints with their correspondent coordinates from the RGB mosaic, which is considered an independent source of higher accuracy. This metric is reported in ground distances to directly compare the results, considering their different spatial resolutions. The accuracy achieved in the tomato experiments throughout the 52 checkpoints and at a 95% confidence level, varies between 9 and 13 cm (Table 3). According to the NSSDA standard, when 50 points are tested, the percentage confidence level allows a maximum of three checkpoints to be above the MAE. This criterion is met for all of the mosaics in the tomato experiments, as shown in Table 3 (last column) and Figure 10 (red points). In the case of the date palm experiment, the accuracy reached throughout the 25 checkpoints at a 95% confidence level, was 18 cm. Following the NSSDA standard for >20 tested points, only one is allowed to be above the MAE, which is a condition achieved by the resulting mosaic (see Figure 9, red points).

An additional spatial quality assessment for the date experiment was performed by comparing a semiautomated georectification with the automated method proposed herein. To do this, matching points between each scanned swath and the RGB-frame reference were manually identified, with a total of 27 stitching points per swath selected (as this number corresponds with the average of inliers retrieved per swath by the automated method; see Table 2). A polynomial affine transformation was performed using these points, achieving an RMSE of 0.102 m, an MAE of 0.096 m, and an accuracy of 0.167 m at a 95% confidence level. Figure 9d,e show the error distribution of the checkpoints achieved for both methods. As anticipated, the error is smaller and more homogenous across the manually-rectified mosaic compared to the automated effort, although the difference in spatial accuracy achieved is around 1 cm.

4.5. Processing Efficiency

Given that the proposed approach can achieve spatial accuracies comparable with those obtained by manually identifying the matching points, one of the key reasons for choosing an automated method will be based on the processing time (i.e., it should be faster and as reliable when compared with manual approaches). The efficiency of an algorithm is usually expressed in terms of its processing time. As such, the computational cost of the proposed automated georectification workflow, coded in Matlab, was measured on a per step basis to allow an intercomparison of the approaches. Some factors, such as 200 GB of RAM memory and 20 processor cores, were set as constant to execute the routines. Table 4 compares the timing measurements per dataset for three general stages: (i) extraction and selection of matching points, (ii) geographic transformation, and (iii) mosaicking. The manual coregistration performed for the date palms imagery was also timed, with an average of 3 min required to manually identify each of the 27 pairs of matching points per swath, from a total of 16 flight lines (i.e., 21.6 h in total). It is noticeable that the time required to execute each stage is correlated with the data size. That is, the larger the data set, the longer the processing time. For the automated solution, nearly 10% of the processing time is used to extract and select the matching points, while another 10% is spent by the geographic transformation. The majority of the time, around 80%, is dedicated to stitching the strips and stacking the bands together into a single hyperspectral mosaic. In contrast, when comparing both approaches (the automated with the semiautomated), a difference of 21.3 h was measured, where 85% of the total time was spent by the handmade selection of points.

Table 4. Georectification processing time by stage.

Crop	Hypers. Mosaic	Mosaic Dimension (Rows × Columns)	Mosaic Size (Giga-bytes)	Matching Points Extraction and Selection (hours)	Geographic Transforma-tion (hours)	Mosaicking Time (hours)	Net Processing Time (hours)
Toma-toes	2017/320	16571 × 16429	220	0.6	0.6	5.3	6.5
	2017/334	16714 × 16143	195	0.5	0.6	5.2	6.3
	2017/340	16571 × 16000	170	0.4	0.5	5.2	6.1
	2018/007	16429 × 16571	220	0.6	0.6	5.3	6.5
Date Palms	Automa-ted	8588 × 7758	17	0.3	0.4	3.0	3.7
	Semiautomated		17	21.6			25

5. Discussion

A range of semiautomated [18,20,22,26,41] and fully automated frameworks [37,38,72–74] have been explored to georectify UAV-based hyperspectral data captured by push-broom cameras. However, challenges related to data collection procedures, quality assessment, and optimization of algorithms require further investigation to expedite data processing and accomplish a standardized positional accuracy of retrieved data. These factors, together with the need for processing large volumes of image time-series, motivated the development of a simplified, expedited, and automated workflow to georectify and mosaic high-spatial-resolution hyperspectral images acquired by UAV-based push-broom spectroradiometers. To address these challenges, an improved coregistration strategy combining SURF feature detector and MLSAC model-fitting algorithm was established to allow robust direct geographic transformation between the hyperspectral scans and an RGB reference orthophoto. An additional novel aspect of the proposed approach is the fact that high positional accuracies can be reached with different percentages of true matches without requiring any additional image treatment and with a limited number of GCPs.

Some considerations relevant to the development and execution of the proposed methodology must be taken into account to assure an effective implementation for multiple applications. For instance, in the data collection stage, it is advised to design a flight plan that allows the simultaneous collection of coincident hyperspectral and RGB frame-based data. Establishing a minimum of requisites, such as atmospheric conditions, side-lap overlaps, flight speed and height, frame rate scanning, and FOV allows the capture of both datasets under similar illumination conditions and to achieve comparable spatial resolutions. However, if different GSD are collected between the RGB reference and the hyperspectral dataset, then the RGB dataset should be resampled to the hyperspectral imagery resolution, to increase the efficiency of the SURF coregistration method. Although SURF is a scale-invariant feature detector, it has been shown elsewhere that the algorithm operates considerably better when comparing similarly scaled images [75]. An alternative to managing the scale difference was proposed by Habib et al. [38], who established a GSD ratio threshold between the spectral scans and the RGB reference to constrain the feature detection in SURF. However, our study demonstrates that resizing the RGB orthomosaic is enough to retrieve hundreds of matches. Another aspect to account for is the flight time, since the coregistration is based on the similarity of the luminance images derived from the hyperspectral swaths and the RGB orthophotos. Both datasets should be consecutively (or simultaneously) collected in order to avoid significant changes in luminance. Theoretically, SURF or any other type of feature detector/descriptor algorithm always retrieves interest points from an image unless it is a constant matrix whose pixel values are all the same [75]. However, the number of features detected can be reduced by the homogeneity of the scene, since the detection is based on local texture analysis. For instance, a poor number of SURF points could be retrieved for an image covering a highly homogeneous and flat desert area. In such a case, the number of true matches between two scenes could be null if these were captured under slightly different illumination conditions, hence requiring ancillary GCPs. Although SURF is also robust under invariant illumination conditions [48], large differences between the images to coregister (e.g., shadows or new elements placed on the ground) can reduce the number of matches and the georectification quality. Considering such factors will not only help to reduce ground-based collection efforts, but it will also make the data more reliable.

Amongst the different approaches used to georectify and mosaic UAV-based hyperspectral data, those using coregistration methods with RGB scenes from frame sensors generally yield better accuracies and products than those based on dense networks of ground control points (GCP) and manual stitching [22,25]. Habib et al. [38] used the same hyperspectral camera and IMU reference employed in this study, with a 17 mm lens and onboard a fixed-wing UAV, to capture 5 cm GSD swaths with 50% side lap over a crop field. Their approach includes a partial rectification of the hyperspectral scans based on a derived DEM from the RGB frame-based dataset and a coregistration strategy based on a modified version of SURF. Their results achieved relative accuracies between 0.5 to 0.9 m RMSE per swath. Considering the comparable date palm study (6 cm GSD) explored here, the relative accuracy achieved for our georectified mosaic (0.1 m RMSE) improved these results by between 67% and 88%. This improvement relies on the use of luminance images and the integration of SURF and MLSAC. In previous approaches [23,38,39], most establish a comparison between the hyperspectral and the RGB data using a single band (often the red band), thereby omitting radiometric differences of both sensors.

In contrast, luminance images are based on a model of a weighted combination of RGB wavelengths that equalizes multiple data sources under a standard metric. By comparing the luminance images derived from the high spectral and the RGB datasets, SURF is able to retrieve thousands of features and hundreds of matching points, as shown in the presented study cases. Furthermore, the strategy of selecting true matches (or inliers) is essential to fit an affine model, especially when the study site has a homogeneous land cover. The alternative proposed by Habib et al. [38] to reduce the number of false matches, was by constraining SURF with some ratios and ranges in the spatial location, scale, and main orientation, achieving a maximum of 350 true matching pairs between consecutive swaths, and fitting an affine model base on them. In contrast, our study implements the MLSAC routine as a strategy to do both, selecting the best matching points or inliers, and fitting the transformation model per swath through a maximum likelihood of the error, where the distance error parameter can be set to be as restrictive as required. In the case of the date palms, only an average of 27 inliers per swath are retrieved, and these are the best points that assure an affine model with an error ≤0.09 m per swath.

One of the aims of automated approaches based purely on computer vision and coregistration algorithms is to reduce field and manual work. Ramirez-Paredes et al. found that navigation and positional data are not required to achieve an alignment line-to-line between the RGB reference and the hyperspectral strips, demonstrating this by combining a light payload sensing system with machine vision algorithms. However, spatial accuracy is the most important factor to evaluate in the georectification and mosaicking process. In order to quantify and minimize the absolute error, GCPs, check points, and onboard navigation sensors are always required. Here, it is demonstrated that an automated method that relies on the RGB reference accuracy, requires just a few well-distributed GCPs (minimum five), high-precision GNSS base stations, and GNSS/IMU sensors integrated with the cameras, to produce high-quality results. Moreover, recent studies [76] have found that a minimum of three GCP/ha are sufficient to assure sub centimeter-level horizontal accuracies when operating similar UAV-based RGB systems at 30 m above the ground approximately. One of our study cases reached absolute accuracies of ~1.5 pixels for RGB orthophotos with 5 mm GSD, and relative accuracies between two and seven pixels for hyperspectral images with millimetric resolution (7 mm). Turner et al. [23] conducted a comparative experiment by using the Headwall Micro-Hyperspec onboard a small multi-rotor UAV, integrated with a dual frequency GNSS antenna, an IMU, and a machine vision camera. Their georectified hyperspectral imagery achieves 2 cm GSD with an absolute accuracy of ~2.5 pixels, by sampling 46 GCPs. Although having a significant level of difference in accuracy, these results support the viability of using an ancillary frame camera and automated coregistration methods in combination with a sufficient quantity of GCPs. Ultimately, the number of required GCPs will depend on the area, the desired accuracy level, the terrain conditions, and the available resources (i.e., equipment, time, people).

In terms of computational efficiency, the robustness of the presented workflow is demonstrated (Table 4) by the parallel implementation of optimized algorithms, following the suggestion of Ramirez-Paredes et al. Although it is not possible to establish a comparison between the automated methods in the literature (since these do not report the process timing and barely describe the computational resources and data size), some aspects can be highlighted regarding the efficiency of some of the adopted algorithms. In comparison with the Habib et al. [38] approach, our method performs the feature detection routine SURF only once, whereas their workflow executes it several times, since there is a feature detection between consecutive swaths, and between the swaths and the RGB orthomosaic. Consequently, under that approach, the computational effort in the extraction and selection of matching points stage could increase considerably as the number of flight lines increases. Another comparison can be established with the geocoding package PARGE [39,77], whose ortho-rectification strategy relies on using navigation data (GNSS/IMU), ancillary sensor information (FOV, scanning frequency), high-resolution digital surface models (DSM), and tens of GCPs, in order to fully reconstruct the geometry of the scanning process. According to Schläepfer et al. [77] the whole processing time that PARGE can take to georectify a typical airborne-based scan of 512 × 2000 pixels at 200 spectral channels, is within about 4 h, achieving submetric accuracies. Based on this performance, it is expected that this approach would require a higher computational and manual effort than the approach proposed herein. Likewise, the SpectralView [40] application provides a quick geometry correction approximation, requiring only a coarse resolution DTM and navigation data to produce georeferenced scans. Based on the preprocessing stage of our study data, one hyperspectral scan of 640 × 2000 pixels with 270 bands can be georeferenced through SpectralView within about 1 h, reaching only a submetric level of accuracy and requiring additional processing (like that proposed herein), in order to obtain consistently high positional accuracies.

Although the presented case studies show this automated approach is a valid, computationally efficient, and accurate alternative to the current variety of georectification methods, some improvements would further strengthen the performance of the methodology. In terms of the extraction and selection of matching points, a further comparative study could explore different possible integrations of new image feature detector methods [75] (like SURF) with model fitting routines [78] (like MLSAC), aiming to strengthen the proposed coregistration strategy. With respect to the spatial accuracy assessment of automated georectification methods, as a best practice, it is suggested to use international spatial quality control tests [69,79] that guide how to decide when the accuracy of the results is sufficient or not, for a specific study purpose. Further work could also involve laying out a dense GCP network over a study site to assess the absolute accuracy of the hyperspectral mosaics, especially for mountainous terrains or nonflat fields. In addition, regarding the computational efficiency of the mosaicking stage, it is advised that efficient stitching and band-stacking strategies that can speed up the creation of the hyperspectral mosaic data cube be explored.

6. Conclusions

In order to address the postprocessing georectification challenges in a timely and computationally efficient manner, a batch processing workflow was presented to produce georectified UAV hyperspectral mosaics captured with push-broom sensors. The approach uses as a reference an auxiliary orthophoto collected with a frame-based camera, which is used to individually coregister each spectral scan. SURF and MLSAC computer vision stitching algorithms were implemented to produce thousands of matching points between the intensity images of the RGB reference and each hyperspectral swath. Affine transformations were estimated to obtain free-distortions scanlines, and to stitch them together as mosaic data cubes. The number of inliers extracted from the matching points is correlated with the accuracy of the results, which demonstrates the importance of the SURF coregistration approach to produce high-quality matches, and the consensus algorithm MLSAC to select the inlier pairs. The methodology was tested with different temporal and high-spatial-resolution scenes collected over two varying landscapes. The hyperspectral mosaics with millimeter spatial resolution (7 mm), achieved centimeter level residual errors, with an RMSE of ~7 cm, MAE of ~ 5 cm, and accuracy of ~9 cm at a 95% confidence level. The hyperspectral dataset with centimetric spatial resolution (6 cm) achieved decimeter level residual errors, with an RMSE of ~11 cm, MAE of ~9 cm, and accuracy of ~18 cm at a 95% confidence level. In terms of the computational complexity of the workflow, SURF and MLSAC provide a robust and highly efficient solution to automate the matching points selection process, assuring enough high-quality points to perform an affine geometric transformation. Additional tests are required for implementing approaches that speed up the mosaicking step, since the composition of a mosaic data cube is computationally intensive. Future work should also focus on testing the proposed approach over different terrains and land surface and atmospheric conditions to further improve the framework.

Author Contributions: Experiments were designed by Y.A., D.T., and S.P., in discussion with M.F.M. and A.L. Data processing was undertaken by Y.A. and Y.M. Exploration and analysis were undertaken by Y.A. The manuscript was drafted by Y.A., with input from M.F.M., D.T., Y.M., and A.L. All authors contributed to the final manuscript production. All authors discussed the results and contributed to the final manuscript production. All authors have read and agreed to the published version of the manuscript.

Funding: This research was supported by the King Abdullah University of Science and Technology (KAUST).

Acknowledgments: The authors thank Matteo Ziliani and Bruno Aragon for their assistance in collecting the UAV data and ancillary measurements.

Conflicts of Interest: The authors declare no conflict of interest.

References

1. Atzberger, C. Advances in Remote Sensing of Agriculture: Context Description, Existing Operational Monitoring Systems and Major Information Needs. *Remote Sens.* **2013**, *5*, 949–981. [CrossRef]
2. McCabe, M.F.; Rodell, M.; Alsdorf, D.E.; Miralles, D.G.; Uijlenhoet, R.; Wagner, W.; Lucieer, A.; Houborg, R.; Verhoest, N.E.; Franz, T.E.; et al. The future of Earth observation in hydrology. *Hydrol. Earth Syst. Sci.* **2017**, *21*, 3879–3914. [CrossRef] [PubMed]
3. Warner, T.A.; Cracknell, A.P. Unmanned aerial vehicles for environmental applications. *Int. J. Remote Sens.* **2017**, *38*, 2029–2036.
4. Manfreda, S.; McCabe, M.F.; Miller, P.E.; Lucas, R.; Pajuelo Madrigal, V.; Mallinis, G.; Ben Dor, E.; Helman, D.; Estes, L.; Ciraolo, G.; et al. On the Use of Unmanned Aerial Systems for Environmental Monitoring. *Remote Sens.* **2018**, *10*, 641. [CrossRef]
5. Aasen, H.; Honkavaara, E.; Lucieer, A.; Zarco-Tejada, P.J. Quantitative Remote Sensing at Ultra-High Resolution with UAV Spectroscopy: A Review of Sensor Technology, Measurement Procedures, and Data Correction Workflows. *Remote Sens.* **2018**, *10*, 1091. [CrossRef]
6. Zhang, C.; Kovacs, J.M. The Application of Small Unmanned Aerial Systems for Precision Agriculture: A Review. *Precis. Agric.* **2012**, *13*, 693–712. [CrossRef]
7. Honkavaara, E.; Saari, H.; Kaivosoja, J.; Pölönen, I.; Hakala, T.; Litkey, P.; Mäkynen, J.; Pesonen, L. Processing and Assessment of Spectrometric, Stereoscopic Imagery Collected using a Lightweight UAV Spectral Camera for Precision Agriculture. *Remote Sens.* **2013**, *5*, 5006–5039. [CrossRef]
8. Roosjen, P.P.J.; Suomalainen, J.M.; Bartholomeus, H.M.; Clevers, J.G.P.W. Hyperspectral Reflectance Anisotropy Measurements Using a Push-broom Spectrometer on an Unmanned Aerial Vehicle—Results for Barley, Winter Wheat, and Potato. *Remote Sens.* **2016**, *8*, 909. [CrossRef]
9. Pádua, L.; Vanko, J.; Hruška, J.; Adão, T.; Sousa, J.J.; Peres, E.; Morais, R. UAS, sensors, and data processing in agroforestry: A review towards practical applications. *Int. J. Remote Sens.* **2017**, *3*, 2349–2391. [CrossRef]
10. Jakob, S.; Zimmermann, R.; Gloaguen, R. The Need for Accurate Geometric and Radiometric Corrections of Drone-Borne Hyperspectral Data for Mineral Exploration: MEPHySTo—A Toolbox for Pre-Processing Drone-Borne Hyperspectral Data. *Remote Sens.* **2017**, *9*, 88. [CrossRef]
11. Adão, T.; Hruška, J.; Pádua, L.; Bessa, J.; Peres, E.; Morais, R.; Sousa, J. Hyperspectral Imaging: A Review on UAV-Based Sensors, Data Processing and Applications for Agriculture and Forestry. *Remote Sens.* **2017**, *9*, 1110. [CrossRef]
12. Burkart, A.; Cogliati, S.; Schickling, A.; Rascher, U. A Novel UAV-Based Ultra-Light Weight Spectrometer for Field Spectroscopy. *IEEE Sens. J.* **2014**, *14*, 62–67. [CrossRef]
13. Turner, D.; Lucieer, A.; Wallace, L. Direct Georeferencing of Ultrahigh-Resolution UAV Imagery. *IEEE Trans. Geosci. Remote Sens.* **2014**, *52*, 2738–2745. [CrossRef]
14. Schickling, A.; Matveeva, M.; Damm, A.; Schween, J.; Wahner, A.; Graf, A.; Crewell, S.; Rascher, U. Combining Sun-Induced Chlorophyll Fluorescence and Photochemical Reflectance Index Improves Diurnal Modeling of Gross Primary Productivity. *Remote Sens.* **2016**, *8*, 574. [CrossRef]
15. Garzonio, R.; Mauro, B.D.; Colombo, R.; Cogliati, S. Surface Reflectance and Sun-Induced Fluorescence Spectroscopy Measurements Using a Small Hyperspectral UAS. *Remote Sens.* **2017**, *9*, 472. [CrossRef]
16. Zeng, C.; Richardson, M.; King, D.J. The impacts of environmental variables on water reflectance measured using a lightweight unmanned aerial vehicle (UAV)-based spectrometer system. *ISPRS J. Photogramm. Remote Sens.* **2017**, *130*, 217–230. [CrossRef]
17. Uto, K.; Seki, H.; Saito, G.; Kosugi, Y.; Komatsu, T. Development of a Low-Cost Hyperspectral Whiskbroom Imager Using an Optical Fiber Bundle, a Swing Mirror, and Compact Spectrometers. *IEEE J. Sel. Top. Appl. Earth Obs. Remote Sens.* **2016**, *9*, 3909–3925. [CrossRef]

18. Zarco-Tejada, P.J.; Diaz-Varela, R.; Angileri, V.; Loudjani, P. Tree height quantification using very high resolution imagery acquired from an unmanned aerial vehicle (UAV) and automatic 3D photo-reconstruction methods. *Eur. J. Agron.* **2014**, *55*, 89–99. [CrossRef]
19. Calderón, R.; Navas-Cortés, J.A.; Lucena, C.; Zarco-Tejada, P.J. High-resolution airborne hyperspectral and thermal imagery for early detection of Verticillium wilt of olive using fluorescence, temperature and narrow-band spectral indices. *Remote Sens. Environ.* **2013**, *139*, 231–245. [CrossRef]
20. Zarco-Tejada, P.J.; Morales, A.; Testi, L.; Villalobos, F.J. Spatio-temporal patterns of chlorophyll fluorescence and physiological and structural indices acquired from hyperspectral imagery as compared with carbon fluxes measured with eddy covariance. *Remote Sens. Environ.* **2013**, *133*, 102–115. [CrossRef]
21. Hyperspectral Imaging Sensors. Available online: https://www.headwallphotonics.com/hyperspectral-sensors (accessed on 6 November 2018).
22. Lucieer, A.; Malenovský, Z.; Veness, T.; Wallace, L. HyperUAS-Imaging Spectroscopy from a Multirotor Unmanned Aircraft System: HyperUAS-Imaging Spectroscopy from a Multirotor Unmanned. *J. Field Robot* **2014**, *31*, 571–590. [CrossRef]
23. Turner, D.; Lucieer, A.; McCabe, M.F.; Parkes, S.; Clarke, I. Push-broom hyperspectral imaging from an Unmanned Aircraft System (UAS)-geometric processing workflow and accuracy assessment. *ISPRS Int. Arch. Photogramm. Remote Sens. Spat. Inf. Sci.* **2017**, *XLII-2/W6*, 379–384. [CrossRef]
24. Malenovský, Z.; Lucieer, A.; King, D.H.; Turnbull, J.D.; Robinson, S.A. Unmanned aircraft system advances health mapping of fragile polar vegetation. *Methods Ecol. Evol.* **2017**, *8*, 1842–1857. [CrossRef]
25. Sankey, T.; Donager, J.; McVay, J.; Sankey, J.B. UAV lidar and hyperspectral fusion for forest monitoring in the southwestern USA. *Remote Sens. Environ.* **2017**, *195*, 30–43. [CrossRef]
26. Suomalainen, J.; Anders, N.; Iqbal, S.; Roerink, G.; Franke, J.; Wenting, P.; Hünniger, D.; Bartholomeus, H.; Becker, R.; Kooistra, L. A Lightweight Hyperspectral Mapping System and Photogrammetric Processing Chain for Unmanned Aerial Vehicles. *Remote Sens.* **2014**, *6*, 11013–11030. [CrossRef]
27. HySpex Mjolnir V-1240. Available online: https://www.hyspex.no/products/mjolnir.php (accessed on 6 November 2018).
28. Hruska, R.; Mitchell, J.; Anderson, M.; Glenn, N.F. Radiometric and Geometric Analysis of Hyperspectral Imagery Acquired from an Unmanned Aerial Vehicle. *Remote Sens.* **2012**, *4*, 2736–2752. [CrossRef]
29. Aasen, H.; Bolten, A. Multi-temporal high-resolution imaging spectroscopy with hyperspectral 2D imagers - From theory to application. *Remote Sens. Environ.* **2018**, *205*, 374–389. [CrossRef]
30. Berni, J.A.J.; Zarco-Tejada, P.J.; Sepulcre-Cantó, G.; Fereres, E.; Villalobos, F. Mapping canopy conductance and CWSI in olive orchards using high resolution thermal remote sensing imagery. *Remote Sens. Environ.* **2009**, *113*, 2380–2388. [CrossRef]
31. Stagakis, S.; González-Dugo, V.; Cid, P.; Guillén-Climent, M.L.; Zarco-Tejada, P.J. Monitoring water stress and fruit quality in an orange orchard under regulated deficit irrigation using narrow-band structural and physiological remote sensing indices. *ISPRS J. Photogramm. Remote Sens.* **2012**, *71*, 47–61. [CrossRef]
32. Torres-Sánchez, J.; López-Granados, F.; Peña, J.M. An automatic object-based method for optimal thresholding in UAV images: Application for vegetation detection in herbaceous crops. *Comput. Electron. Agric.* **2015**, *114*, 43–52. [CrossRef]
33. Pérez-Ortiz, M.; Peña, J.M.; Gutiérrez, P.A.; Torres-Sánchez, J.; Hervás-Martínez, C.; López-Granados, F. A semi-supervised system for weed mapping in sunflower crops using unmanned aerial vehicles and a crop row detection method. *Appl. Soft Comput.* **2015**, *37*, 533–544. [CrossRef]
34. Hagen, N.; Kester, R.T.; Gao, L.; Tkaczyk, T.S. Snapshot advantage: A review of the light collection improvement for parallel high-dimensional measurement systems. *Opt. Eng.* **2012**, *51*, 111702. [CrossRef] [PubMed]
35. Hagen, N.; Kudenov, M.W. Review of snapshot spectral imaging technologies. *Opt. Eng.* **2013**, *52*, 090901. [CrossRef]
36. Yuan, H.; Yang, G.; Li, C.; Wang, Y.; Liu, J.; Yu, H.; Feng, H.; Xu, B.; Zhao, X.; Yang, X. Retrieving Soybean Leaf Area Index from Unmanned Aerial Vehicle Hyperspectral Remote Sensing: Analysis of RF, ANN, and SVM Regression Models. *Remote Sens.* **2017**, *9*, 309. [CrossRef]
37. Ramirez-Paredes, J.P.; Lary, D.J.; Gans, N.R. Low-altitude Terrestrial Spectroscopy from a Push-broom Sensor. *J. Field Robot* **2015**, *33*, 837–852. [CrossRef]

38. Habib, A.; Xiong, W.; He, F.; Yang, H.L.; Crawford, M. Improving Orthorectification of UAV-Based Push-Broom Scanner Imagery Using Derived Orthophotos from Frame Cameras. *IEEE J. Sel. Top. Appl. Earth Obs. Remote Sens.* **2017**, *10*, 262–276. [CrossRef]
39. Schläpfer, D.; Schaepman, M.E.; Itten, K.I. PARGE: Parametric geocoding based on GCP-calibrated auxiliary data. *Proc. SPIE 3438 Imaging Spectrom. IV.* **1998**, *3438*, 334–344.
40. SpectralView. Available online: http://www.headwallphotonics.com/software (accessed on 5 May 2019).
41. Habib, A.; Zhou, T.; Masjedi, A.; Zhang, Z.; Evan Flatt, J.; Crawford, M. Boresight Calibration of GNSS/INS-Assisted Push-Broom Hyperspectral Scanners on UAV Platforms. *IEEE J. Sel. Top. Appl. Earth Obs. Remote Sens.* **2018**, *11*, 1734–1749. [CrossRef]
42. Chen, Y.H.; Lin, H.Y.S.; Su, C.W. Full-Frame Video Stabilization Via SIFT Feature Matching. In Proceedings of the Tenth International Conference on Intelligent Information Hiding and Multimedia Signal Processing, Kitakyushu, Japan, 27–29 August 2014; pp. 361–364.
43. Shene, T.N.; Sridharan, K.; Sudha, N. Real-Time SURF-Based Video Stabilization System for an FPGA-Driven Mobile Robot. *IEEE Trans. Ind. Electron.* **2016**, *63*, 5012–5021. [CrossRef]
44. Jeon, S.; Yoon, I.; Jang, J.; Yang, S.; Kim, J.; Paik, J. Robust Video Stabilization Using Particle Keypoint Update and l1-Optimized Camera Path. *Sensors* **2017**, *17*, 337. [CrossRef]
45. Anand, R.; Veni, S.; Aravinth, J. Big Data Challenges in Airborne Hyperspectral Image for Urban Landuse Classification. In Proceedings of the International Conference on Advances in Computing, Communications and Informatics (ICACCI), Udupi, India, 13–16 September 2017; pp. 1808–1814.
46. Shi, Y.; Thomasson, J.A.; Murray, S.C.; Pugh, N.A.; Rooney, W.L.; Shafian, S.; Rajan, N.; Rouze, G.; Morgan, C.L.; Neely, H.L.; et al. Unmanned Aerial Vehicles for High-Throughput Phenotyping and Agronomic Research. *PLoS ONE* **2016**, *11*, e0159781. [CrossRef] [PubMed]
47. Hassaballah, M.; Abdelmgeid, A.A.; Alshazly, H.A. Image Features Detection, Description and Matching. In *Image Feature Detectors and Descriptors. Studies in Computational Intelligence*; Awad, A., Hassaballah, M., Eds.; Springer International Publishing: Warsaw, Poland, 2016; Volume 630, pp. 11–45.
48. Bay, H.; Ess, A.; Tuytelaars, T.; Van Gool, L. Speeded-Up Robust Features (SURF). *Comput. Vis. Image Underst.* **2008**, *110*, 346–359. [CrossRef]
49. Torr, P.H.S.; Zisserman, A. MLESAC: A New Robust Estimator with Application to Estimating Image Geometry. *Comput. Vis. Image Underst.* **2000**, *78*, 138–156. [CrossRef]
50. Johansen, K.; Morton, M.J.; Malbeteau, Y.M.; Aragon, B.; Al-Mashharawi, S.K.; Ziliani, M.G.; Angel, Y.; Fiene, G.M.; Negrão, S.S.; Mousa, M.A.; et al. Unmanned Aerial Vehicle-based Phenotyping using Morphometric and Spectral Analysis can Quantify Responses of Wild Tomato Plants to Salinity Stress. *Front. Plant Sci.* **2019**, *10*, 370. [CrossRef] [PubMed]
51. Aaftab, A.; Burhan, A.; Muhammad, N.; Ihsanullah, D.; Sidra, R. Effectiveness of silicon and irrigation scheduling for mitigating drought stress in sorghum (Sorghum bicolor L.) in arid region of Saudi Arabia. *Int. J. Biosci.* **2018**, *12*, 266–278.
52. Aly, A.A.; Al-Omran, A.M.; Sallam, A.S.; Al-Wabel, M.I.; Al-Shayaa, M.S. Vegetation cover change detection and assessment in arid environment using multi-temporal remote sensing images and ecosystem management approach. *Solid Earth.* **2016**, *7*, 713–725. [CrossRef]
53. Matrice 600. Available online: https://www.dji.com/matrice600 (accessed on 6 November 2018).
54. Matrice 100. Available online: https://www.dji.com/matrice100 (accessed on 6 May 2019).
55. Zenmuse X3. Available online: https://www.dji.com/zenmuse-x3 (accessed on 6 December 2017).
56. Universal Ground Control Station. Available online: https://www.ugcs.com (accessed on 6 November 2018).
57. Leica Viva GS15—Smart Antenna. Available online: https://leica-geosystems.com/products/gnss-systems/smart-antennas/leica-viva-gs15 (accessed on 17 May 2019).
58. Leica AS10 GNSS—Compact Antenna. Available online: https://leica-geosystems.com/products/gnss-reference-networks/antennas/leica-as10 (accessed on 6 November 2018).
59. Leica Geo Office. Available online: https://leica-geosystems.com/products/total-stations/software/leica-geo-office (accessed on 16 December 2017).
60. Turner, D.; Lucieer, A.; Watson, C. An Automated Technique for Generating Georectified Mosaics from Ultra-High Resolution Unmanned Aerial Vehicle (UAV) Imagery, Based on Structure from Motion (SfM) Point Clouds. *Remote Sens.* **2012**, *4*, 1392–1410. [CrossRef]
61. Agisoft PhotoScan. Available online: https://www.agisoft.com (accessed on 6 November 2018).

62. Verhoeven, G. Taking computer vision aloft—Archaeological three-dimensional reconstructions from aerial photographs with photoscan. *Archaeol. Prospect.* **2011**, *18*, 67–73. [CrossRef]
63. Ziliani, M.G.; Parkes, S.D.; Hoteit, I.; McCabe, M.F. Intra-Season Crop Height Variability at Commercial Farm Scales Using a Fixed-Wing UAV. *Remote Sens.* **2018**, *10*, 2007. [CrossRef]
64. Kanan, C.; Cottrell, G.W. Color-to-Grayscale: Does the Method Matter in Image Recognition? *PLoS ONE* **2012**, *7*, e29740. [CrossRef]
65. Recommendation ITU-R BT.601-7. Available online: http://www.itu.int/dms_pubrec/itu-r/rec/bt/R-REC-BT.601-7201103-I!!PDF-E.pdf (accessed on 6 November 2018).
66. Fischler, M.A.; Bolles, R.C. Random sample consensus: A paradigm for model fitting with applications to image analysis and automated cartography. *Commun. ACM* **1981**, *24*, 381–395. [CrossRef]
67. Hartley, R.; Zisserman, A. *Multiple View Geometry in Computer Vision*, 2nd ed.; Cambridge University Press: New York, NY, USA, 2004; pp. 25–64.
68. Sammut, C.; Webb, G.I. Mean Absolute Error. In *Encyclopedia of Machine Learning*; Springer: Boston, MA, USA, 2011.
69. Geospatial Positioning Accuracy Standards. Part 3: National Standard for Spatial Data Accuracy, FGDC-STD-007.3-1998. Available online: https://www.fgdc.gov/standards/projects/FGDC-standards-projects/accuracy/part3/chapter3 (accessed on 6 November 2018).
70. ISO 19157:2013 Geographic Information-Data Quality. Available online: https://www.iso.org/standard/32575.html (accessed on 6 November 2018).
71. Sedaghat, A.; Ebadi, H. Distinctive Order Based Self-Similarity descriptor for multi-sensor remote sensing image matching. *ISPRS J. Photogramm. Remote Sens.* **2015**, *108*, 62–71. [CrossRef]
72. Goncalves, H.; Corte-Real, L.; Goncalves, J.A. Automatic Image Registration through Image Segmentation and SIFT. *IEEE Trans. Geosci. Remote Sens.* **2011**, *49*, 2589–2600. [CrossRef]
73. Sedaghat, A.; Ebadi, H. Remote Sensing Image Matching Based on Adaptive Binning SIFT Descriptor. *IEEE Trans. Geosci. Remote Sens.* **2015**, 5283–5293. [CrossRef]
74. Liu, J.; Gong, J.; Guo, B.; Zhang, W. A Novel Adjustment Model for Mosaicking Low-Overlap Sweeping Images. *IEEE Trans. Geosci. Remote Sens.* **2017**, *55*, 4089–4097. [CrossRef]
75. Tareen, S.A.K.; Saleem, Z. A comparative analysis of SIFT, SURF, KAZE, AKAZE, ORB, and BRISK. In Proceedings of the International Conference on Computing, Mathematics and Engineering Technologies (iCoMET), Sukkur, Pakistan, 3–4 March 2018; pp. 1–10.
76. Oniga, V.-E.; Breaban, A.-I.; Statescu, F. Determining the Optimum Number of Ground Control Points for Obtaining High Precision Results Based on UAS Images. *Proceedings* **2018**, *2*, 352. [CrossRef]
77. Schläpfer, D.; Hausold, A.; Richter, R. A Unified Approach to Parametric Geocoding and Atmospheric/Topographic Correction for Wide FOV Airborne Imagery. Part 1: Parametric Ortho-Rectification Process. In Proceedings of the EARSeL Workshop on Imaging Spectroscopy, Enschede, The Netherlands, 11–13 July 2000.
78. Choi, S.; Taemin, K.; Wonpil, Y. Performance Evaluation of RANSAC Family. In Proceedings of the British Machine Vision Conference, Newcastle, UK, 3–6 September 2009; pp. 81.1–81.12.
79. Mesas-Carrascosa, F.J.; Rumbao, I.C.; Berrocal, J.A.B.; Porras, A.G.F. Positional Quality Assessment of Orthophotos Obtained from Sensors Onboard Multi-Rotor UAV Platforms. *Sensors* **2014**, *14*, 22394–22407. [CrossRef]

MDPI
St. Alban-Anlage 66
4052 Basel
Switzerland
Tel. +41 61 683 77 34
Fax +41 61 302 89 18
www.mdpi.com

Remote Sensing Editorial Office
E-mail: remotesensing@mdpi.com
www.mdpi.com/journal/remotesensing

www.ingramcontent.com/pod-product-compliance
Lightning Source LLC
LaVergne TN
LVHW071033170726
843515LV00006B/1271

* 9 7 8 3 0 3 6 5 0 8 7 8 8 *